Jennifer Moore C20
Manch...

POLYMER PHYSICS

POLYMER PHYSICS

Ulf W. Gedde

Associate Professor of Polymer Technology
Department of Polymer Technology
Royal Institute of Technology
Stockholm, Sweden

CHAPMAN & HALL
London · Glasgow · Weinheim · New York · Tokyo · Melbourne · Madras

Published by Chapman & Hall, 2–6 Boundary Row, London SE1 8HN, UK

Chapman & Hall, 2–6 Boundary Row, London SE1 8HN, UK

Blackie Academic & Professional, Wester Cleddens Road, Bishopbriggs, Glasgow G64 2NZ, UK

Chapman & Hall GmbH, Pappelallee 3, 69469 Weinheim, Germany

Chapman & Hall USA, 115 Fifth Avenue, New York, NY 10003, USA

Chapman & Hall Japan, ITP-Japan, Kyowa Building, 3F, 2-2-1 Hirakawacho, Chiyoda-ku, Tokyo 102, Japan

Chapman & Hall Australia, 102 Dodds Street, South Melbourne, Victoria 3205, Australia

Chapman & Hall India, R. Seshadri, 32 Second Main Road, CIT East, Madras 600 035, India

First edition 1995

© 1995 Chapman & Hall

Typeset in 10/12 pt Palatino by Techset Composition Ltd, Salisbury, Wiltshire
Printed in England by Clays Ltd St Ives plc

ISBN 0 412 59020 4 (HB)   0 412 62640 3 (PB)

Apart from any fair dealing for the purposes of research or private study, or criticism or review, as permitted under the UK Copyright Designs and Patents Act, 1988, this publication may not be reproduced, stored, or transmitted, in any form or by any means, without the prior permission in writing of the publishers, or in the case of reprographic reproduction only in accordance with the terms of the licences issued by the Copyright Licensing Agency in the UK, or in accordance with the terms of licences issued by the appropriate Reproduction Rights Organization outside the UK. Enquiries concerning reproduction outside the terms stated here should be sent to the publishers at the London address printed on this page.

The publisher makes no representation, express or implied, with regard to the accuracy of the information contained in this book and cannot accept any legal responsibility or liability for any errors or omissions that may be made.

A catalogue record for this book is available from the British Library

Library of Congress Catalog Card Number: 94–74687

∞ Printed on permanent acid-free text paper, manufactured in accordance with ANSI/NISO Z39.48-1992 and ANSI/NISO Z39.48-1984 (Permanence of Paper).

To Maria, Alexander and Raija

CONTENTS

Preface xi

1 A brief introduction to polymer science 1
    1.1 Fundamental definitions 1
    1.2 Configurational states 2
    1.3 Homopolymers and copolymers 5
    1.4 Molecular architecture 6
    1.5 Common polymers 6
    1.6 Molar mass 6
    1.7 Polymerization 12
    1.8 Thermal transitions and physical structures 13
    1.9 Polymer materials 15
    1.10 A short history of polymers 15
    1.11 Summary 18
    1.12 Exercises 18
    1.13 References 18
    1.14 Suggested further reading 18

2 Chain conformations in polymers 19
    2.1 Introduction 19
    2.2 Experimental determination of dimensions of chain molecules 21
    2.3 Characteristic dimensions of 'random coil' polymers 23
    2.4 Models for calculating the average end-to-end distance for an ensemble of statistical chains 24
    2.5 Random-flight analysis 33
    2.6 Chains with preferred conformation 35
    2.7 Summary 36
    2.8 Exercises 37
    2.9 References 38

3 The rubber elastic state 39
    3.1 Introduction 39
    3.2 Thermo-elastic behaviour and thermodynamics: energetic and entropic elastic forces 41
    3.3 The statistical mechanical theory of rubber elasticity 44
    3.4 Swelling of rubbers in solvents 48
    3.5 Deviations from classical statistical theories 48
    3.6 Small-angle neutron scattering data 51
    3.7 The theory of Mooney and Rivlin 51
    3.8 Summary 51
    3.9 Exercises 52
    3.10 References 53
    3.11 Suggested further reading 53

4 Polymer solutions — 55
- 4.1 Introduction — 55
- 4.2 Regular solutions — 55
- 4.3 The Flory–Huggins theory — 58
- 4.4 Concentration regimes in polymer solutions — 65
- 4.5 The solubility parameter concept — 66
- 4.6 Equation-of-state theories — 68
- 4.7 Polymer–polymer blends — 70
- 4.8 Summary — 73
- 4.9 Exercises — 74
- 4.10 References — 75
- 4.11 Suggested further reading — 75

5 The glassy amorphous state — 77
- 5.1 Introduction to amorphous polymers — 77
- 5.2 The glass transition temperature — 78
- 5.3 Non-equilibrium features of glassy polymers and physical ageing — 82
- 5.4 Theories for the glass transition — 87
- 5.5 Mechanical behaviour of glassy, amorphous polymers — 89
- 5.6 Structure of glassy, amorphous polymers — 95
- 5.7 Summary — 96
- 5.8 Exercises — 97
- 5.9 References — 97
- 5.10 Suggested further reading — 98

6 The molten state — 99
- 6.1 Introduction — 99
- 6.2 Fundamental concepts in rheology — 99
- 6.3 Measurement of rheological properties of molten polymers — 104
- 6.4 Flexible-chain polymers — 105
- 6.5 Liquid-crystalline polymers — 109
- 6.6 Summary — 127
- 6.7 Exercises — 128
- 6.8 References — 129
- 6.9 Suggested further reading — 129

7 Crystalline polymers — 131
- 7.1 Background and a brief survey of polymer crystallography — 131
- 7.2 The crystal lamella — 137
- 7.3 Crystals grown from the melt and the crystal lamella stack — 147
- 7.4 Supermolecular structure — 151
- 7.5 Methods of assessing supermolecular structure — 155
- 7.6 Degree of crystallinity — 157
- 7.7 Relaxation processes in semicrystalline polymers — 162
- 7.8 Summary — 164
- 7.9 Exercises — 165
- 7.10 References — 166
- 7.11 Suggested further reading — 167

8 Crystallization kinetics — 169
- 8.1 Background — 169
- 8.2 The equilibrium melting temperature — 171
- 8.3 The general Avrami equation — 175
- 8.4 Growth theories — 178
- 8.5 Molecular fractionation — 189
- 8.6 Orientation-induced crystallization — 194
- 8.7 Summary — 195
- 8.8 Exercises — 197
- 8.9 References — 198
- 8.10 Suggested further reading — 198

9 Chain orientation — 199
- 9.1 Introduction — 199
- 9.2 Definition of chain orientation — 199
- 9.3 Methods for assessment of uniaxial chain orientation — 203
- 9.4 Methods for assessment of biaxial chain orientation — 208
- 9.5 How chain orientation is created — 208
- 9.6 Properties of oriented polymers — 211
- 9.7 Summary — 214
- 9.8 Exercises — 215
- 9.9 References — 216
- 9.10 Suggested further reading — 216

10 Thermal analysis of polymers — 217
- 10.1 Introduction — 217
- 10.2 Thermo-analytical methods — 218
- 10.3 Thermal behaviour of polymers — 226
- 10.4 Summary — 234
- 10.5 Exercises — 234
- 10.6 References — 236
- 10.7 Suggested further reading — 237

11 Microscopy of polymers — 239
- 11.1 Introduction — 239
- 11.2 Optical microscopy (OM) — 241
- 11.3 Electron microscopy — 244
- 11.4 Preparation of specimens for microscopy — 247
- 11.5 Applications of polymer microscopy — 252
- 11.6 Summary — 256
- 11.7 Exercises — 256
- 11.8 References — 257
- 11.9 Suggested further reading — 257

12 Spectroscopy and scattering of polymers — 259
  12.1 Introduction — 259
  12.2 Spectroscopy — 260
  12.3 Scattering and diffraction methods — 269
  12.4 Summary — 273
  12.5 Exercises — 273
  12.6 References — 273
  12.7 Suggested further reading — 273

13 Solutions to problems given in exercises — 275
  Chapter 1 — 275
  Chapter 2 — 275
  Chapter 3 — 278
  Chapter 4 — 279
  Chapter 5 — 281
  Chapter 6 — 282
  Chapter 7 — 283
  Chapter 8 — 286
  Chapter 9 — 287
  Chapter 10 — 289
  Chapter 11 — 291
  Chapter 12 — 291
  References — 292

Index — 293

PREFACE

This book is the result of my teaching efforts during the last ten years at the Royal Institute of Technology. The purpose is to present the subject of polymer physics for undergraduate and graduate students, to focus the fundamental aspects of the subject and to show the link between experiments and theory. The intention is not to present a compilation of the currently available literature on the subject. Very few reference citations have thus been made. Each chapter has essentially the same structure: starting with an introduction, continuing with the actual subject, summarizing the chapter in 300–500 words, and finally presenting problems and a list of relevant references for the reader. The solutions to the problems presented in Chapters 1–12 are given in Chapter 13. The theme of the book is essentially polymer science, with the exclusion of that part dealing directly with chemical reactions. The fundamentals in polymer science, including some basic polymer chemistry, are presented as an introduction in the first chapter. The next eight chapters deal with different phenomena (processes) and states of polymers. The last three chapters were written with the intention of making the reader think practically about polymer physics. How can a certain type of problem be solved? What kinds of experiment should be conducted?

This book would never have been written without the help of my friend and adviser, Dr Anthony Bristow, who has spent many hours reading through the manuscript, criticizing the content, the form and the presentation. I also wish to thank my colleagues at the Department, Maria Conde Braña, Kristian Engberg, Anders Gustafsson, Mikael Hedenqvist, Anders Hult, Jan-Fredrik Jansson, Håkan Jonsson, Joanna Kiesler, Sari Laihonen, Bengt Rånby, Patrik Roseen, Fredrik Sahlén, Marie-Louise Skyff, Bengt Stenberg, Björn Terselius, Toma Tränkner, Göran Wiberg and Jens Viebke, who have provided help of different kinds, ranging from criticism of the manuscript to the provision of micrographs, etc. Special thanks are due to Dr Richard Jones, Cavendish Laboratory, University of Cambridge, UK, who read through all the chapters and made some very constructive criticisms. I also want to thank friends and collaborators from other departments/companies: Profs Richard Boyd, University of Utah, USA; Andrew Keller, University of Bristol, UK; David Bassett, University of Reading, UK; Clas Blomberg, Royal Institute of Technology, Sweden; Josef Kubat, Chalmers University of Technology, Sweden; Torbjörn Lagerwall, Chalmers University of Technology, Sweden, and Mats Ifwarson, Studsvik Material AB, Sweden. I wish to emphasize, however, that I alone have responsibility for the book's shortcomings.

I am also indebted to Chapman & Hall for patience in waiting for the manuscript to arrive and for performing such an excellent job in transforming the manuscript to this pleasant shape. More than anything, I am grateful to my family for their support during the almost endless thinking and writing process.

Ulf W. Gedde
Stockholm

A BRIEF INTRODUCTION TO POLYMER SCIENCE

1.1 FUNDAMENTAL DEFINITIONS

Polymers consist of large molecules, i.e. macromolecules. According to the basic IUPAC definition (Metanomski 1991):

> A polymer is a substance composed of molecules characterized by the multiple repetition of one or more species of atoms or groups of atoms (constitutional repeating units) linked to each other in amounts sufficient to provide a set of properties that do not vary markedly with the addition of one or a few of the constitutional repeating units.

The word **polymer** originates from the Greek words 'poly' meaning many and 'mer' meaning part. Figure 1.1 shows the structure of polypropylene, an industrially important polymer. The constitutional repeating units, which are also called simply 'repeating units', are linked by covalent bonds, and the atoms of the repeating unit are also linked by covalent bonds. A molecule with only a few constitutional repeating units is called an **oligomer**. The physical properties of an oligomer vary with the addition or removal of one or a few constitutional repeating units from its molecules. A **monomer** is the substance that the polymer is made from, which in the case of polypropylene is propylene (propene) (Fig. 1.1). The process that converts a monomer to a polymer is called **polymerization**.

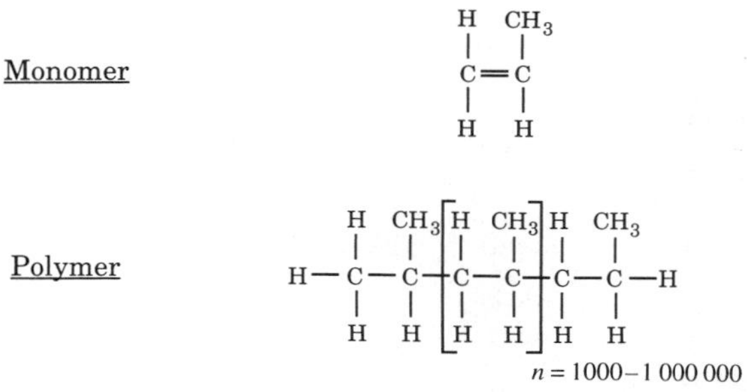

The polymers dealt with in this book are exclusively organic carbon-based polymers. Other common elements in the organic polymers are hydrogen, oxygen, nitrogen, sulphur and silicon. Table 1.1 presents some typical bond energies and bond lengths of different covalent and secondary bonds. When assessing the stability of primary and secondary bonds, these energies are compared with the thermal energy, i.e. RT, where R is the gas constant and T is the absolute temperature given in kelvin. The thermal energy is approximately 2.5 kJ mol^{-1} at 300 K and approximately 4 kJ mol^{-1} at 500 K.

The large difference in dissociation energy and bond force constant ('stiffness') between the covalent bonds (so-called primary bonds) and the weak secondary bonds between different molecules is of great importance for polymer properties. The identity of the molecules, i.e. the entities linked by covalent bonds, is largely preserved during melting. There are many examples of polymers that degrade early at low temperatures but that involve only a few of the existing primary bonds. Melting involves mainly the rupture and re-establishment of a great many secondary bonds.

Polymer crystals show very direction-dependent (anisotropic) properties. The Young's modulus of polyethylene at room temperature is approximately 300 GPa in the chain-axis direction and only 3 GPa in the transverse directions (Fig. 1.2). This considerable difference in modulus is due to the presence of two

Figure 1.1 The structure of a monomer (propylene) and a polymer (polypropylene). The constitutional repeating unit is shown between the brackets.

Table 1.1 Dissociation energy and length of different bonds

Bond type	Energy (kJ mol^{-1})	Bond length (nm)
Covalent bond	300–500	0.15 (C–C; C–N; C–O)
		0.11 (C–H)
		0.135 (C=C)
van der Waals bond	10	0.4
Dipole–dipole bond	>10	0.4
Hydrogen bond	10–50	0.3

types of bond connecting the different atoms in the crystals: strong and stiff bonds along the chain axis and weak and soft secondary bonds acting in the transverse directions (Fig. 1.2). A whole range of other properties, e.g. the refractive index, also show strong directional dependence. The orientation of the polymer molecules in a material is enormously important. The Young's modulus of a given polymer can be changed by a factor of 100 by changing the degree of chain orientation. This is an important topic discussed in Chapter 9.

1.2 CONFIGURATIONAL STATES

The term **configuration** refers to the 'permanent' stereostructure of a polymer. The configuration is defined by the polymerization method, and a polymer preserves its configuration until it reacts chemically. A change in configuration requires the rupture of chemical bonds. Different configurations exist in polymers with stereocentres (tacticity) and double bonds (*cis* and *trans* forms). A polymer with the constitutional repeating unit –CH$_2$–CHX– exhibits two different stereoforms (configurational base units) for each constitutional repeating unit (Fig. 1.3). The following convention is adapted to distinguish between the two stereoforms. One of the chain ends is first selected as the near one. The selected asymmetric carbon atom (the one with the attached X atom (group of atoms) should be pointing upwards. The term **d form** is given to the arrangement with the X group pointing to the right (from the observer at the near end). The **l form** is the mirror-image of the d form, i.e. the X group points in this case to the left. This convention is, however, not absolute. If the near and far chain ends are reversed, i.e. if the chain is viewed from the opposite direction, the d and l notation is reversed for a given chain. When writing a single chain down on paper, the convention is that the atoms shown on the left-hand side are assumed to be nearer to the observer than the atoms on the right-hand side.

Tacticity is the orderliness of the succession of configurational base units in the main chain of a polymer molecule. An **isotactic** polymer is a regular polymer consisting only of one species of configurational base unit, i.e. only the d or the l form (Fig. 1.4).

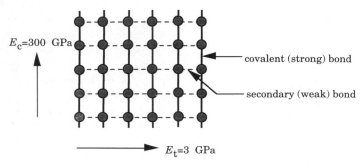

Figure 1.2 Schematic representation of a polymer crystal illustrating its anisotropic nature. The moduli for polyethylene parallel (E_c) and transverse (E_t) to the chain axis are shown.

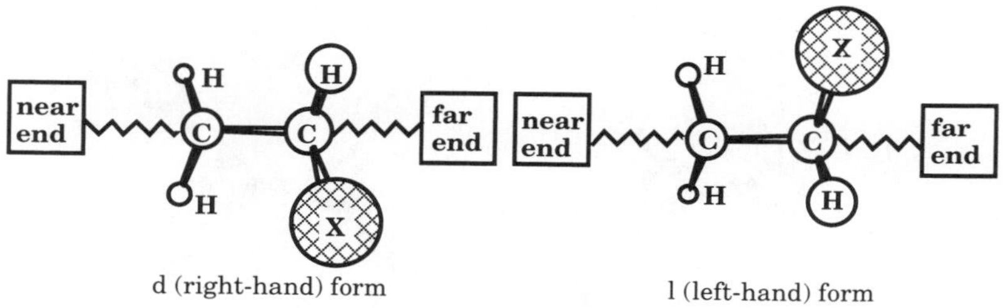

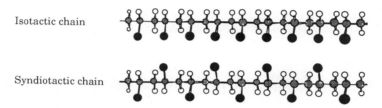

Figure 1.3 Configurational base units of a polymer with the constitutional repeating unit –CH$_2$–CHX–. The 'X' is an atom or a group of atoms different from hydrogen.

Isotactic chain

Syndiotactic chain

Figure 1.4 Regular tactic chains of [–CH$_2$–CHX–]$_n$, where X is indicated by a filled circle.

These can be converted into each other by simple rotation of the whole molecule. Thus, in practice there is no difference between an all-d chain and an all-l chain. A small mismatch in the chain ends does not alter this fact. A **syndiotactic** polymer consists of an alternating sequence of the different configurational base units, i.e. ...dldldldldldl... (Fig. 1.4). An **atactic** polymer has equal numbers of randomly distributed configurational base units.

Carbon-13 nuclear magnetic resonance (NMR) is the most useful method of assessing tacticity. By C-13 NMR it is possible to assess the different sequential distributions of adjacent configurational units that are called **dyads**, **triads**, **tetrads** and **pentads**. The two possible dyads are shown in Fig. 1.5. A chain with 100% **meso** dyads is perfectly isotactic whereas a chain with 100% **racemic** dyads is perfectly syndiotactic. A chain with a 50/50 distribution of meso and racemic dyads is atactic.

Triads express the sequential order of the configurational base units of a group of three adjacent constitutional repeating units. The following triads are possible: mm, mr and rr (where meso = m; racemic = r). Tetrads include four repeating units and the following six sequences are possible: mmm, mmr, mrm, mrr, rmr and rrr. Sequences with a length of up to five constitutional repeating units, called pentads, can be distinguished by C-13 NMR. The following ten different pentads are possible: mmmm, mmmr, mmrm, mmrr, mrrm, rmmr, rmrm, mrrr, mrrr and rrrr.

Polymers with double bonds in the main chain, e.g. polydienes, show different stereostructures. Figure 1.6 shows the two stereoforms of 1,4-polybutadiene: *cis* and *trans*. The double bond is rigid and allows no torsion, and the *cis* and *trans* forms are not transferable into each other. Polyisoprene is another well-known example: natural rubber consists almost exclusively of the *cis* form whereas gutta-percha is composed of the *trans* form. Both polymers are synthesized by

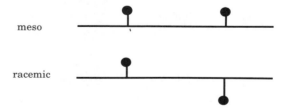

Figure 1.5 Dyads of a vinyl polymer with the constitutional repeating unit –CH$_2$–CHX–, where the X group is indicated by a filled circle.

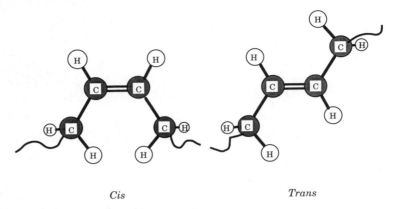

Cis   *Trans*

Figure 1.6 Stereoforms of 1,4-polybutadiene showing only the constitutional repeating unit with the rigid central double bond.

'nature', and this shows that stereoregularity was achieved in nature much earlier than the discovery of coordination polymerization by Ziegler and Natta.

The polymerization of diene monomers may involve different addition reactions (Fig. 1.7). 1,2 addition yields a polymer with the double bond in the pendant group whereas 1,4 addition gives a polymer with the unsaturation in the main chain.

Vinyl polymers (–CH$_2$–CHX–) may show different configurations with respect to the head (CHX) and tail (CH$_2$): head-to-head, with –CHX bonded to CH$_2$– and head-to-head–tail-to-tail, with –CHX bonded to –CHX followed by –CH$_2$ bonded to –CH$_2$ (Fig. 1.8).

The configurational state determines the low-temperature physical structure of the polymer. A polymer with an irregular configuration, e.g. an atactic polymer, will never crystallize and freezes to a glassy structure at low temperatures, whereas a polymer with a regular configuration, e.g. an isotactic polymer, may crystallize at some temperature to form a semicrystalline material. There is a spectrum of intermediate cases in which crystallization occurs but to a significantly reduced level. This topic is further discussed in Chapters 5 and 7.

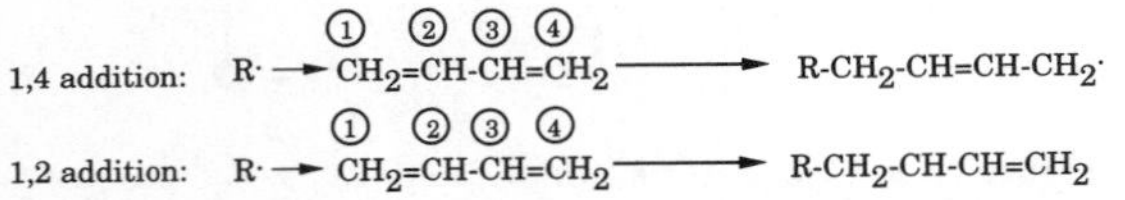

A conformational state refers to the stereostructure of a molecule defined by its sequence of bonds and torsion angles. The change in shape of a given molecule due to torsion about single (sigma) bonds is referred to as a change of **conformation** (Fig. 1.9). Double and triple bonds, which in addition to the rotationally symmetric sigma bond also consist of one or two rotationally asymmetric pi bonds, permit no torsion. There are only small energy barriers, from a few to 10 kJ mol^{-1}, involved in these torsions. The conformation of polymers is the subject of Chapter 2. The multitude of conformations in polymers is very important for the behaviour of polymers. The rapid change in conformation is responsible for the sudden extension of a rubber polymer on loading and the extraordinarily high ultimate extensibility of the network. This is the topic of Chapter 3. The high segmental flexibility of the molecules at high temperatures and the low flexibility at low temperatures is a very useful signature of a polymer.

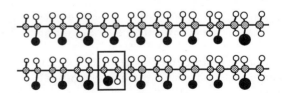

Figure 1.8 Head-to-tail configuration (upper chain) and a chain with a head-to-head junction followed by a tail-to-tail sequence (lower chain).

1,4 addition: R· → CH$_2$=CH-CH=CH$_2$ ① ② ③ ④   → R-CH$_2$-CH=CH-CH$_2$·

1,2 addition: R· → CH$_2$=CH-CH=CH$_2$ ① ② ③ ④   → R-CH$_2$-CH-CH=CH$_2$

Figure 1.7 Different additions of butadiene.

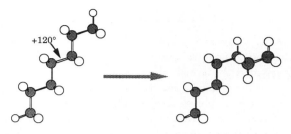

Figure 1.9 Examples of conformational states of a few repeating units of a polyethylene chain. The right-hand form is generated by 120° torsion about the single bond indicated by the arrow.

1.3 HOMOPOLYMERS AND COPOLYMERS

A **homopolymer** consists of only one type of constitutional repeating unit (A). A **copolymer**, on the other hand, consists of two or more constitutional repeating units (A, B, etc). Several classes of copolymer are possible: **block** copolymers, **alternating** copolymers, **graft** copolymers and **statistical** copolymers (Fig. 1.10).

The different copolymers with constitutional repeating units A and B are named according to the source-based nomenclature rules as follows: unspecified type, poly(A-*co*-B); statistical copolymer, poly(A-*stat*-B); alternating copolymer, poly(A-*alt*-B); graft copolymer, poly(A-*graft*-B). Note that the constitutional repeating unit of the backbone chain of the graft copolymer is specified first.

A **random** copolymer is a special type of statistical copolymer. The probability of finding a given constitutional repeating unit at any given site in a random copolymer is independent of the nature of the adjacent units at that position. A statistical copolymer may, however, obey known statistical laws, e.g. Markovian statistics. The term 'random copolymer' is occasionally used for polymers with the additional restriction that the constitutional repeating units are present in equal amounts. The notation for a random copolymer is poly(A-*ran*-B).

Copolymerization provides a route for making polymers with special, desired property profiles. A statistical copolymer consisting of units A and B, for instance, has in most cases properties in between those of the homopolymers (polyA and polyB). An important deviation from this simple rule arises if either polyA or polyB is semicrystalline. The statistical copolymer (polyA-*stat*-B)) is for most compositions fully amorphous. Block and graft copolymers form in most cases a two-phase morphology and the different phases obey properties similar to those of the respective homopolymers.

Di-block (A-B) and **tri-block** (A-B-A) copolymers are made by so-called living polymerization (section 1.7). These polymers have found applications as thermoplastic elastomers and as compatibilizers to increase the adhesion between the phases in polymer blends.

Terpolymers consist of three different repeating units: A, B and C.

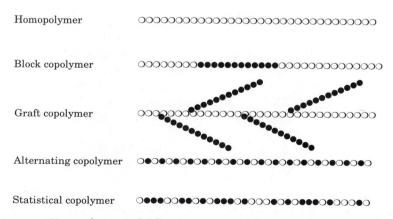

Figure 1.10 Homopolymers and different classes of copolymers. Unit A: ○; unit B: ●.

1.4 MOLECULAR ARCHITECTURE

Molecular architecture deals with the shape of a polymer molecule. Examples of polymers with different molecular architecture are shown in Fig. 1.11. A **short-chain branch** has an oligomeric nature whereas a **long-chain branch** is of polymeric length. A **network** polymer consists of many interconnected chain segments and many different molecular paths exist between any two atoms. A **dendrimer** or **hyperbranched** polymer, not shown in Fig. 1.11, consists of a constitutional repeating unit including a branching group. The number of branches increases thus according to a power law expression with the number of the 'generation' of the polymer.

The molecular architecture is important for many properties. Short-chain branching tends to reduce crystallinity. Long-chain branches have profound effects on rheological properties. Typical of ladder polymers is a high strength and a high thermal stability. Hyperbranched polymers consist of molecules with an approximately spherical shape, and it has been shown that their melt viscosity is significantly lower than that of their linear analogue with the same molar mass. Crosslinked polymers are thermosets, i.e. they do not melt. They also show little creep under constant mechanical loading.

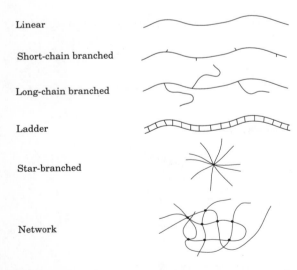

Figure 1.11 Schematic representation of structures of polymers with different molecular architecture.

1.5 COMMON POLYMERS

A list of common polymers with their constitutional repeating units is presented in Table 1.2. Most polymers are named **polyx**. If 'x' is a single word, the name of the polymer is written out directly, as in the case of polystyrene. However, if 'x' consists of two or more words, parentheses should be used, as in the case of poly(methyl methacrylate). There are two different systems for naming polymers. The **structure-based** names are rigorous but seldom used in practice. They are simply given as poly(constitutional repeating unit) and the rules of nomenclature for the constitutional repeating unit are no different from those of any other organic substance. The **source-based** names attach 'poly' to the name of the monomer. Polystyrene is a source-based name. Its structure-based equivalent is poly(1-phenylethylene). Polyethylene (source-based name) is denoted polymethylene according to the structure-based system. The names of polymers treated in this book are almost exclusively related to the source-based system. Most polymers have abbreviated names, which are also presented in Table 1.2. It is clearly acceptable to use abbreviations in scientific papers and technical reports, but the full name of the polymer should be given the first time it appears in the text. There are numerous other names of polymers used frequently. 'Nylon', 'Kevlar' and 'Vectra' are a few well-known examples. The source-based or structure-based names are preferred over these truly trivial names.

1.6 MOLAR MASS

The enormous size of polymer molecules gives them unique properties. Figure 1.12 shows the influence of molar mass on the melting point of polyethylene. Low molar mass substances (oligomers) show a strong increase in melting point with increasing molar mass, whereas a constant melting point is approached in the polymer molar mass range. Other polymers show a similar behaviour.

Other properties such as fracture toughness and Young's modulus show a similar molar mass dependence, with constant values approached in the high molar mass region. Polymer properties are often obtained in the molar mass range from 10 000 to 30 000 g mol^{-1}. Rheological properties such as melt

Table 1.2 Constitutional repeating units of common polymers

Polymer name, abbreviation	Constitutional repeating unit	Polymer name, abbreviation	Constitutional repeating unit
Polyethylene, PE	—CH$_2$—CH$_2$—	Poly(1,4-butadiene)	—CH$_2$—CH=CH—CH$_2$—
Polypropylene, PP	—CH$_2$—CH(CH$_3$)—	Polyisoprene	—CH$_2$—C(CH$_3$)=CH—CH$_2$—
Polyvinylchloride, PVC	—CH$_2$—CHCl—	Polyacrylonitrile, PAN	—CH$_2$—CH(CN)—
Polystyrene, PS	—CH$_2$—CH(C$_6$H$_5$)—	Polymethylmethacrylate, PMMA	—CH$_2$—C(CH$_3$)(COOCH$_3$)—
Poly(vinyl alcohol), PVAL	—CH$_2$—CH(OH)—	Poly(n-alkylmethacrylate)	—CH$_2$—C(CH$_3$)(CO—O—(CH$_2$)$_{n-1}$—CH$_3$)—
Polyvinylacetate, PVAC	—CH$_2$—CH(O—CO—CH$_3$)—	Poly(n-alkylacrylate)	—CH$_2$—CH(CO—O—(CH$_2$)$_{n-1}$—CH$_3$)—
Poly(4-methyl-1-pentene)	—CH$_2$—CH(CH$_2$—CH(CH$_3$)$_2$)—	Poly(ethylene terephthalate), PETP	—O—CH$_2$—CH$_2$—O—CO—C$_6$H$_4$—CO—

Table 1.2 continued

Polymer name, abbreviation	Constitutional repeating unit	Polymer name, abbreviation	Constitutional repeating unit
Poly(butylene terephthalate), PBTP	—O—C(H)(H)—C(H)(H)—C(H)(H)—C(H)(H)—O—C(=O)—C$_6$H$_4$—C(=O)—	Polyoxymethylene, POM	—C(H)(H)—O—
Polytetrafluoroethylene, PTFE	—C(F)(F)—C(F)(F)—	Polyethyleneoxide, PEO	—C(H)(H)—C(H)(H)—O—
Polyamide 6, PA 6	—C(=O)—(CH$_2$)$_5$—N(H)—	Poly(vinylidene dichloride), PVDC	—C(H)(Cl)—C(H)(Cl)—
Polyamide n, PA n	—C(=O)—(CH$_2$)$_{n-1}$—N(H)—	Poly(vinylidene difluoride), PVDF	—C(H)(F)—C(H)(F)—
Polyamide 6,10, PA 6,10	—N(H)—(CH$_2$)$_6$—N(H)—C(=O)—(CH$_2$)$_8$—C(=O)—		

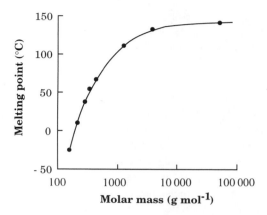

Figure 1.12 Molar mass dependence of the equilibrium melting point of oligo- and polyethylene. Drawn after data collected by Boyd and Phillips (1993).

viscosity show a progressive strong increase with increasing molar mass even at high molar masses (see Chapter 6). High molar mass polymers are therefore difficult to process but, on the other hand, they have very good mechanical properties in the final products.

There is currently no polymerization method available that yields a polymer with only one size of molecules. Variation in molar mass among the different molecules is characteristic of all synthetic polymers. They show a broad distribution in molar mass. The molar mass distribution ranges over three to four orders of magnitude in many cases. The full representation of the molar mass distribution is currently only achieved with size exclusion chromatography, naturally with a number of experimental limitations. Other methods yield different averages. The most commonly used averages are defined as follows. The **number** average is given by:

$$\bar{M}_n = \frac{\sum_i N_i M_i}{\sum_i N_i} = \sum_i n_i M_i \quad (1.1)$$

where N_i is the number of molecules of molar mass M_i, and n_i is the numerical fraction of those molecules. The **mass** or **weight** average is given by:

$$\bar{M}_w = \frac{\sum_i N_i M_i^2}{\sum_i N_i M_i} = \frac{\sum_i W_i M_i}{\sum_i W_i} = \sum_i w_i M_i \quad (1.2)$$

where W_i is the mass of the molecules of molar mass M_i, and w_i is the mass fraction of those molecules. The **Z** average is given by:

$$\bar{M}_z = \frac{\sum_i N_i M_i^3}{\sum_i N_i M_i^2} \quad (1.3)$$

while the **viscosity** average

$$\bar{M}_v = \left(\frac{\sum_i N_i M_i^{1+i}}{\sum_i N_i M_i} \right)^{1/a} \quad (1.4)$$

where a is a constant that takes values between 0.5 and 0.8 for different combinations of polymer and solvent. The viscosity average is obtained by viscometry. The intrinsic viscosity is given by:

$$[\eta] = \lim_{c \to 0} \left(\frac{\eta - \eta_0}{c \eta_0} \right)$$

where c is the concentration of polymer in the solution, η_0 is the viscosity of the pure solvent and η is the viscosity of the solution. The viscosities are obtained from the flow-through times (t and t_0) in the viscometer:

$$\frac{\eta}{\eta_0} \approx \frac{t}{t_0}$$

and the intrinsic viscosity is converted to the viscosity average molar mass according to the Mark–Houwink (1938) viscosity equation:

$$[\eta] = K \bar{M}_v^a \quad (1.5)$$

where K and a are the Mark–Houwink parameters. These constants are unique for each combination of polymer and solvent and can be found tabulated in the appropriate reference literature. The Mark–Houwink parameters given are in most cases based on samples with a narrow molar mass distribution. If eq. (1.5) is used for a polymer sample with a broad molar mass distribution, the molar mass value obtained is indeed the viscosity average.

All these averages are equal only for a perfectly monodisperse polymer. In all other cases, the averages are different: $\bar{M}_n < \bar{M}_v < \bar{M}_w < \bar{M}_z$. The viscosity average is often relatively close to the weight average.

Let us now show that the mass average is always greater than the number average:

$$\sum_i N_i(M_i - \bar{M}_n)^2 \geq 0 \quad (1.6a)$$

$$\sum_i N_i M_i^2 + \sum_i N_i \bar{M}_n^2 - 2\sum_i N_i M_i \bar{M}_n \geq 0 \quad (1.6b)$$

If eq. (1.6b) is divided by $\sum N_i$, the following expression is obtained:

$$\frac{\sum_i N_i M_i^2}{\sum_i N_i} + \bar{M}_n^2 - \frac{2\sum_i N_i M_i \bar{M}_n}{\sum_i N_i} \geq 0$$

and thus

$$\frac{\sum_i N_i M_i^2}{\sum_i N_i} \geq \bar{M}_n^2 \quad (1.7)$$

The following simplifications lead to the desired result:

$$\frac{\sum_i N_i M_i^2}{\sum_i N_i} \times \frac{1}{\bar{M}_n} \geq \bar{M}_n \Rightarrow \frac{\sum_i N_i M_i^2}{\sum_i N_i} \times \frac{\sum_i N_i}{\sum_i N_i M_i} \geq \bar{M}_n$$

and

$$\bar{M}_w \geq \bar{M}_n \quad (1.8)$$

Equality occurs only when a sample is truly monodisperse, i.e. when all molecules are of the same molar mass. It can be shown that the breadth of the molar mass distribution, expressed as its standard deviation (σ), is related to the ratio $\bar{M}_w/\bar{M}_n$ (polydispersity index) as follows:

$$\frac{\sigma}{\bar{M}_n} = \sqrt{\frac{\bar{M}_w}{\bar{M}_n} - 1} \quad (1.9)$$

The standard deviation takes the value zero for $\bar{M}_w/\bar{M}_n = 1$. The polydispersity index takes a high value for a sample with a broad molar mass distribution, i.e. a high σ value.

A wide range of methods can be used for the assessment of molar mass (Table 1.3). Some of the methods require no calibration and may be referred to as absolute, whereas other methods are relative. The latter require calibration with samples of known molar mass.

The concentration of **end groups** in a given sample provides direct information about the number of polymer molecules per gram, i.e. the molar mass. Infrared spectroscopy, NMR and titration of acid end groups in polyesters have been used for end-group analysis. One drawback of these methods is that they can only be used on low molar mass polymers.

Colligative properties are those properties of a solution which depend only upon the number of solute species present in a certain volume, and not on the nature of the solute species. It is thus logical that measurement of the colligative properties makes determination of $\bar{M}_n$ possible. The important colligative effects that are used for molar mass determination are boiling point elevation (ebulliometry), freezing point depression (cryoscopy) and

Table 1.3 Experimental techniques for molar mass determination

Method	Result	Comments
End-group analysis	$\bar{M}_n$	Absolute method, restricted to low molar mass
Colligative methods: ebulliometry, cryoscopy and osmometry	$\bar{M}_n$	Absolute methods, ebulliometry/cryoscopy, restricted to low molar mass
Light scattering	$\bar{M}_w$	Absolute method
Viscometry	$\bar{M}_v$	Relative method, easy to use
Size exclusion chromatography (SEC)	Molar mass distribution	Relative method, requires calibration

osmotic pressure (membrane osmometry). Ebulliometry and cryoscopy are restricted to samples with low molar masses, typically less than 10 000 g mol^{-1}. The number average molar mass is obtained according to the following general equation:

$$\left(\frac{\Delta T_x}{c}\right)_{c \to 0} = \left(\frac{V_1 R T_0^2}{\Delta H_x}\right) \times \frac{1}{\bar{M}_n} \quad (1.10)$$

where ΔT_x is the change in transition temperature (freezing point or boiling point), c is the concentration of polymer in the solution, V_1 is the molar volume of the solvent, R is the gas constant, T_0 is the transition temperature for the pure solvent and ΔH_x is the transition enthalpy. Membrane osmometry is useful for samples of $\bar{M}_n \leq 100\,000$ g mol^{-1}, and the number average molar mass is obtained from the following expression:

$$\left(\frac{\Pi}{c}\right)_{c \to 0} = \frac{RT}{\bar{M}_n} \quad (1.11)$$

where Π is the osmotic pressure.

Size exclusion chromatography (SEC), often referred to as gel permeation chromatography (GPC), gives the whole molar mass distribution. A dilute solution of the polymer is injected into a gel column. The flow-through times of the different molar mass species depend on their hydrodynamic volumes, i.e. on the size of the molecular coil. Large molecules have little accessibility to the pores of the gel and they are eluated after only a short period of time. Smaller molecules can penetrate into a much larger volume of the porous gel, and they remain in the column for a longer period of time. The concentration of polymer passing through the column is recorded continuously as a function of time by measurement of refractive index or infrared light absorption. SEC is a relative method. Calibration with narrow molar mass fractions of the polymer studied is necessary. It is also possible to use standards of another polymer and then by calculation, using the Mark–Houwink parameters of the polymers, to convert the molar mass scale of the calibrant polymer to that of the polymer studied. This procedure is known as 'universal calibration'. The hydrodynamic volume is proportional to the product $[\eta]M$. If calibration is done with polymer (calibrant index 2), the universal calibration procedure

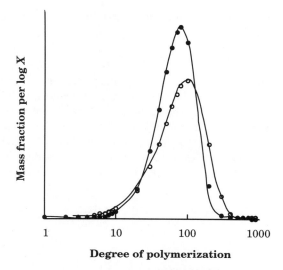

Figure 1.13 Theoretical chain length distribution curves based on: ● the Schultz distribution; ○ the Schultz–Flory distribution. Both distributions have $X_n = 50$.

is carried out according to eq. (1.12), derived as follows:

$$[\eta]_1 M_1 = [\eta]_2 M_2$$
$$K_1 M_1^{1+a_1} = K_2 M_2^{1+a_2}$$
$$M_1 = \left[\frac{K_2}{K_1} M_2^{1+a_2}\right]^{1/1+a_1} \quad (1.12)$$

where index 1 refers to the polymer studied.

Details of the light scattering method are given in Chapters 2 and 12.

An alternative way of describing the molecular size is by using the degree of polymerization (X) which is related to the molar mass (M) as follows:

$$X = \frac{M}{M_{rep}} \quad (1.13)$$

where M_{rep} is the molar mass of the constitutional repeating unit. It is useful to define the same kind of averages for X as are used for molar mass. Different polymerization methods yield polymers with different molar mass distributions. A few illustrative examples are shown in Fig 1.13. The following chain-length distribution was originally derived by Flory for step-growth polymerization (section 1.7):

$$n_i = \frac{1}{\bar{X}_n}\left(1 - \frac{1}{\bar{X}_n}\right)^{i-1} \quad (1.14)$$

where n_i is the number fraction of molecules of $X = i$ and $\bar{X}_n$ is the number average of the degree of polymerization. This distribution is called the **most probable** (or **Schultz–Flory**) distribution. For chain-growth radical polymerization (section 1.7) with recombination exclusively through recombination, the X distribution is described by the **Schultz distribution**:

$$n_i = \frac{4i}{(\bar{X}_n - 1)^2}\left[\frac{1}{1 + \dfrac{2}{\bar{X}_n - 1}}\right]^i \quad (1.15)$$

1.7 POLYMERIZATION

The polymerization process can in simple terms be divided into **step-growth** and **chain-growth** polymerization.

A typical example of step-growth polymerization is the formation of a polyester from a hydroxycarboxylic acid:

2 HO–R–COOH ⇌
   HO–R–CO–O–R–COOH + H$_2$O

HO–R–COOH + HO–R–CO–O–R–COOH ⇌
   HO–R–CO–O–R–CO–O–R–COOH + H$_2$O

etc.

The kinetics of polymerization is not affected by the size of the reacting species. The number of reactive groups, in this case hydroxyl groups and acid groups, decreases with increasing length of the molecules. At any given moment, the system will consist of a mixture of growing chains and water. One difficult problem is that all reactions are reversible, with an equilibrium being established for each reaction. The consumption of reactive groups and the formation of a high molar mass polymer requires the removal of water from the system. The degree of polymerization X_n is equal to $1/(1 - p)$, where p is the degree of consumption of reactive groups (yield). As yield varies, the following X_n values are obtained:

Yield	0.10	0.9	0.99	0.999	0.9999
X_n	1.1	10	100	1000	10 000

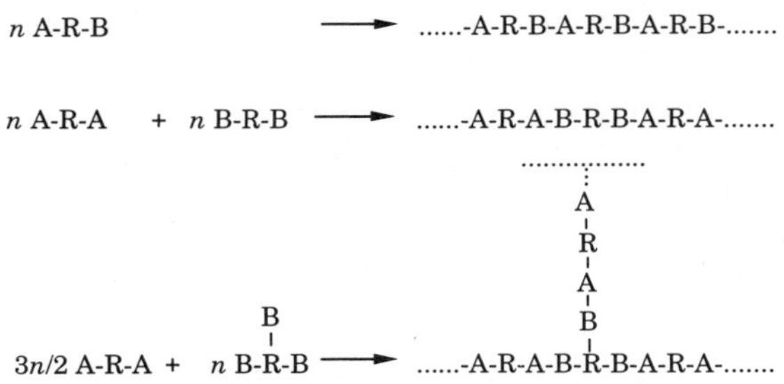

n A-R-B ⟶ ....-A-R-B-A-R-B-A-R-B-.......

n A-R-A + n B-R-B ⟶ ....-A-R-A-B-R-B-A-R-A-.......

$3n/2$ A-R-A + n B-R-B ⟶ ....-A-R-A-B-R-B-A-R-A-.......
with branches: A–R–A–B at B positions

Figure 1.14 Different molecular architectures arising from different combinations of monomers of different functionalities.

Step-growth polymerization is involved in the formation of, e.g., polyesters and polyamides. Different techniques are available for obtaining a high yield and high molar mass. If an acid chloride is used instead of the carboxylic acid, HCl is formed instead of water. The former is more easily removed from the system, and higher yields and molar masses are obtained. The long reaction time needed to reach a high yield is a considerable disadvantage with step-growth polymerization. Polymers with different molecular architectures can be made using monomers of different functionality (Fig. 1.14). Tri-functional monomers yield branched and ultimately crosslinked polymers.

Chain-growth polymerization involves several consecutive stages: initiation, propagation and termination. Each chain is individually initiated and grows very rapidly to a high molar mass until its growth is terminated. At a given time, there are essentially only two types of molecule present: monomer and polymer. The number of growing chains is always very low. Chain-growth polymerization is divided into several subgroups depending on the mechanism: radical, anionic, cationic or coordination polymerization. A generalized scheme for radical polymerization is shown below.

The initiation is accomplished by thermal or UV-initiated degradation of an organic peroxide or similar unstable compound (initiator). Free radicals are generated which attack the double bond of the unsaturated monomer (typically a vinyl monomer: CH$_2$=CHX) and the radical centre is moved to the end of the 'chain'.

$$\text{ROOR} \rightarrow 2\,\text{RO}\cdot$$

$$\text{RO}\cdot + \text{M} \rightarrow \text{ROM}\cdot$$

Propagation is a chain reaction involving very rapid addition of monomer to the radicalized chain.

$$\text{ROM}\cdot + \text{M} \rightarrow \text{ROMM}\cdot$$

$$\text{ROMM}\cdot + \text{M} \rightarrow \text{ROMMM}\cdot$$

etc.

It is possible that a radicalized chain abstracts a hydrogen from an adjacent polymer molecule and that the reactive site is then moved from one molecule to another (chain transfer). This leads in most cases to the formation of a molecule with a long-chain branch. The chain transfer reaction may also be intramolecular, which is a well-known mechanism giving branches in high-pressure polyethylene.

$$\begin{array}{c} X \\ | \\ R_1\text{-CH}_2\text{-C}\cdot \\ | \\ H \end{array} + \begin{array}{c} X \\ | \\ R_2\text{-CH}_2\text{-C-R}_3 \\ | \\ H \end{array} \rightarrow$$

$$\begin{array}{c} X \\ | \\ R_1\text{-CH}_2\text{-C-H} \\ | \\ H \end{array} + \begin{array}{c} X \\ | \\ R_2\text{-CH}_2\text{-C-R}_3 \\ | \\ \cdot \end{array}$$

$$\begin{array}{c} X \\ | \\ R_2\text{-CH}_2\text{-C-R}_3 \\ | \\ \cdot \end{array} \xrightarrow{+nM} \begin{array}{c} X \\ | \\ R_2\text{-CH}_2\text{-C-R}_3 \\ | \\ R_4 \end{array}$$

The propagation is stopped either by combination or by disproportionation

$$R_1M\cdot + \cdot MR_2 \rightarrow R_1MMR_2 \quad \text{(combination)}$$

$$\begin{array}{c} X \\ | \\ R_1\text{-CH}_2\text{-C}\cdot \\ | \\ H \end{array} + \begin{array}{c} X \\ | \\ \cdot\text{C-CH}_2\text{-R}_2 \\ | \\ H \end{array} \rightarrow$$

$$\begin{array}{c} X \\ | \\ R_1\text{-CH}_2\text{-C-H} \\ | \\ H \end{array} + \begin{array}{c} X \\ | \\ \text{C=CH-R}_2 \\ | \\ H \end{array}$$

(disproportionation)

Both anionic and cationic polymerization include both initiation and chain-wise propagation but with one important difference from radical polymerization: in neither case is there any natural termination reaction. In both cases it is possible to achieve the conditions for living polymerization, where all chains are constantly growing until all monomer is consumed and the molar mass distribution of the resulting polymer is narrow. It is also possible to add a new monomer and to prepare exact block copolymers. The presence of small traces of an impurity, e.g. water, leads to chain transfer reactions and a termination of the growing polymer chains.

Coordination polymerization was discovered in the 1950s by Ziegler (1955) and Natta (1959). This technique makes it possible to produce stereoregular polymers such as isotactic polypropylene and linear polyethylene. A Ziegler–Natta catalyst requires a combination of the following substances: (i) a transition metal compound from groups IV–VIII; (ii) an organometallic compound from groups I–III; and (iii) a dry, oxygen-free, inert hydrocarbon solvent. Commonly used systems have involved aluminium alkyls and titanium halides. The polymerization is often rapid and exothermic, requiring external cooling. The polymer normally precipitates around the catalyst suspension. The initial work of Ziegler and Natta involved unsupported catalysts, whereas much of the commercial polymer made by coordination polymerization is achieved by supported catalysts. From a mechanistic point of view they are of the same class as the Ziegler–Natta catalysts. The pioneering work was due to Hogan and Banks using activated chromic oxides on silica supports. This development led to the polymerization of linear polyethylene and linear low-density polyethylene (copolymers of ethylene and higher 1-alkenes).

1.8 THERMAL TRANSITIONS AND PHYSICAL STRUCTURES

It is useful to divide the polymers into two main classes: the **fully amorphous** and the **semicrystalline**. The fully amorphous polymers show no sharp, crystalline Bragg reflections in the X-ray diffractograms taken at any temperature. The reason why these polymers are unable to crystallize is commonly their irregular chain structure. Atactic polymers, statistical copolymers and highly branched polymers belong to this class of polymers (Chapter 5).

The semicrystalline polymers show crystalline Bragg reflections superimposed on an amorphous background. Thus, they always consist of two components differing in degree of order: a crystalline component composed of thin (10 nm) lamella-shaped crystals and an amorphous component. The degree of crystallinity can be as high as 90% for certain low molar mass polyethylenes and as low as 5% for polyvinlychloride. Chapters 7 and 8 deal with the semicrystalline polymers.

A third, recently developed group of polymers, is the **liquid-crystalline** polymers showing orientational order but not positional order. They are thus intermediates between the amorphous and the crystalline polymers. A detailed discussion of liquid crystalline polymers is given in Chapter 6.

The differences in crystallinity lead to differences in physical properties. Figure 1.15 shows, for example, the temperature dependence of the relaxation modulus for different polystyrenes. The relaxation modulus is defined as the stress divided by the strain as recorded after 10 s of constant straining, a so-called stress relaxation experiment.

At 100°C, the fully amorphous polystyrenes show a drop in modulus by a factor of 1000. This 'transition' is called the **glass transition**. All fully amorphous polymers show a similar modulus–temperature curve around the glass transition temperature (T_g). The material is said to be 'glassy' at temperatures below T_g (region I). Under these conditions, they are hard plastics with a modulus close to 3 GPa. The glassy polymer is believed to show very little segmental mobility. Conformational changes are confined to small groups of atoms. The deformation is predominantly due to stretching of secondary bonds and bond angle deformation ('frozen spaghetti deformation').

The glass transition shows many kinetic peculiarities and it is not a true thermodynamic phase transition like melting of a crystal. This is one of the subjects of Chapter 5. In region II, the transitional region, the polymer shows damping. Such materials are referred to as 'leatherlike'.

At temperatures above T_g, the materials are rubber-like with a modulus of a few megapascals (region III). Above the glass transition temperature relatively large groups of atoms, of the order of 100 main chain atoms, can change their conformation. Crosslinked materials show elastic properties in this temperature region. The rate at which the conformational changes occur is so high that the strain response to a step stress is instantaneous. This rubber elastic behaviour is treated in Chapter 3. Uncrosslinked polymers show a pronounced drop in modulus at higher temperatures. The temperature region at which the modulus remains practically constant, the so-called **rubber plateau**, is much longer for the high molar mass material than for the low molar mass material (Fig. 1.15). This indicates that the low modulus characteristic of materials in region V is due to sliding motions of molecules which occur more readily in low molar mass polymers with few chain entanglements. Crosslinked polymers show no region V behaviour because the crosslinks prevent the sliding motion.

Semicrystalline polystyrene shows a weak glass transition at 100°C (Fig. 1.15), due to a softening of the amorphous component of this two-phase polymer. The fraction of crystalline component was not reported by Tobolski, but it was probably about 20%. The crystalline component remains unchanged by the glass transition. The crystallites act as crosslinks and the rubber modulus of the amorphous component is higher than that of the wholly amorphous polystyrene. The glass transition is hardly visible in high-crystalline polymers such as polyethylene. The pronounced drop

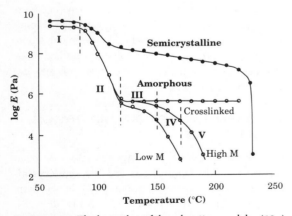

Figure 1.15 The logarithm of the relaxation modulus (10 s) as a function of temperature for semicrystalline (isotactic) polystyrene and fully amorphous (atactic) polystyrene in three 'versions': low molar mass uncrosslinked, and high molar mass uncrosslinked and crosslinked. Drawn after data from Tobolski (1960, p. 75).

in modulus occurring at 230°C is due to the melting of the crystalline component. The melting and crystallization of semicrystalline polymers is an important part of polymer physics and is treated in Chapters 7 and 8.

1.9 POLYMER MATERIALS

It is common to divide plastic materials into **thermoplastics** and **thermosets**. Thermoplastics are composed of linear or branched polymer molecules, and for that reason they melt. Thermoplastics are first synthesized and then at a later stage moulded. Thermosets are crosslinked polymers that do not melt. An uncrosslinked prepolymer is given the desired final shape and the polymer is crosslinked at a later stage while it is kept in the mould.

The properties of a polymer material are determined by the structure of the polymers used, the additives and the processing methods and conditions. It is possible to make an extremely stiff and strong fibrous material from polyethylene. Conventionally processed polyethylene has a stiffness of only about 1 GPa, whereas fibrous polyethylene may exhibit a longitudinal modulus of 100 GPa. Some polymer materials are almost pure with only a small content of additives, whereas others consist of predominantly non-polymeric constituents. Composites consist of reinforcing fibres and the function of the polymer is merely to provide the shape of the product and to transfer forces from one fibre to another. The reinforcing fibres give the material its high strength and stiffness.

This book deals primarily with the polymers, but nevertheless additives play an important role. Polymers would not be used to the extent they are if it were not for the additives. Some polymers such as polyethylene may only contain a small portion of antioxidant to prevent the polymer from oxidizing. Other polymers, particularly for rubbers, contain both large numbers and large amounts of additives, e.g. antioxidants, carbon black, oil, fillers, reinforcing fibres, initiators, an accelerator for vulcanization, and an inhibitor in order to avoid early crosslinking. An increasingly important field is the prevention of fire without the use of halogen-containing polymers. This is accomplished with the use of small-molecule fire retardants. Some polymers such as polyvinylchloride are used with plasticizers, i.e. miscible low molar mass liquids. Numerous polymeric materials contain large fractions of fillers. The purpose of using fillers can be cost reduction, reduced mould shrinkage, promotion of nucleation and improvement of mechanical properties. The list of additives used in polymers is extended but, for our purposes, the list presented here is sufficient.

1.10 A SHORT HISTORY OF POLYMERS

It is intended in this section to give only a very short presentation of the development of polymer materials and ideas in polymer science. A detailed presentation of this field is given by Morawetz (1985).

The first polymers used were all obtained from natural products. Natural rubber from *Hevea* trees was being used by the American Indians when Columbus arrived in 1492. Cellulose in different forms, starch and collagen in leather are other examples of natural polymers used. Modification of native polymers started in the mid-nineteenth century and the first wholly synthetic polymer was made at the beginning of the twentieth century. The science of polymers began in the 1920s.

The development of polymer science and technology has occurred primarily during the last 60–70 years and the commercial introduction of new polymers has proceeded through three time stages giving rise to three generations of polymers.

The first generation was introduced before 1950 and includes polystyrene, polyvinylchloride, low-density polyethylene, polyacrylates, polymethacrylates, glass-fibre reinforced polyesters, aliphatic polyamides, styrene-butadiene rubber and the first synthetic paints (alkyds).

The second generation of polymers was introduced during 1950–65 and includes a number of engineering plastics such as high-density polyethylene, isotactic polypropylene, polycarbonates, polyurethanes, epoxy resins, polysulphones and aromatic polyesters, also used for films and fibres. New rubber materials, acrylic fibres made of polyacrylonitrile and latex paint were also introduced.

The third generation, introduced since 1965, consists mainly of speciality polymers with a more complex chemical structure. These polymers were characterized by very high thermal and chemical stability and high strength/stiffness. Examples are poly(phenylene sulphide) (Ryton®), polyaryletherketone (PEEK®), polyimides (Kapton®), aromatic polyesters (Ekonol® and Vectra®), aromatic polyamides (Nomex® and Kevlar®), and fluor-containing polymers (Teflon® and Viton®). Parallel to this development of new polymers, existing polymers such as polyethylene have undergone significant improvement. Crosslinked polyethylene and new fracture-tough thermoplastic polyethylenes are examples of the more recently introduced materials.

A 'polymer calendar' is presented below with the important breakthroughs indicated. Many important accomplishments have been omitted to keep the list reasonably short.

1844: Charles Goodyear discovered that sulphur-containing natural rubber turned elastic after heat treatment. Vulcanization was discovered and utilized.

1862: Alexander Parks modified cellulose with nitric acid to form cellulose nitrate and, by mixing this polymer with a plasticizer, he made a material named Parkesine. A few years later, a similar material named celluloid (cellulose nitrate plasticized with camphor) was patented by John and Isaiah Hyatt.

1905: Leo Baekeland made Bakelite, the first wholly synthetic polymer. Bakelite is a thermoset made from phenol and formaldehyde.

1920: The macromolecular concept was formulated by Hermann Staudinger. The idea had been presented by Staudinger at a lecture in 1917. However, the concept of large molecules was not new at that time. Peter Klason, a Swedish chemist, had reported in 1897 that lignin in wood was formed mainly from coniferyl alcohol units connected to 'large molecules', mainly by ether bonds. During the 1920s the macromolecular idea was under debate with Staudinger in favour and a relatively large group of scientists against the new idea. Staudinger received the Nobel Prize in 1953.

1930s: Werner Kuhn, Herman Mark and Eugene Guth found evidence that polymer chains in solution were flexible and that the viscosity in solution was related to the molar mass of the polymer.

1934– : The statistical mechanical theory for rubber elasticity was first qualitatively formulated by Werner Kuhn, Eugene Guth and Herman Mark. The entropy-driven elasticity was explained on the basis of conformational states. The initial theory dealt only with single molecules, but later development by these pioneers and by other scientists formulated the theory also for polymer networks. The first stress–strain equation based on statistical mechanics was formulated by Eugene Guth and Hubert James in 1941.

1930s: Wallace Carothers, a research chemist at DuPont, USA, studied polycondensation reactions, synthesizing first aliphatic polyesters, and later and more importantly polychloroprene and polyamide 6,6 (Nylon). Carothers's research not only supported the macromolecular concept but also showed the industrial importance of synthetic polymers.

1930s: Paul Flory derived and experimentally confirmed the Gaussian molar mass distribution for polymers made by step-growth polymerization. Later Flory showed that polymers made by anionic polymerization adapted to the narrower Poisson distribution of chain lengths. Flory also postulated the existence of chain-transfer reactions in chain-growth polymerization. Flory received the Nobel Prize for chemistry in 1974 for these and later fundamental achievements in the physical chemistry of macromolecules.

1933: Styrene-butadiene rubber was made in Germany.

1936: Epoxy resins were made by Pierre Castan (Switzerland).

1938: Silicone rubbers were made by Eugene Rochow (USA).

1939: Polytetrafluoroethylene (Teflon®) was made by Roy Plunkett (USA).

1942: Paul Flory and Maurice Huggins presented, independently, the thermodynamics theory for polymer solutions.

1940s: Flory was very active in many areas during this decade. He made his contribution to rubber elasticity together with Rehner, developed a theory for gelation by which the gel point can be predicted from the degree of 'conversion', and developed a theory for the excluded volume effect of polymer molecules in solution. Flory introduced the theta solvent concept. Under theta conditions, the polymer molecules have unperturbed dimensions. Flory also predicted that the shape of molecules in a pure melt should be the same as under theta conditions.

1940s: Low-density polyethylene was made by Eric William Fawcett (UK).

1940s: Glass-fibre reinforced polyester was made in Germany.

1950s: Karl Ziegler (Germany) and Guilio Natta (Italy) discovered that polymerization in the presence of certain metal-organic catalysts yielded stereoregular polymers. Both were awarded the Nobel Prize for chemistry in 1963. Their work led to the development of linear polyethylene and isotactic polypropylene. This discovery of Ziegler and Natta's was, however, preceded by the preparation of isotactic polypropylene by Paul Hogen (Phillips Petroleum Co., USA) and linear polyethylene at DuPont, USA.

1949–56: Theories for liquid crystals of rod-like polymers were proposed by Lars Onsager in 1949 and by Paul Flory in 1956.

1956: Michael Szwarc, USA, discovered living anionic polymerization. This technique permitted the preparation of narrow molar mass fractions and 'exact' di- and tri-block copolymers. Thermoplastic elastomers, such as Kraton® (Shell Chemical Co., USA), are prepared by living anionic polymerization.

1957: Andrew Keller, Bristol, UK, found that polymer molecules were folded at the large surfaces of lamella-shaped single crystals of polyethylene. The general shape of the single crystals and the chain axis orientation (but not the explicit expression for chain folding) was also reported by Erhart Fischer and Paul Till in 1957. The first suggestion of chain folding goes, however, back to Keith Storks in 1938 dealing with gutta-percha but it passed largely unnoticed by the scientific society.

1960: Polyoxymethylene (Delrin®) was made by DuPont, USA.

1961: Aromatic polyamide (Nomex®) was made by DuPont, USA.

1962: Blends of poly(phenylene oxide) (PPO) and polystyrene with the commercial name Noryl® (General Electric Co., USA) were first made.

1970–85: Ultra-oriented polyethylene with mechanical properties approaching those of metals was made by solid-state processes, in some cases combined with solution processes. A number of scientists were active in the field: Ian Ward (UK), Roger Porter (USA), Albert Pennings and Piet Lemstra (Netherlands). The pioneering work of elucidating the mechanisms of transformation from the isotropic to the fibrous polymer is due to Anton Peterlin (USA).

1971: Pierre-Gilles de Gennes, a French physicist, who was awarded the Nobel Prize for physics in 1992, presented the reptation model which describes the diffusion of chain molecules in a matrix of similar chain molecules. The reptation model was later further developed by Masao Doi and Sam Edwards.

1972– : The first melt-processable (later categorized as thermotropic liquid-crystalline) polymer, based on p-hydroxybenzoic acid and biphenol terephthalate, was reported by Steven Cottis in 1972. This polymer is now available on the market as Xydar®. In 1973, the first well-characterized thermotropic polymer, a copolyester of p-hydroxybenzoic acid and ethylene terephthalate, was patented by Herbert Kuhfuss and W. Jerome Jackson (Eastman-Kodak Co., USA). They reported the discovery of liquid-crystalline behaviour in this polymer in 1976. At the beginning of the 1980s, the Celanese Company developed a family of processable thermotropic liquid crystalline polymers based on hydroxybenzoic acid and hydroxynaphthoic acid, later named Vectra®.

Mid-1970s– : Theories for the crystallization of polymers were introduced by John Hoffman and coworkers, and later in the 1980s by David Sadler, University of Bristol (UK).

1977: Stefanie Kwolek and Paul Morgan, research chemists at DuPont, reported that solutions of poly(phenylene terephthalamide) could be spun to superstrong and stiff fibres. They showed that the solutions possessed liquid-crystalline order. The fibres were later commercialized under the name Kevlar®.

1977: The first electrically conductive polymer was prepared by doping of polyacetylene by Alan MacDiarmid, Alan Heeger and Hideka Shirikawa.

1978: The German scientists Heino Finkelmann, Michael Happ, Michael Portugall and Hellmut Ringsdorf suggested that a decoupling of the main chain and the mesogen motions was possible through the insertion of a flexible spacer in side-chain liquid-crystalline polymers. Since this breakthrough, an abundance of side-chain polymers have been synthesized and the combinations of main chains, spacers and mesogens appear to be infinite.

1980s: The molecular interpretation of relaxation processes in polymers developed strongly using molecular mechanics modelling particularly due to the work of Richard Boyd, USA. David Bassett and associates at the University of Reading (UK) introduced the permanganic etching technique, which led to a strong development in the understanding of the morphology of polymers.

1.11 SUMMARY

This chapter should be considered as a basic introduction to polymer science necessary for the understanding of the following chapters. Fundamental concepts are introduced. The difference between configuration and conformation is explained. Polymer synthesis is briefly described. Molar mass averages are defined and the methods used for their measurement are briefly presented. The properties of polymer materials are determined not only by their polymer constituents but also by the low molar mass additives and by the processing methods used. The major thermal transitions are briefly described. The use of (native) polymers is many thousands of years old, but it took until the beginning of the twentieth century for the first wholly synthetic polymer (Bakelite) to be made. The history of polymer science began in 1917 with Staudinger's introduction of the macromolecular concept. Finally, a list of important subsequent accomplishments in polymer science and technology is presented.

1.12 EXERCISES

1.1. When was the first synthetic polymer made? What polymer was it?

1.2. Write the constitutional repeating unit structures of the following polymers: PE, PP, PMMA and PA 8.

1.3. Explain briefly the difference between the concepts of configuration and conformation.

1.4. Why cannot atactic polystyrene crystallize?

1.5. What are the main differences between step-growth and chain-growth polymerization?

1.6. Is it possible to make isotactic polystyrene by radical polymerization?

1.7. What is the name of the technique that reveals the entire molar mass distribution?

1.8. Explain why measurement of colligative properties yields the number average molar mass.

1.13 REFERENCES

Boyd, R. H. and Phillips, P. J. (1993) *The Science of Polymer Molecules*. Cambridge University Press, Cambridge.

Mark, H. (1938) *Der Feste Korper*, 103.

Metanomski, W. V. (ed.) (1991) *Compendium of Macromolecular Nomenclature*. Blackwell Scientific, Oxford.

Morawetz. H. (1985) *Polymers: The Origins and Growth of a Science*. Wiley, New York.

Natta, G. and Pasguan, I. (1959) in *Advances in Catalysis and Related Subjects* (Eley, D. D., Sellwood, P. W. and Weisz, B., eds) **II**, p. 2, Academic Press, New York.

Tobolski, A. V. (1960) *Properties and Structure of Polymers*. Wiley, New York.

Ziegler, K., Holzkamp, E., Breil, H. and Martin, H. (1955) *Angew. Chem.* **67**, 541.

1.14 SUGGESTED FURTHER READING

Rånby, B. (1993) 'Background – polymer science before 1977', in *Conjugate Polymers and Related Structures*, Nobel Symposium 81. Oxford University Press, Oxford.

Young, R. J. and Lowell, P. A. (1990) *Introduction to Polymers*, 2nd edn. Chapman & Hall, London.

CHAIN CONFORMATIONS IN POLYMERS

2.1 INTRODUCTION

A polymer molecule can take many different shapes (**conformations**) primarily due its degree of freedom for rotation about σ bonds. Studies of the heat capacity of ethane (CH_3–CH_3) indicate that the bond linking the carbon atoms is neither completely rigid nor completely free to rotate. Figure 2.1 shows the different rotational positions of ethane as viewed along the C–C bond. The hydrogen atoms repel each other, causing energy maxima in the **eclipsed** position and energy minima in the stable **staggered** position. The torsion angle may be defined as in Fig. 2.1, $\phi = 0$ for the eclipsed position and $\phi > 0$ for clockwise rotation round the further carbon atom. Some authors set $\phi = 0$ for the staggered position. In both cases, the value of ϕ is independent of the viewing direction (turning the whole molecule round).

Figure 2.2 shows the conformational energy plotted as a function of the torsion angle and the energy difference between the stable staggered position. The energy barrier (eclipsed) is equal to 11.8 kJ mol^{-1} which may be compared with the thermal energy at room temperature, $RT \approx 8.31 \times 300$ J mol$^{-1} \approx 2.5$ kJ mol^{-1}.

The alkane with additional two carbon atoms, n-butane (CH_3–CH_2–CH_2–CH_3), has different **stable** conformational states, referred to as **trans** (T) and **gauche** (G and G'), as shown in Fig. 2.3. The conformational energy 'map' of n-butane is shown in Fig. 2.4. The energy difference between the trans and gauche states is 2.1 ± 0.4 kJ mol^{-1}. Calculations and experiments have shown that there is an angular displacement by 5–10° of the gauche states from their 120° angle towards the trans state, i.e. the gauche states are located at 110–115° from the trans state. The energy barrier between the trans and the gauche states is 15 kJ mol^{-1}. The energy barrier between the two gauche states is believed to be very high, but its actual value is not precisely known.

Normal pentane has two rotational bonds and hence potentially nine combinations, but only six of them are distinguishable: TT, TG, TG', GG, G'G' and GG'. The conformations GT, G'T and G'G are identical with TG, TG' and GG'. Two pairs of mirror-images are present, namely TG and TG' and GG and G'G'. The energy for the conformation GG' is much greater than predicted from the data presented in Fig. 2.4 because of strong steric repulsion of the two CH_3 groups separated by three CH_2 groups (Fig. 2.5). The dependence of the potential energy of one σ bond on the actual torsion angle of the nearby bonds is referred to as a **second-order interaction**.

The **rotational isomeric state approximation**, which is a convenient procedure for dealing with the conformational states of polymers, was introduced by Flory. Each molecule is treated as existing only in

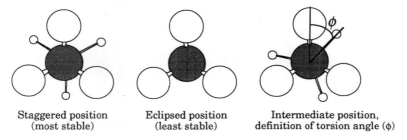

Staggered position (most stable)   Eclipsed position (least stable)   Intermediate position, definition of torsion angle (φ)

Figure 2.1 Rotational isomers of ethane from a view along the C–C bond: carbon – shaded; hydrogen – white.

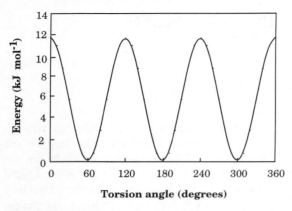

Figure 2.2 Conformational energy of ethane as a function of torsion angle.

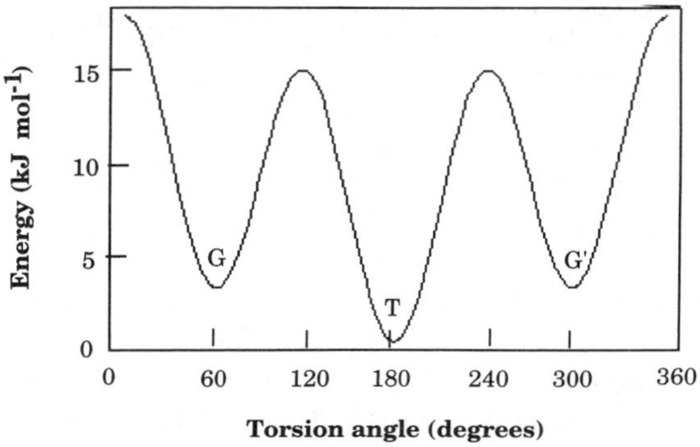

Figure 2.4 Conformational energy of n-butane as a function of torsion angle of the central carbon–carbon bond. The outer carbon–carbon bonds are assumed to be in their minimum energy states (staggered positions).

discrete torsional angle states corresponding to the potential energy minima, i.e. to different combinations of T, G and G'. Fluctuations about the minima are ignored. This approximation means that the continuous distribution over the torsional angle space ϕ is replaced by a distribution over many discrete states. This approximation is well established for those bonds with barriers substantially greater than the thermal energy (RT).

Let us now consider an alkane with n carbons. The question is how many different conformations this molecule can take. The molecule with n carbons has $n - 1$ σ main-chain bonds. The two end bonds do not contribute to different conformations, whereas each of the other carbon–carbon bonds is in one of the three rotational states T, G and G'. The number of different conformations following this simple scheme is 3^{n-3}. A typical polymer molecule may have 10 000 carbons and thus $3^{9997} \approx 10^{4770}$ conformations, i.e. an enormously large number of states. However, this treatment is oversimplified. First, due to symmetry, the number of distinguishable conformations is less than 3^{9997}, although it is correct in order of magnitude. Second, the energy of certain conformations is very high, e.g. those containing GG', giving them a very low statistical weight. The energy map shown in Fig. 2.4 is of limited applicability for predicting the probability of conformations in polyethylene. The

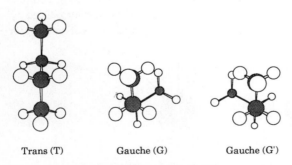

Trans (T)   Gauche (G)   Gauche (G')

Figure 2.3 Conformational states of n-butane. Note that the views of the gauche conformers are along the middle carbon–carbon bond. Carbon – shaded; hydrogen – white.

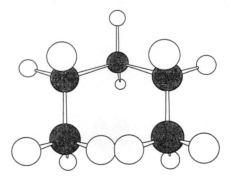

Figure 2.5 Illustration of the steric repulsion in the high-energy GG' conformer in n-pentane: carbon – shaded; hydrogen – white.

interdependence of the torsion angle potentials, as demonstrated in the high-energy GG' sequence, has to be considered. These **higher-order interactions** are discussed later in this chapter.

Polymer chains exhibit in several cases a **random** chain conformation, i.e. a random distribution of trans and gauche states. Figure 2.6 illustrates the idea of a random (so-called **Gaussian**) chain. Random macromolecular chains are found in solutions of polymers in good solvents, in polymer melts and probably also in glassy amorphous polymers. Crystalline polymers, on the other hand, consist of long sequences of bonds with a regular arrangement of energetically preferred chain conformations interrupted by chain folds or by statistical sequences.

The first part of this chapter deals with the statistics of the Gaussian chain. Expressions for the characteristic dimension of the random chain (average **end-to-end distance** or **radius of gyration**) are derived as a function of molar mass, chain flexibility and temperature.

The particularly simple relationships between the average end-to-end distance of the random coil and the chain length that are derived in section 2.4 are valid under the ideal solution conditions referred to as theta conditions. The dimension of the unperturbed polymer chain is only determined by the short-range effects and the chain behaves as a 'phantom' chain that can intersect or cross itself freely. It is important to note that these conditions are also met in the pure polymer melt, as was first suggested by Flory (the so-called Flory theorem) and as was later experimentally confirmed by small-angle neutron scattering.

The statistical variation of the end-to-end distance is considered in so-called random flight analysis. This analysis forms the basis for one of the most prominent theories in polymer physics, the theory of rubber elasticity, which is presented in Chapter 3. The second part of Chapter 2 introduces the concept of 'preferred chain conformation', i.e. the state of the molecule in a crystal.

It is essential to understand that an ensemble of random chains can only be described by means of a spatial distribution function. Two different measures of the random chain are commonly used: (a) the end-to-end distance (r); and (b) the radius of gyration (s). The former is simply the distance between the chain ends. The radius of gyration is defined as the root-mean-square distance of the collection of atoms from their common centre of gravity:

$$s^2 = \frac{\sum_{i=1}^{n} m_i \bar{r}_i^2}{\sum_{i=1}^{n} m_i} \quad (2.1)$$

where $\bar{r}_i$ is the vector from the centre of gravity to atom i. Debye showed many years ago that for large values of n, i.e. for polymers, the following relationship holds between the second moment of the mean values:

$$\langle s^2 \rangle = \frac{\langle r^2 \rangle}{6} \quad (2.2)$$

Random coils can thus be characterized by either of the two dimensions, the average end-to-end distance or the radius of gyration. The average end-to-end distance is used in the rest of this chapter to characterize the random chain.

2.2 EXPERIMENTAL DETERMINATION OF DIMENSIONS OF CHAIN MOLECULES

The size of random chain coils, predominantly in solutions of polymers, can be obtained by several experimental techniques. The size of the molecular coil of a particular polymer is dependent on the solvent (Fig. 2.7). A good solvent expands the coil. A poor solvent, on the other hand, causes shrinkage. In between these two extremes, so-called theta solvents are found (see Chapter 4). Typical of these

Figure 2.6 Gaussian chain.

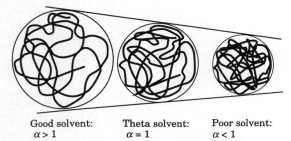

Good solvent: $\alpha > 1$ Theta solvent: $\alpha = 1$ Poor solvent: $\alpha < 1$

Figure 2.7 Random coils in solvents of different solvent power; α is the linear coil expansion factor which according to the definition is equal to 1 in a theta solvent.

are that **inter**molecular and **intra**molecular interactions are similar in magnitude. The further discussion of chain dimensions refers to theta conditions. Light scattering and viscometry are probably the two most commonly used methods for determining the coil size. The scattering of light is caused by the difference in refractive index between solvent and solute (polymer) and its angular dependence is a function of coil size. The viscosity is sensitively dependent on the hydrodynamic volume of the coil, and is briefly presented. The arrangements of macromolecules in concentrated solutions or in the solid state have more recently been studied by means of small-angle neutron scattering (SANS). The conformation, i.e. r or s, of deuterated polymer chains in host protonated chains is determined by SANS.

The following equation describes the scattering of light by solutions of polymers as a function of the size of the random coil, described here by the average of the square of the end-to-end distance (r) and the mass average molar mass $(\bar{M}_w)$:

$$\frac{Kc}{R_\theta} = \frac{1}{\bar{M}_w} + \frac{1}{\bar{M}_w} \frac{16}{3} \pi^2 \frac{\langle r^2 \rangle}{6\lambda^2} \sin^2 \frac{\theta}{2} + 2A_2 c \quad (2.3)$$

where

$$K = \begin{cases} \dfrac{4\pi^2 n_1^2}{N_A \lambda_0^4} \left(\dfrac{\partial n}{\partial c}\right)_T f & \text{(vertical polarization)} \\ \dfrac{4\pi^2 n_1^2}{N_A \lambda_0^4} \left(\dfrac{\partial n}{\partial c}\right)_T f \cos^2 \theta & \text{(horizontal polarization)} \\ \dfrac{4\pi^2 n_1^2}{N_A \lambda_0^4} \left(\dfrac{\partial n}{\partial c}\right)_T f \left(\dfrac{1+\cos^2 \theta}{2}\right) & \text{(unpolarized)} \end{cases}$$

n_1 is the refractive index of the solvent, c is the concentration of polymer, N_A is the Avogadro number, λ_0 is the wavelength *in vacuo*, λ is the wavelength in the medium, θ is the scattering angle, f is a correction factor, A_2 is a constant and R_θ is the reduced scattered intensity defined as

$$R_\theta = \frac{r^2 I_\theta}{I_0}$$

where r is the distance between the sample and the point at which the intensity I_θ is recorded and I_0 is the intensity of the incoming light.

Equation (2.3) permits the separate determination of $\bar{M}_w$ and $\langle r^2 \rangle$ which is schematically shown in the so-called Zimm plot (Fig. 2.8). They are obtained simply from:

$$\lim_{\theta \to 0} \frac{Kc}{R_\theta} = \frac{1}{\bar{M}_w} + 2A_2 c \quad (2.4)$$

and

$$\lim_{c \to 0} \frac{Kc}{R_\theta} = \frac{1}{\bar{M}_w} + \frac{1}{\bar{M}_w} \frac{16}{3} \pi^2 \frac{\langle r^2 \rangle}{6\lambda^2} \sin^2 \frac{\theta}{2} \quad (2.5)$$

The viscometry relies on the relationship:

$$[\eta] = \frac{\Phi(\langle r^2 \rangle)^{3/2}}{M} = \left(\frac{\eta_r - 1}{c}\right)_{c \to 0} \quad (2.6)$$

where $[\eta]$ is the limiting viscosity number with η_r being the relative viscosity. Φ is a constant which is the same for all systems; an experimental and theory-based value for near-theta conditions is $\Phi = 2.6 \times 10^{21}$ with r in centimetres and $[\eta]$ in dl g^{-1}.

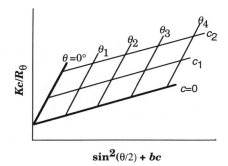

Figure 2.8 Schematic Zimm plot.

2.3 CHARACTERISTIC DIMENSIONS OF 'RANDOM COIL' POLYMERS

The dimensions of different polymer chains depend strongly on the type of environment and temperature. A good solvent expands the coil. A poor solvent, on the other hand, causes shrinkage. Between these two extremes, so-called theta solvents are found (illustrated in Fig. 2.7; see Chapter 4). Typical of these is that intermolecular and intramolecular interactions are similar in magnitude. All polymers dissolved in theta solvents adapt to the following equation:

$$\langle r^2 \rangle_0 = Cnl^2 \quad (2.7)$$

where the subscript signifies theta conditions, and C is a constant which depends on the nature of the polymer. It is possible to make a distinction between short-range and long-range interactions. The short-range interaction is described by the examples presented in section 2.1, where $\langle r^2 \rangle_0$ is determined only by the short-range effects. The long-range intramolecular interaction, illustrated in Fig. 2.9, causes expansion of the coil, which can be expressed by a linear expansion factor α:

$$\langle r^2 \rangle = \alpha^2 \langle r^2 \rangle_0 \quad (2.8)$$

The expansion factor (α) is also affected by temperature and type of solvent. It has been shown both theoretically and experimentally that α is a function of the molar mass in good solvents:

$$\alpha^5 - \alpha^3 = C\psi\sqrt{n}\left(1 - \frac{\theta}{T}\right) \quad (2.9)$$

where C is a polymer-related constant, ψ is the interaction entropy (see Chapter 4) and θ is the theta temperature or Flory temperature. At theta conditions ($T = \theta$), α becomes 1, whereas in a good solvent ($T > \theta$), α is proportional to $n^{1/10}$. By combining eqs (2.8) and (2.9), the following expression is obtained,

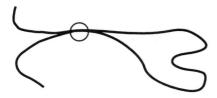

Figure 2.9 Steric interference between two chain segments widely separated in sequence along the chain.

where C_1 and C_2 are constants:

$$\langle r^2 \rangle = C_1 n^{1/5} \langle r^2 \rangle_0 = C_2 n^{1/5} n = C_2 n^{6/5} \quad (2.10)$$

i.e. $r \propto n^{3/5}$. These arguments provide an exponent which is very close to the experimental values reported for solutions of polymers in good solvents, $r \propto n^{0.59}$. It must be stressed that at $T = \theta$, $\langle r^2 \rangle = \langle r^2 \rangle_0$ and the latter shows a molar mass dependence according to eq. (2.7). At this temperature, the enthalpic and entropic contributions from solvent–polymer and polymer–polymer interactions do not appear. The volume expansion caused by the long-range polymer–polymer interactions is exactly compensated for by the solvent–polymer interaction under theta conditions. The segments of a molecule under theta conditions are arranged in a way which indicate that they do not 'sense' the other segments of the same molecule. The molecules behave like 'ghosts' or 'phantoms' and are sometimes also referred to as phantom chains.

Flory proposed that the polymer molecules in the molten state are unperturbed (phantom-like) as they are in a theta solvent and that the same simple equation, eq. (2.7) relating $\langle r^2 \rangle_0$ with chain length (n), should hold also for molten polymers. Small-angle neutron scattering data were available many years later, and indeed they support the Flory theorem. A particularly elegant reasoning explaining the Flory theorem is found in de Gennes (1979). Consider a dense system of identical chains. One of the chains is focused. Let us call this molecule M1. The repeating units of M1 are subjected to a repulsive potential (denoted U) created by the excluded volume (long-range) effects from its own repeating units. This potential would lead to an expansion of the coil of M1 from the unperturbed molecule size if it were not for the surrounding identical molecules M2, M3, etc. The latter generate a counteracting, attractive potential acting inward on molecule M1 of exactly the same size as the repulsive potential. Chain M1 is thus subjected to no net force and remains unperturbed.

Table 2.1 presents data for polymers under theta conditions and it turns out that polymers with flexible backbones exhibit low C values. The flexible ether groups in polyethyleneoxide and the bulky pendant phenyl groups of polystyrene explain the shifts in C

Table 2.1 C values for some polymers under theta conditions

Polymer	$C (M = \infty)^a$
Polyethylene	6.7
Polyethyleneoxide	4.0
Polystyrene, atactic	10.0

Source: Flory (1989)
[a] See eq. (2.7).

value of these polymers with respect to that of polyethylene.

The temperature expansion of the unperturbed chain dimensions $d(\ln\langle r^2\rangle_0)/dT$ is preferably determined by measuring the force in a sample of a lightly crosslinked network polymer as a function of elongation at different temperatures. The fundamental theoretical aspects are explored further in Chapter 4. Table 2.2 presents data collected by Mark (1976).

The decrease in coil dimension with increasing temperature found for polyethylene is expected; for computational details, see the first part of section 2.4.3. The high-energy gauche states are more populated at higher temperatures than at low temperatures, causing a decrease in end-to-end distance with increasing temperature. Atactic polystyrene exhibits expanding coils with increasing temperature The bulky phenyl group causes the extended all-trans conformation to be less energetically favourable than angular states and the 'extended' states become more populated at elevated temperatures. The preferred conformation of poly(dimethyl siloxane) is the all-trans state. This conformation is, however, not extended because the Si–O–Si and O–Si–O bond angles are different. More extended conformations are obtained by inclusion of gauche states along the chain. The latter are more frequent at higher temperatures.

2.4 MODELS FOR CALCULATING THE AVERAGE END-TO-END DISTANCE FOR AN ENSEMBLE OF STATISTICAL CHAINS

2.4.1 THE FREELY JOINTED CHAIN

Figure 2.10 illustrates the model: a chain consisting of n segments (main-chain bonds), each bond having a length l. The end-to-end vector ($\bar{r}$) is the sum of the individual bond vectors according to:

$$\bar{r} = \sum_{i=1}^{n} \bar{r}_i \qquad (2.11)$$

The square end-to-end distance (r^2) becomes:

$$r^2 = \sum_{i=1}^{n} \bar{r}_i \sum_{j=1}^{n} \bar{r}_j = \sum_{i=1}^{n} \bar{r}_i^2 + 2 \sum_{i=1}^{n-1} \sum_{j=i+1}^{n} \bar{r}_i \bar{r}_j \qquad (2.12)$$

Equation 2.12 is valid for any polymer chain regardless of structure. We have to consider an ensemble of N chains each comprising n segments. The average of the squared end-to-end distance, $\langle r^2\rangle$, is equal to:

$$\langle r^2\rangle = \frac{1}{N} \sum_{k=1}^{N} \bar{r}_{k^*}^2 = \sum_{i=1}^{n} \langle \bar{r}_i^2\rangle + 2 \sum_{i=1}^{n-1} \sum_{j=i+1}^{n} \langle \bar{r}_i \bar{r}_j\rangle$$

$$= \begin{bmatrix} \langle \bar{r}_1\bar{r}_1\rangle + \langle \bar{r}_1\bar{r}_2\rangle + \cdots + \langle \bar{r}_1\bar{r}_n\rangle + \\ \langle \bar{r}_2\bar{r}_1\rangle + \langle \bar{r}_2\bar{r}_2\rangle + \cdots + \langle \bar{r}_2\bar{r}_n\rangle + \\ \cdots \quad \cdots \quad \cdots \quad\quad \cdots\cdots\cdots \quad + \\ \cdots \quad \cdots \quad \cdots \quad\quad \cdots\cdots\cdots \quad + \\ \langle \bar{r}_n\bar{r}_1\rangle + \langle \bar{r}_n\bar{r}_2\rangle + \cdots + \langle \bar{r}_n\bar{r}_n\rangle \end{bmatrix}$$

$$(2.13)$$

The first term $\sum_{i=1}^{n} \langle \bar{r}_i^2\rangle$ is the sum of the diagonal elements in the square array, whereas

$$\sum_{i=1}^{n-1} \sum_{j=i+1}^{n} \langle \bar{r}_i \bar{r}_j\rangle$$

Table 2.2 Temperature coefficients of unperturbed chain dimensions

Polymer	Temp. (°C)	$10^3 \, d(\ln\langle r^2\rangle)/dT \, (K^{-1})$
Polyethylene	140–190	−1.1
Atactic polystyrene	120–170	+0.3
Polydimethylsiloxane	40–100	+0.8

Source: Mark (1976).

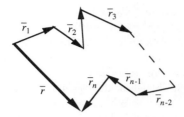

Figure 2.10 Definition of quantities in a jointed-chain model.

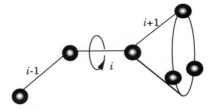

Figure 2.11 Three different rotational isomers are generated by torsion about bond i. If the three states are equally populated, the average vector of bond $i+1$ has no component perpendicular to a vector parallel to the ith bond.

constitutes the sum of the elements above the diagonal of the array.

The scalar product of the arbitrary segment vectors $\bar{r}_i$ and $\bar{r}_j$ is:

$$\langle \bar{r}_i \bar{r}_j \rangle = l^2 \langle \cos \theta_{ij} \rangle \quad (2.14)$$

where θ_{ij} is the angle between the two bond vectors. The following relationship is obtained by combining eqs (2.13) and (2.14):

$$\langle r^2 \rangle = nl^2 + 2l^2 \sum_{i=1}^{n-1} \sum_{j=i+1}^{n} \langle \cos \theta_{ij} \rangle$$

$$= \begin{bmatrix} l^2 & + l^2 \langle \cos \theta_{12} \rangle + \cdots + l^2 \langle \cos \theta_{1n} \rangle + \\ l^2 \langle \cos \theta_{21} \rangle + & l^2 & + \cdots + l^2 \langle \cos \theta_{2n} \rangle + \\ \cdots & \cdots & \cdots & \cdots & \cdots & + \\ \cdots & \cdots & \cdots & \cdots & \cdots & + \\ l^2 \langle \cos \theta_{n1} \rangle + l^2 \langle \cos \theta_{n2} \rangle + \cdots + & l^2 \end{bmatrix} \quad (2.15)$$

Equation (2.15) is still a general formulation and is valid for any continuous polymer chain.

The **freely jointed chain** consists of a chain of bonds: the orientation of the different bonds is completely uncorrelated. No direction is preferred.

Thus, insertion of $\langle \cos \theta_{ij} \rangle = 0$ for $i \neq j$ in eq. (2.15) leads to:

$$\langle r^2 \rangle = nl^2 \quad (2.16)$$

The molar mass dependence of the end-to-end distance follows a square-root law.

2.4.2 THE FREELY ROTATING CHAIN

The model of the **freely rotating chain** assumes that the bond angle (τ) is constant. No particular chain conformation is preferred and the average projection of bond $i+1$ along a direction perpendicular to bond i is zero (Fig. 2.11). In this case, $\langle \cos \theta_{ij} \rangle \neq 0$ also for $i \neq j$, and

$$\langle \bar{r}_i \bar{r}_{i+1} \rangle = l^2 \cos(180 - \tau)$$
$$\langle \bar{r}_i \bar{r}_{i+2} \rangle = l^2 \cos^2(180 - \tau)$$
$$\langle \bar{r}_i \bar{r}_j \rangle = l^2 [\cos(180 - \tau)]^{j-i}$$

which after combination with eqs (2.13) and (2.15) gives:

$$\langle r^2 \rangle = nl^2 + 2l^2 \sum_{i=1}^{n-1} \sum_{j=i+1}^{n} [\cos(180 - \tau)]^{j-i}$$

$$= \begin{bmatrix} l^2 & + l^2 \cos(180 - \tau) + \cdots + l^2 [\cos(180 - \tau)]^{n-1} + \\ l^2 \cos(180 - \tau) + & l^2 & + \cdots + l^2 [\cos(180 - \tau)]^{n-2} + \\ \cdots & \cdots & \cdots & \cdots & + \\ \cdots & \cdots & \cdots & \cdots & + \\ l^2 [\cos(180 - \tau)]^{n-1} + & \cdots & \cdots & l^2 \end{bmatrix} \quad (2.17)$$

The summation can be performed over a single variable (k) by substituting $j - i$ by k:

$$\langle r^2 \rangle = nl^2 \left[1 + \frac{2}{n} \sum_{k=1}^{n-1} (n-k)\alpha^k \right] \quad (2.18)$$

where $\alpha = \cos(180 - \tau)$.

Equation (2.18) can be simplified as follows:

$$\langle r^2 \rangle = nl^2 \left[1 + \frac{2}{n} \sum_{k=1}^{n-1} (n-k)\alpha^k \right]$$

$$= nl^2 \left[1 + 2 \sum_{k=1}^{n-1} \alpha^k - \frac{2}{n} \sum_{k=1}^{n-1} k\alpha^k \right]$$

$$= nl^2 \left[1 + \frac{2(\alpha - \alpha^n)}{1 - \alpha} - \frac{2}{n} \left(\frac{\alpha(1-\alpha)^n}{(1-\alpha)^2} - \frac{n\alpha^n}{1-\alpha} \right) \right]$$

$$= nl^2 \left[1 + \frac{2\alpha}{1 - \alpha} - \frac{2\alpha}{n} \frac{(1-\alpha)^n}{(1-\alpha)^2} \right]$$

For infinitely long chains ($n = \infty$):

$$\langle r^2 \rangle = nl^2 \left[1 + \frac{2\alpha}{1-\alpha} \right] = nl^2 \left[\frac{1+\alpha}{1-\alpha} \right]$$

$$= nl^2 \left[\frac{1 + \cos(180 - \tau)}{1 - \cos(180 - \tau)} \right] \quad (2.19)$$

Insertion of the bond angle (τ) value for an sp^3 hybridized carbon of 110° gives the following degree-of-polymerization dependence of a methylene chain with free rotation:

$$\langle r^2 \rangle \approx 2nl^2$$

2.4.3 THE CHAIN OF HINDERED ROTATION

Chain with independent torsion angle potential

The freely rotating chain is a good approximation for a real polymer chain at high temperatures when the energy difference between trans and gauche states is small compared to RT. At lower temperatures, the low-energy states become more populated than the high-energy states, which for a polyethylene chain leads to a decrease in the low-energy trans population with increasing temperature. The low-temperature conformation of polymethylene is more extended than is predicted by eq. (2.19).

Initially, an expression for a chain with independent torsion angle potentials is derived, e.g. only first-order interactions are considered. In order to derive a relationship between the second moment of the end-to-end vector and molar mass, it is useful to consider the individual bonds as different Cartesian coordinate systems (CS_i) – see Fig. 2.12.

The following equation can be used to transform a vector $\bar{V}$ expressed in CS_{i+1} to its representation $\bar{V}'$ in CS_i:

$$\bar{V}' = \mathbf{T}_i \bar{V} \quad (2.20)$$

where $\mathbf{T}_i$ is the following second-order Cartesian tensor:

$$\mathbf{T}_i = \begin{bmatrix} \cos \theta_i & \sin \theta_i & 0 \\ \sin \theta_i \cos \phi_i & -\cos \theta_i \cos \phi_i & \sin \phi_i \\ \sin \theta_i \sin \phi_i & -\cos \theta_i \sin \phi_i & -\cos \phi_i \end{bmatrix} \quad (2.21)$$

The transformation tensor can be understood by considering that the unit vector along x_{i+1} has the following coordinates in CS_i:

$$\begin{bmatrix} x_i \\ y_i \\ z_i \end{bmatrix} = \begin{bmatrix} \cos \theta_i \\ \sin \theta_i \cos \phi_i \\ \sin \theta_i \sin \phi_i \end{bmatrix}$$

The following notation in CS_i is valid for (0, 1, 0) and (0, 0, 1) in CS_{i+1}:

$$\begin{bmatrix} x_i \\ y_i \\ z_i \end{bmatrix} = \begin{bmatrix} \sin \theta_i \\ -\cos \theta_i \cos \phi_i \\ -\cos \theta_i \sin \phi_i \end{bmatrix} \quad \text{for (0, 1, 0)}$$

$$\begin{bmatrix} x_i \\ y_i \\ z_i \end{bmatrix} = \begin{bmatrix} 0 \\ \sin \phi_i \\ -\cos \phi_i \end{bmatrix} \quad \text{for (0, 0, 1)}$$

Any vector $\bar{V} = (v_x, v_y, v_z)$ in CS_{i+1} has the following coordinates in CS_i:

$$v'_x = v_x \cos \theta_i + v_y \sin \theta_i + v_z 0$$

$$v'_y = v_x \sin \theta_i \cos \phi_i - v_y \cos \theta_i \cos \phi_i + v_z \sin \phi_i$$

$$v'_z = v_x \sin \theta_i \sin \phi_i - v_y \cos \theta_i \sin \phi_i - v_z \cos \phi_i$$

that is to say

$$\bar{V}' = \mathbf{T}_i \bar{V}$$

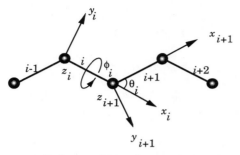

Figure 2.12 Definition of CS_i and CS_{i+1}; the torsion angle ϕ is set to zero for the planar trans conformation. Axis y_i is in the plane of bonds i and $i-1$, and z_i completes the right-handed Cartesian coordinate system.

The scalar product $\bar{r}_i\bar{r}_j$ can be strictly formulated using the following matrix notations:

$$\bar{r}_i\bar{r}_j = \bar{r}_i^T(T_i \ldots T_{j-1})\bar{r}_j \qquad (2.22)$$

where

$$\bar{r}_i^T = (1, 0, 0)$$

and

$$\bar{r}_j = \begin{pmatrix} 1 \\ 0 \\ 0 \end{pmatrix}$$

The scalar product $\bar{r}_i\bar{r}_j$ thus becomes:

$$\bar{r}_i\bar{r}_j = l^2(T_i \ldots T_{j-1})_{11}$$

where the subscript denotes the (1, 1)th element of the product tensor. For a chain with free rotation ($\langle\cos\phi_i\rangle = 0$ and $\langle\sin\phi_i\rangle = 0$), the average T_i tensor is simplified to:

$$T_i^* = \begin{bmatrix} \cos\theta_i & \sin\theta_i & 0 \\ 0 & 0 & 0 \\ 0 & 0 & 0 \end{bmatrix} \qquad (2.23)$$

The scalar product $\bar{r}_i\bar{r}_j$ becomes:

$$\bar{r}_i\bar{r}_j = l^2(T_i^* \ldots T_{j-1}^*)_{11} = l^2(\cos\theta)^{j-i} \qquad (2.24)$$

which was derived in section 2.4.2. A chain with restricted rotation has a preference for certain rotational isomers, e.g. a polyethylene chain prefers the trans state ($\phi = 0$ according to the earlier definition). For a symmetric molecule, it may be argued that isomers of torsion angle $\phi = +x$ are of the same population as those of $\phi = -x$ which implies that $\langle\sin\phi_i\rangle = 0$. The transformation matrix becomes:

$$T_i = \begin{bmatrix} \cos\theta & \sin\theta & 0 \\ \sin\theta\langle\cos\phi\rangle & -\cos\theta\langle\cos\phi\rangle & 0 \\ 0 & 0 & -\langle\cos\phi\rangle \end{bmatrix}$$

$$(2.25)$$

We may then use eq. (2.25) and implement the matrices in the following equation:

$$\langle r^2 \rangle = nl^2 + 2\sum_{i=1}^{n-1}\sum_{j=i+1}^{n} \langle\bar{r}_i\bar{r}_j\rangle$$

$$= nl^2 + 2l^2(1, 0, 0)\left[\sum_{k=1}^{n-1}(n-k)T_i^k\right]\begin{bmatrix} 1 \\ 0 \\ 0 \end{bmatrix}$$

$$(2.26)$$

This series of matrices can be treated as scalar quantities and the series converges just as was shown for scalar quantities (see eq. (2.19)) to:

$$\langle r^2 \rangle = nl^2\left[\frac{E + \langle T\rangle}{E - \langle T\rangle}\right]_{11} \qquad (2.27)$$

where E is the unit matrix:

$$E = \begin{bmatrix} 1 & 0 & 0 \\ 0 & 1 & 0 \\ 0 & 0 & 1 \end{bmatrix} \qquad (2.28)$$

By combining eqs (2.25), (2.27) and (2.28), an expression valid for a chain with **hindered rotation** with independent torsion angle potentials is obtained:

$$\langle r^2 \rangle = nl^2\left[\frac{1 + \cos(180 - \tau)}{1 - \cos(180 - \tau)}\right]\left[\frac{1 + \langle\cos\phi\rangle}{1 - \langle\cos\phi\rangle}\right]$$

$$(2.29)$$

The temperature dependence of $\langle r^2 \rangle$ originates from the temperature dependence of $\langle\cos\phi\rangle$, as may be illustrated by the following example. It is here assumed that the concentrations of the three possible rotational isomers T, G and G' of the polyethylene chain are dependent only on the energy levels of the three rotational isomers and that the energy level of the rotational state of a certain bond is not influenced by the torsion angles of the surrounding bonds, i.e. only first-order interactions are considered.

The rotational partition function z is a measure of the number of rotational states which the system can adopt at the temperature of interest, which at $T = 0$ K is equal to 1 and increases with increasing temperature. For n-butane it is equal to the sum of the statistical weights (u_η, see definition in eq. 2.36) of the possible conformations, i.e.:

$$z = 1 + \sigma + \sigma = 1 + 2\sigma \quad (2.30)$$

where

$$\sigma = e^{-E_g/RT} \quad (2.31)$$

E_g being the energy difference between the gauche and the trans states. Provided that the rotational potential of the bonds is independent of the actual torsion angles of the nearby bond, the average $\langle f \rangle$ of any function $f(\phi)$ is given by

$$\langle f \rangle = \frac{\sum_\eta u_\eta f(\phi_\eta)}{z} \quad (2.32)$$

which may be applied to $\cos \phi$:

$$\langle \cos \phi \rangle = \frac{\sum_\eta u_\eta \cos \phi_\eta}{z}$$

$$= \frac{1 + \sigma \cos(120°) + \sigma \cos(-120°)}{1 + \sigma + \sigma}$$

$$= \frac{1 - \sigma}{1 + 2\sigma} \quad (2.33)$$

which is inserted in eq. (2.29) to give:

$$\langle r^2 \rangle = nl^2 \left[\frac{1 + \cos(180 - \tau)}{1 - \cos(180 - \tau)} \right] \left[\frac{2 + \sigma}{3\sigma} \right] \quad (2.34)$$

At 140°C, using $E_g = 2.1$ kJ mol^{-1}, $\sigma = 0.54$; the second moment of the end-to-end distance becomes:

$$\langle r^2 \rangle = nl^2 \times 2 \times \frac{2 + 0.54}{3 \times 0.54} = 3.4 nl^2$$

which is lower than the experimentally obtained $(6.7 \pm 0.1)nl^2$ (Table 2.1). Agreement with experimental data is obtained by also considering higher-order interactions. Flory showed that an analysis using second-order interactions brings the predicted data closer to the experimental data.

Equation (2.34) predicts that the end-to-end distance is a function of both chain flexibility (controlled by E_g) and the temperature (T). The last factor in eq. (2.34) approaches unity at high temperatures, and eq. (2.34) then becomes identical with eq. (2.19). Polyethylene with an extended low-energy state becomes increasingly more coiled with increasing temperature. Figure 2.13 shows that the end-to-end distance predicted from eq. (2.34) of polyethylene ($E_g = 2.1$ kJ mol^{-1}) decreases with increasing temperature and approaches the value of the chain of free rotation. It is also shown in the same graph that the end-to-end distance increases with increasing energy difference between the trans and gauche states.

Figure 2.14 demonstrates the disordering of the polymer chain which occurs at elevated temperatures. Figure 2.15 illustrates the effect of a change in the energy difference between trans and gauche states on the chain conformation, i.e. the static chain flexibility. A novel type of polymer, the **liquid-crystalline**

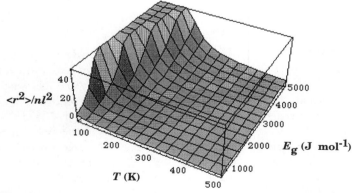

Figure 2.13 Normalized end-to-end distance $\langle r^2 \rangle / nl^2$ as a function of temperature and energy difference between G and T (E_g).

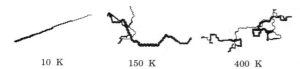

Figure 2.14 Simulated polyethylene chains ($E_g = 2.1$ kJ mol^{-1}; $n = 200$; temperature as shown in figure 1 using software developed by Nairn (1990)).

Figure 2.15 Simulated chains ($T = 300$ K; $n = 200$) using software developed by Nairn (1990). Energy difference values between straight and angular states are given in graph.

polymers, consists of very rigid segments exhibiting very low chain flexibility. The rigid-rod molecules self-align locally along a certain common **director** at certain temperatures and/or certain polymer concentrations. These polymers are isotropic overall provided that no external, aligning field has been applied.

Chain with interdependent bonds: statistical weight matrices and end-to-end distances for interdependent bonds

This presentation essentially follows Flory's (1989) treatment. The conformation of a chain with n bonds, each bond having three possible torsion angles, may be specified by $n - 2$ components according to:

$$2113131\ldots$$

It is here assumed that the potential of a given bond i depends only on the states of the adjacent bonds $i - 1$ and $i + 1$. The total conformation energy of this chain sequence is then:

$$E\{\phi\} = E_2 + E_{21} + E_{11} + E_{13} + E_{31} + E_{13} + E_{31} + \cdots$$

which can be expressed in more general terms as:

$$E\{\phi\} = \sum_{i=2}^{n-1} E_i(\phi_{i-1}, \phi_i) = \sum_{i=2}^{n-2} E_{\zeta\eta;i} \quad (2.35)$$

where ζ is the state of bond $i - 1$ and η is the state of bond i. The first term is indexed by η only. The interaction between bonds i and $i + 1$ is taken into consideration in the next term.

A few important aspects of eq. (2.35) should be noted. The energy $E_{\zeta\eta;i}$ is determined by the preceding bond being in its assigned state, ϕ_i, and the succeeding bond being tentatively in its trans state. Thus interactions due to bonds $i - 1$ and $i + 1$, these being in states ϕ_ζ and ϕ_T (energy = 0), are included in the term $E_{\zeta\eta;i}$. The effect of the procedure of setting bond $i + 1$ to its trans state, is removed when the next term of the series is evaluated. Adoption of this procedure makes $E_{\zeta\eta;i} = 0$ if η is trans, irrespective of ζ.

For polyethylene, the following energies are obtained:

$$E_{\zeta T} = 0 \text{ for } \zeta = T, G, G'$$

$$E_{TG} = E_{TG'} = E_{GG} = E_{G'G'} = 2.1 \text{ kJ mol}^{-1}$$

$$E_{GG'} = E_{G'G} = 12.4 \text{ kJ mol}^{-1}$$

The following problem demonstrates the method. The problem is to calculate the conformational energy (in kilojoules per mole) for conformations of n-pentane.

$$E = \sum_{i=2}^{3} E_{\zeta\eta;i} = E_{\eta;2} + E_{\zeta\eta;3}$$

TT: $E = E_{T;2} + E_{TT;3} = 0 + 0 = 0$

TG: $E = E_{T;2} + E_{TG;3} = 0 + 2.1 = 2.1$

TG': $E = E_{T;2} + E_{TG';3} = 0 + 2.1 = 2.1$

GG: $E = E_{G;2} + E_{GG;3} = 2.1 + 2.1 = 4.2$

G'G': $E = E_{G';2} + E_{G'G';3} = 2.1 + 2.1 = 4.2$

GG': $E = E_{G;2} + E_{GG';3} = 2.1 + 12.4 = 14.5$

The statistical weight $u_{\zeta\eta;i}$ corresponding to the energy $E_{\zeta\eta;i}$ of a certain conformation $\zeta\eta$ is defined by:

$$u_{\zeta\eta;i} = \exp(-E_{\zeta\eta;i}/RT) \quad (2.36)$$

The statistical weights of the nine states originating from the rotation about two adjacent bonds are conveniently expressed in the **statistical weight matrix**

$$\mathbf{U}_i = [u_{\zeta\eta}]_i \quad (2.37)$$

with states ζ for bond $i - 1$ indexing rows and η for bond i indexing columns. The statistical weight matrix of polyethylene is:

$$U = \begin{array}{c} \\ T \\ G \\ G' \end{array} \begin{array}{c} T \quad G \quad G' \end{array} \begin{bmatrix} 1 & \sigma & \sigma \\ 1 & \sigma & 0 \\ 1 & 0 & \sigma \end{bmatrix} \quad (2.38)$$

The statistical weight $\exp(-E_{GG'}/RT)$ is approximated by zero. Generalization to describe any symmetric chain, i.e. a chain containing no asymmetric carbons, with a three-fold rotational symmetry gives the following statistical weight matrix:

$$U = \begin{bmatrix} 1 & \sigma & \sigma \\ 1 & \sigma\psi & \sigma\omega \\ 1 & \sigma\omega & \sigma\psi \end{bmatrix} \quad (2.39)$$

where $\sigma\psi$ describes the GG (or G'G') interaction and $\sigma\omega$ describes the GG' (or G'G) interaction.

The symmetry of the chain requires that:

$$u_{12} = u_{13}; \ u_{21} = u_{31}; \ u_{22} = u_{33}; \ u_{23} = u_{32}$$

If all succeeding bonds are in a trans state:

$$u_{11} = u_{12} = u_{13}$$

Are higher-order interactions important? GG'G has a very serious steric overlap but since it contains GG' its low frequency is considered in the 'two-bond model' GGG'G' is also a 'bad' conformation with steric overlap, but again it contains the GG' pair.

The statistical weight of a certain conformation in the chain molecule is:

$$\Omega_{\{\phi\}} = \prod_{i=2}^{n-1} u_{\zeta\eta;i} \quad (2.40)$$

and the conformation partition function for a chain with n bonds is:

$$Z = \sum_{\{\phi\}} \Omega_{\{\phi\}} = \sum_{\{\phi\}} \prod_{i=2}^{n-1} u_{\zeta\eta;i} \quad (2.41)$$

This evaluation of Z by summing all possible conformations is a gigantic task and can only be accomplished for very short chains. Other methods must be used. Matrix multiplication and a treatment which was used for the Ising ferromagnet can be used. Let us consider a chain with n bonds, each in two rotational states (α or β). The statistical matrices become:

$$U_2 = \begin{bmatrix} u_\alpha & 0 \\ 0 & u_\beta \end{bmatrix} \quad (2.42)$$

$$U = \begin{bmatrix} u_{\alpha\alpha} & u_{\alpha\beta} \\ u_{\beta\alpha} & u_{\beta\beta} \end{bmatrix} \quad (2.43)$$

The partition function is the sum of all elements in the product tensor, i.e.

$$Z = [1, 1]U_2 U^{n-3} \begin{bmatrix} 1 \\ 1 \end{bmatrix} \quad (2.44)$$

The same result is obtained if U is rewritten as

$$U_2 = \begin{bmatrix} u_\alpha & u_\beta \\ 0 & 0 \end{bmatrix} \quad (2.45)$$

and

$$Z = [1, 0]U_2 U^{n-3} \begin{bmatrix} 1 \\ 1 \end{bmatrix} \quad (2.46)$$

which can be simplified to

$$Z = [1, 0]U^{n-2} \begin{bmatrix} 1 \\ 1 \end{bmatrix} \quad (2.47)$$

In a more general form, with more rotational states per bond:

$$Z = J^* \left[\prod_{i=2}^{n-1} U_i \right] J \quad (2.48)$$

where

$$J^* = [1 \ 0 \ \ldots \ 0] \text{ and } J = \begin{bmatrix} 1 \\ 1 \\ \vdots \\ 1 \end{bmatrix} \quad (2.49)$$

For a chain with only one type of bond, this equation simplifies to

$$Z = J^* U^{n-2} J \quad (2.50)$$

This expression can be simplified and converted to an algebraic expression by the similarity expression, by which U is transformed to the diagonal tensor Λ with eigenvalues λ_η of U as elements:

$$A^{-1}UA = \begin{bmatrix} \lambda_1 & 0 & 0 \\ 0 & \lambda_2 & 0 \\ 0 & 0 & \lambda_3 \end{bmatrix} \equiv \Lambda \quad (2.51)$$

This equation may be rewritten as

$$\mathbf{BUA} = \mathbf{\Lambda} \quad (2.52)$$

where $\mathbf{B} = \mathbf{A}^{-1}$, the two matrices being related as:

$$\mathbf{AB} = \begin{bmatrix} 1 & 0 & 0 \\ 0 & 1 & 0 \\ 0 & 0 & 1 \end{bmatrix} \equiv \mathbf{E} \quad (2.53)$$

If eq. (2.52) is premultiplied by $\mathbf{A}$:

$$\mathbf{ABUA} = \mathbf{A}\mathbf{\Lambda}$$
$$\mathbf{EUA} = \mathbf{A}\mathbf{\Lambda} \quad (2.54)$$
$$\mathbf{UA} = \mathbf{A}\mathbf{\Lambda}$$

The latter can be separated into three vector equations:

$$\mathbf{UA}_k = \mathbf{A}_k \lambda_k \quad k = 1, 2, 3 \quad (2.55)$$

where $\mathbf{A}_k$ are the column eigenvectors of $\mathbf{U}$:

$$\mathbf{A}_k = \begin{bmatrix} A_{1k} \\ A_{2k} \\ A_{3k} \end{bmatrix} \quad k = 1, 2, 3 \quad (2.56)$$

If eq. (2.52) is postmultiplied by $\mathbf{B}$:

$$\mathbf{BUAB} = \mathbf{\Lambda B}$$
$$\mathbf{BUE} = \mathbf{\Lambda B} \quad (2.57)$$
$$\mathbf{BU} = \mathbf{\Lambda B}$$

which can be written in the form

$$\mathbf{B}_k^* \mathbf{U} = \lambda_k \mathbf{B}_k^* \quad (2.58)$$

where $\mathbf{B}_k^* = [B_{k1}, B_{k2}, B_{k3}]$ are the eigenrows of $\mathbf{U}$. Since $\mathbf{AB} = \mathbf{E}$, we can write:

$$\mathbf{B}_j^* \mathbf{A}_k = \delta_{jk} \quad (2.59)$$

where δ_{jk} is the Kronecker delta (equal to 1 for $j = k$ and equal to 0 for $j \neq k$). Equation (2.55) can be rewritten as:

$$(\mathbf{U} - \lambda_k \mathbf{E})\mathbf{A}_k = 0 \quad (2.60)$$

which has solution

$$|\mathbf{U} - \lambda_k \mathbf{E}| = 0 \quad (2.61)$$

The expressions

$$\mathbf{BUA} = \mathbf{\Lambda}$$
$$\mathbf{ABUABU} = \mathbf{A}\mathbf{\Lambda B}$$
$$\mathbf{EUE} = \mathbf{A}\mathbf{\Lambda B}$$
$$\mathbf{U} = \mathbf{A}\mathbf{\Lambda B}$$

can be combined with eq. (2.50) to yield:

$$Z = \mathbf{J}^* \mathbf{A}\mathbf{\Lambda}^{n-2} \mathbf{BJ}$$

or

$$Z = [A_{11}\lambda_1^{n-2} \quad A_{12}\lambda_2^{n-2} \quad \ldots \quad A_{1v}\lambda_v^{n-2}] \begin{bmatrix} \sum_{\eta=1}^{v} B_{1\eta} \\ \vdots \\ \sum_{\eta=1}^{v} B_{v\eta} \end{bmatrix} \quad (2.62)$$

The partition function can be written:

$$Z = \sum_{\zeta=1}^{v} \Gamma_\zeta \lambda_\zeta^{n-2} \quad (2.63)$$

where

$$\Gamma_\zeta = A_{1\zeta} \sum_{\zeta=1}^{v} B_{\zeta\eta}$$

which for large values of n can be approximated by

$$Z \cong \Gamma_1 \lambda_1^{n-2} \quad (2.64)$$

where λ_1 is the largest eigenvalue.

At even larger n values, eq. (2.64) simplifies to:

$$Z \cong \lambda_1^{n-2} \quad (2.65)$$

Conformation partition function for chains of three-fold symmetry

If the statistical weight matrix $\mathbf{U}$ is inserted in eq. (2.61):

$$\mathbf{U} = \begin{bmatrix} 1 & \sigma & \sigma \\ 1 & \sigma\psi & \sigma\omega \\ 1 & \sigma\omega & \sigma\psi \end{bmatrix}$$

The following expression is obtained:

$$[\lambda - \sigma(\psi - \omega)]$$
$$\times [\lambda^2 - \lambda(1 + \sigma\psi + \sigma\omega) + \sigma(\psi + \omega - 2)]$$
$$= 0$$

with solutions:

$$\lambda_{1,2} = \tfrac{1}{2}[1 + \sigma(\psi + \omega) \pm \sqrt{[1 - \sigma(\psi + \omega)]^2 + 8\sigma}]$$
$$\lambda_3 = \sigma(\psi - \omega)$$

$$(2.66)$$

with the eigenvector **A** and eigenrow matrices **B** = **A**$^{-1}$ equal to:

$$\mathbf{A} = \begin{bmatrix} 1 - \lambda_2 & -(\lambda_1 - 1) & 0 \\ 1 & 1 & -1 \\ 1 & 1 & 1 \end{bmatrix} \quad (2.67)$$

$$\mathbf{B} = \frac{1}{\lambda_1 - \lambda_2} \begin{bmatrix} 1 & \dfrac{\lambda_1 - 1}{2} & \dfrac{\lambda_1 - 1}{2} \\ -1 & \dfrac{1 - \lambda_2}{2} & \dfrac{1 - \lambda_2}{2} \\ 0 & \dfrac{-(\lambda_1 - \lambda_2)}{2} & \dfrac{\lambda_1 - \lambda_2}{2} \end{bmatrix} \quad (2.68)$$

Combination of eqs (2.63), (2.67) and (2.60) yields:

$$Z = \left[\frac{1 - \lambda_2}{\lambda_1 - \lambda_2}\right]\lambda_1^{n-1} + \left[\frac{\lambda_1 - 1}{\lambda_1 - \lambda_2}\right]\lambda_2^{n-1} \quad (2.69)$$

It is interesting to compare the gauche and trans contents based on interdependent torsion angle potentials with those based on independent torsion angles.

Independent torsion angles: $\Delta E_g = 2.1$ kJ mol^{-1}, $T = 413$ K (section 2.4.3.1). $p_g = \sigma/(1 + 2\sigma) = 0.54/(1 + 2 \times 0.54) = 0.260$; $p_t = 1 - 0.260 \times 2 = 0.480$.

Interdependent torsion angles: $\sigma = 0.54$; $\psi = 1$; $\omega = 0.03$ (12.4 kJ mol^{-1}) inserted in eqs (2.66)–(2.69): $p_t = 0.62$: $P_g = 0.19$.

The first model underestimates the trans content because it does not consider the two bond interactions GG' (G'G).

Flory derived an expression for the second moment of the end-to-end distance but, due to the significant mathematical complexity, the derivation is not presented here. The statistical weights of the conformations $\Omega_{\{\phi\}}$ are introduced in eq. (2.26), which is repeated here:

$$\langle r^2 \rangle = nl^2 + 2l^2 (1, 0, 0) \left[\sum_{k=1}^{n-1} (n-k) \mathbf{T}_i^k \right] \begin{bmatrix} 1 \\ 0 \\ 0 \end{bmatrix}$$

and, after complex matrix mathematics, an expression of the following type is obtained:

$$\langle r^2 \rangle = C(\mathbf{U}) n l^2 \quad (2.70)$$

where C is a constant for a given polymer that depends on the statistical weight matrix (**U**). Flory was able to fit eq. (2.70) to experimental data, ($C = 6.7 \pm 0.2$) using realistic values in **U**.

2.4.4 THE EQUIVALENT CHAIN

The close resemblance between the experimentally established relationship (eq. (2.7)) and the various derived equations indicates that the ideas implemented in the analysis are basically correct. The proportionality constant C is due to short-range interactions. Flexible polymers have only short sequences of bonds with orientational dependence whereas stiff polymers have significantly longer segments of correlated bonds. Hence, a real chain of sufficient length may be represented by an **equivalent chain** comprising n' hypothetical bonds each of length l' connected by **free** joints (Fig. 2.16). The values of n' and l' are obtained by considering that

$$r_{\max} = n' l' \quad (2.71)$$

$$\langle r^2 \rangle = n' l'^2 \quad (2.72)$$

Thus, for polyethylene with $r_{\max} = 0.83 nl$ and $\langle r^2 \rangle_0 = 6.7 nl^2$ (Table 2.1), it follows that $n/n' \approx 10$ real bonds per equivalent segment.

2.4.5 LONG-RANGE INTERACTION

The theoretical analysis presented above has considered only short-range interaction, involving

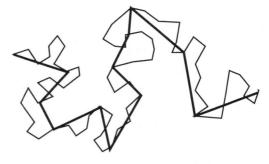

Figure 2.16 Schematic representation of the equivalent chain.

only the spatial, energetic limitations arising from torsions about two adjacent bonds. This simplification is valid, as was pointed out in section 2.3, under theta conditions.

The long-range interactions become apparent in good solvents. Flory showed early that the experimental molar mass dependence of the second moment of the end-to-end distance, $\langle \bar{r}^2 \rangle \propto n^{0.59}$, can be derived according to the following scheme.

The repulsive energy (H_{rep}) is proportional to the volume of the molecular sphere and the number of pairs of bonds:

$$H_{rep} = c_1 c^2 r^3 = c_1 \frac{n^2}{r^6} r^3 = c_1 \frac{n^2}{r^3} \quad (2.73)$$

where c_1 is a constant and c is the concentration of polymer. The conformational entropy (S) is given by:

$$S = c_2 - \frac{r^2}{c_3} \quad (2.74)$$

where c_2 and c_3 are constants. The free energy (G) becomes:

$$G = H - TS = c_1 \left(\frac{n^2}{r^3}\right) + T\left(\frac{r^2}{c_3}\right) + c_4 \quad (2.75)$$

where c_4 is a constant. If the minimum of G is sought by conventional methods, the following equilibrium radius (r^*) is obtained:

$$r^* \propto n^{3/5} \quad (2.76)$$

2.5 RANDOM-FLIGHT ANALYSIS

In section 2.4 expressions were derived for the average end-to-end distance. Random-flight analysis yields an expression for the **distribution** of the end-to-end distance.

It is assumed here that the polymer chain takes discrete steps in three dimensions. For simplicity it is assumed that only *two* types of step exist in each direction: forward and backward. The first task is to determine the average step length. The origin of the chain segment is located in the centre of the sphere shown in Fig. 2.17. The other end of the segment is located on the surface of the sphere and it is assumed that the chain segment may have any direction. All parts of the sphere are equally probable. If we cut the

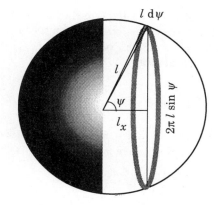

Figure 2.17 The distribution of bonds in real space: x coordinate.

sphere in the middle and consider the probability ($P(l_x)$) that the segment takes a step in the x direction with a length of l_x:

$$P(l_x)dl_x = \frac{2\pi l(\sin \psi) l \, d\psi}{2\pi l^2} = \sin \psi \, d\psi \quad (2.77)$$

$$= \left(\frac{1}{2}\right)^{2k} \frac{(2k)!}{k!k!}$$

$$\langle l_x^2 \rangle = \int_0^1 l_x^2 P(l_x) dl_x = \int_0^{\pi/2} l^2 \cos^2 \psi \sin \psi \, d\psi \quad (2.78)$$

By substitution in eq. (2.78) of $t = \cos \psi$ and $dt = -\sin \psi \, d\psi$

$$\langle l_x^2 \rangle = l^2 \int_0^{-1} -t^2 \, dt = \frac{l^2}{3}$$

$$(\langle l_x^2 \rangle)^{1/2} = \frac{l}{\sqrt{3}} \quad (2.79)$$

The average forward (and backward) step length in any direction (x, y or z) is $l/\sqrt{3}$. The next task is to derive the statistics of positive and negative steps. The end-to-end distance is proportional to the net balance of forward (+) and backward (−) steps. In one dimension (x):

$$x = (n_+ - n_-) \frac{l}{\sqrt{3}} \quad (2.80)$$

where n_+ and n_- are respectively the number of forward and backward steps. The probability of the occurrence of a certain combination of n_+/n_- is binomially distributed:

$$P(n_+, n_-) = \left(\frac{1}{2}\right)^n \frac{n!}{n_+! n_-!} \quad (2.81)$$

The difference between forward and backward steps is denoted m, where $m = n_+ - n_-$:

$$P(n, m) = \left(\frac{1}{2}\right)^n \frac{n!}{\left(\frac{n+m}{2}\right)! \left(\frac{n-m}{2}\right)!} \quad (2.82)$$

The maximum probability occurs for $m = 0$:

$$P(n, 0) = \left(\frac{1}{2}\right)^n \frac{n!}{\left(\frac{n}{2}\right)! \left(\frac{n}{2}\right)!}$$

This equation can be simplified by applying Stirling's approximation ($x! \approx x^x e^x \sqrt{2\pi x}$):

$$P(n, 0) \approx \left(\frac{1}{2}\right)^n \frac{n^n e^n \sqrt{2\pi n}}{\left(\left(\frac{n}{2}\right)^{n/2} e^{n/2} \sqrt{\pi n}\right)^2} = \sqrt{\frac{2}{\pi n}} \quad (2.83)$$

Substitution of n and m in eq. (2.82) according to $2k = n$ and $x = m/2$ leads to the following expression:

$$P(k, x) = \left(\frac{1}{2}\right)^{2k} \frac{(2k)!}{(k+x)!(k-x)!} \quad (2.84)$$

The maximum probability (P_0; for $x = 0$) is given by:

$$P_0 = \left(\frac{1}{2}\right)^{2k} \frac{(2k)!}{k! \, k!} \quad (2.85)$$

The ratio of $P(k, x)$ to P_0 is:

$$\frac{P(k, x)}{P_0} = \frac{k! \, k!}{(k+x)!(k-x)!}$$

$$= \frac{k(k-1)(k-2) \ldots (k-x+1)}{(k+1)(k+2) \ldots (k+x)}$$

$$= \frac{\left(1 - \frac{1}{k}\right)\left(1 - \frac{2}{k}\right) \ldots \left(1 - \frac{(x-1)}{k}\right)}{\left(1 + \frac{1}{k}\right)\left(1 + \frac{2}{k}\right) \ldots \left(1 + \frac{(x-1)}{k}\right)\left(1 + \frac{x}{k}\right)}$$

$$= \frac{\prod_{i=1}^{x-1}\left(1 - \frac{i}{k}\right)}{\prod_{i=1}^{x}\left(1 + \frac{i}{k}\right)} \quad (2.86)$$

Taking the logarithm of eq. (2.86):

$$\ln\left[\frac{P(k, x)}{P_0}\right] = \sum_{i=1}^{x-1} \ln\left(1 - \frac{i}{k}\right) - \sum_{i=1}^{x} \ln\left(1 + \frac{i}{k}\right) \quad (2.87)$$

The Maclaurin expansion,

$$\ln(1 + z) \approx z - \frac{z^2}{2} + \frac{z^3}{3} - \frac{z^4}{4} \approx z \quad (\text{if } z \ll 1)$$

is applied to eq. (2.87):

$$\ln\left[\frac{P(k, x)}{P_0}\right]$$

$$= \sum_{i=1}^{x-1}\left(-\frac{i}{k}\right) - \sum_{i=1}^{x}\left(\frac{i}{k}\right)$$

$$= -\frac{1}{k}\left(\sum_{i=1}^{x-1} i + \sum_{i=1}^{x} i\right)$$

$$= -\frac{1}{k}(2(1 + 2 + 3 + 4 + \cdots + x - 1) + x)$$

$$= -\frac{2}{k}\left(1 + 2 + 3 + 4 + \cdots + x - 1 + \frac{x}{2}\right)$$

$$= -\left(\frac{x(x-1)}{k} - \frac{x}{k}\right) = -\frac{x^2}{k} \quad (2.88)$$

Combining eqs (2.83) and (2.87) gives:

$$P(n, m) = \sqrt{\frac{2}{\pi n}} \exp(-m^2/2n) \quad (2.89)$$

The step length in the x-direction is $2l/\sqrt{3}$ since when n_+ increases by one, n_- has to decrease by one. The relationship between $P(n, m)$ and $P(x)$ becomes:

$$P(n, m) = P(x)\Delta x$$

where $\Delta x = 2l/\sqrt{3}$ and $m = \sqrt{3}\,x/l$ which are inserted in eq. (2.89) giving:

$$P(x)dx = \sqrt{\frac{3}{2\pi}} \left(\frac{1}{\sqrt{n}\,l}\right) \exp(-3x^2/2nl^2)dx \quad (2.90)$$

The same types of expression can be derived for the distribution function in both the y and z directions. It is here assumed that the actual location of the chain end in, for instance, y space does not affect the location of the chain end in the other two dimensions, i.e. $P(x) = f(x)$ only. The probability of finding the chain end in the point (x, y, z) in a chain originating at the origin with the other chain end is given by:

$$P(x, y, z)dx\,dy\,dz = \left(\frac{3}{2\pi nl^2}\right)^{3/2}$$
$$\times \exp(-3(x^2 + y^2 + z^2)/2nl^2)dx\,dy\,dz \quad (2.91)$$

An alternative way of writing eq. (2.91) is:

$$P(x, y, z)dx\,dy\,dz = \left(\frac{3}{2\pi nl^2}\right)^{3/2}$$
$$\times \exp(-3r^2/2nl^2)dx\,dy\,dz \quad (2.92)$$

$P(x, y, z)$ decreases monotonically with increasing r. The most likely occurrence is to find the other chain end at the starting point. By considering that the second moment of the end-to-end distance for a freely jointed chain is equal to nl^2, eq. (2.92) may also be rewritten in the form:

$$P(x, y, z)dx\,dy\,dz = \left(\frac{3}{2\pi \langle r^2 \rangle_0}\right)^{3/2}$$
$$\times \exp(-3r^2/2\langle r^2 \rangle_0)dx\,dy\,dz \quad (2.93)$$

This equation can be applied to any real polymer under theta conditions considering that any such chain can be represented by a hypothetical equivalent chain with n' freely jointed links, each link being of length l' (section 2.4.4). Equation (2.93) is, according to the Flory theorem, also applicable

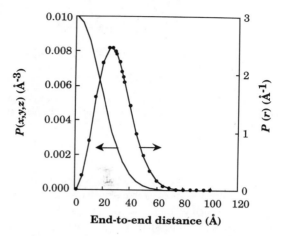

Figure 2.18 Schematic representation of distribution functions: $P(x, y, z)$ and $P(r)$.

to molecules in a molten polymer and to molecules in a rubber polymer. This is important and the basis for the statistical mechanical theory of rubber elasticity presented in Chapter 3.

The radial distribution function $P(r)$ is obtained by multiplying $P(x, y, z)$ with the area of the surface of the sphere with radius r:

$$P(r)dr = P(x, y, z)dx\,dy\,dz\left(\frac{4\pi r^2\,dr}{dx\,dy\,dz}\right)$$
$$= 4\pi r^2 \left(\frac{3}{2\pi \langle r^2 \rangle_0}\right)^{3/2} \exp(-3r^2/2\langle r^2 \rangle_0)dr \quad (2.94)$$

Figure 2.18 shows the shape of the distribution functions. The radial distribution function, $P(r)$, is the product of the monotonically falling $P(x, y, z)$ and the parabolic function r^2.

2.6 CHAINS WITH PREFERRED CONFORMATION

Polymer molecules are found in a preferred conformational state in crystals. The experimental techniques for determining the preferred conformation are mainly X-ray and electron diffraction. The difficult determination of the crystal unit cell must be followed by further molecular mechanical modelling to establish the exact chain conformation.

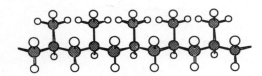

Figure 2.19 Isotactic polypropylene in all-trans conformation showing the steric problem associated with the pendant methyl groups.

Polyethylene obviously shows the simplest polymer structure. The all-trans conformation is energetically the most stable conformation and has been established by numerous diffraction experiments. For polymers with pendant side groups, e.g. isotactic polypropylene (iPP) and isotactic polystyrene (iPS), the extended all-trans conformation is of high energy due to steric repulsion of the side groups (Fig. 2.19).

For iPP, two sequences, /TG/TG/TG/TG/ and /G'T/G'T/G'T/G'T/, have the same minimum conformational energy. Both conformations produce helices. Three polymer repeating units produce a repeating unit in one turn of the helix. This kind of helix is denoted 3_1. The two conformations have different pitches. A view along the helical axis is shown in Fig. 2.20. Other isotactic polymers, e.g. iPS, also prefer the 3_1 TG helix for the same reason as does iPP.

Polyoxymethylene (POM) has no large side groups. The repeating unit of POM is [–CH$_2$–O–]. The lowest-energy conformation is an all-gauche sequence, i.e. ...GGGGGGG... or

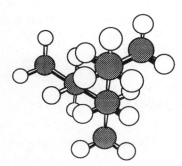

Figure 2.20 View along helical axis of 3_1 helix of isotactic polypropylene. The cross-section of the backbone part of the molecule is triangular and the pendant methyl groups are directed out from the corners of the triangle.

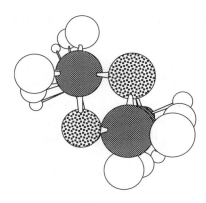

Figure 2.21 View along helical axis of POM (2_1 helix): carbon – shaded; hydrogen – white; oxygen – dotted.

...G'G'G'G'G'G'G'.... This generates a 2_1 helix, i.e. two repeating units complete one helical period in exactly one turn. The conformation of POM is not **exactly** all-gauche; the conformation is somewhat distorted from G and the helix is 9/5 (9 repeating units in 5 turns). Figure 2.21 shows the view along the helical axis of POM.

The reason why G is preferred over T in POM is not fully understood but it is related to the polar character of the C–O bond. In the eclipsed position, the electrostatic attractions between the positive carbon and the negative oxygen are at a maximum. This should contribute to the stabilization of the gauche state over the trans state. The energy difference between T and G is 7 kJ mol^{-1} which is greater than expected from the electrostatic attraction.

2.7 SUMMARY

A polymer molecule can adopt many different shapes primarily due to its degree of freedom for torsion about σ bonds. These states are referred to as conformations. A polymer molecule in a solution, in the molten state and probably also in the glassy, amorphous state, can be characterized as a random coil.

It has been experimentally shown that the second moment of the end-to-end distance ($\langle r^2 \rangle_0$) of unperturbed polymer chains, which only appear under so-called theta conditions, is proportional to the number of bonds (n) and the square of the length of each bond (l):

$$\langle r^2 \rangle_0 = Cnl^2 \quad (2.95)$$

where C is a constant, which depends on the segmental flexibility of the polymer.

This kind of relationship can be derived on the basis of very simple models. For polyethylene, the values for C are for a freely jointed chain $C = 1$, for a freely rotating chain $C = 2$, and for a chain with hindered rotation $C > 2$. Flory showed that it was sufficient to consider the energetics of the torsion about two nearby bonds to obtain agreement between predicted and experimental values ($C = 6.8$ at 410 K for polyethylene in the theta state).

The unperturbed state, i.e. the state of a polymer under theta conditions, is characterized by the absence of long-range interactions. The segments of a molecule under theta conditions are arranged in a way which indicates that they do not 'sense' the other segments of the same molecule. The molecules behave like 'ghosts' or 'phantoms' and are sometimes also referred to as phantom chains. Flory proposed that the spatial extension of polymer molecules in the molten state is the same as in the theta solvent and that the same simple equation (eq. (2.95)) between $\langle r^2 \rangle_0$ and chain length (n) should hold. Small-angle neutron scattering data were available many years later and supported the Flory theorem. In good solvents where, in addition to short-range interactions, long-range interactions also play a role, the perturbed state can be described by the following equation:

$$\langle r^2 \rangle \propto n^{6/5} \quad (2.96)$$

The phantom (unperturbed) polymer chain can be represented by a hypothetical chain with $n' = n/C$ freely jointed segments each of length $l' = Cl$. If n and l are replaced by n' and l' in the equation for the freely jointed chains, eq. (2.95) is obtained.

For phantom chains the distribution function of the end-to-end distance is Gaussian, taking the form:

$$P(x, y, z)dx\,dy\,dz = \left(\frac{3}{2\pi \langle r^2 \rangle_0}\right)^{3/2}$$
$$\times \exp(-3r^2/2\langle r^2 \rangle_0)dx\,dy\,dz \quad (2.97)$$

This Gaussian expression is fundamental to the statistical mechanical theory of rubber elasticity.

Polymer chains in crystals take their preferred conformation, i.e. their low-energy state. Linear polymers with small pendant groups, e.g. polyethylene, exhibit an extended all-trans conformation. Isotactic polymers with the repeating unit –CH_2–CHX– exhibit a helical structure if the X group is sufficiently large (i.e. if X is a methyl group or larger). Even linear polymers with no large pendant group may, due to electrostatic repulsion between nearby dipoles, form a helical structure. Polyoxymethylene belongs to this category.

2.8 EXERCISES

2.1. Calculate the average end-to-end distance for polyethylene with $M = 10^7$ g mol^{-1} at 140°C under theta conditions. Compare this value with the contour length of these molecules.

2.2. Write the different, distinguishable conformations of n-hexane. Calculate the conformational energy of each of them and calculate their statistical weights at 20, 100 and 400 K.

2.3. Compare n-pentane in the GG' state with isotactic PP in an all-trans conformation. Build the molecules using a molecular model and make the comparison.

2.4. Build the preferred conformation of iPP and POM. Use a molecular model.

2.5. Calculate the trans and gauche contents in PE at 20, 100, 200, 300, 400 and 600 K. First, consider only first-order interaction, i.e. independent torsion angle energies. Then consider second-order interactions as well, i.e. the interdependence of torsion about two adjacent bonds. Calculate for each case the constant C in the equation $\langle r^2 \rangle_0 = Cnl^2$.

2.6. Crystalline polymers consist of alternating thin lamellar crystals, typically of a thickness of 10–20 nm, and amorphous interlayers. What is the consequence of the fact that all chains leaving the crystals perform a random walk in the amorphous interlayer before re-entering the crystals?

2.7. Size exclusion chromatography (SEC) is used to determine the molar mass distribution of polymers. SEC is not an absolute method. It requires calibration. Narrow molar mass fractions of atactic polystyrene are commonly used for calibration. The time for a given molecular species to flow through the column is dependent on the hydrodynamic volume of the molecule. It has been shown that the hydrodynamic volume is proportional to the product of the intrinsic viscosity $[\eta]$ and the molar mass M. The Mark–Houwink equation relates the two quantities according to:

$$[\eta] = K \cdot M^a \quad (2.98)$$

where K and a are constants unique for a given combination of polymer, solvent and temperature. Derive the relationship between the molar masses of the polymer studied and of atactic polystyrene for a given eluation time.

2.9 REFERENCES

Boyd, R. H. and Phillips, P. J. (1993) *The Science of Polymer Molecules*. Cambridge University Press, Cambridge.

de Gennes, P.-G. (1979) *Scaling Concepts in Polymer Physics*. Cornell University Press, Ithaca, NY, and London.

Flory, P. J. (1953) *Principles of Polymer Chemistry*. Cornell University Press, Ithaca, NY, and London.

Flory, P. J. (1989) *Statistical Mechanics of Chain Molecules*. Hanser, Munich, Vienna and New York.

Mark, J. E. (1976) *Rubber Chemistry and Technology*, **48**, 495.

Nairn, J. A. (1990) Lattice 4.0™, A Macintosh Application; Salt Lake City, Dept. of Materials Science & Engineering, University of Utah.

THE RUBBER ELASTIC STATE

3.1 INTRODUCTION

Natural rubber is obtained as a latex from a tree called *Hevea Braziliensis*. It consists predominantly of *cis*-1,4-polyisoprene (Fig. 3.1). The word 'rubber' is derived from the ability of this material to remove marks from paper, which was noted by Priestley in 1770. Rubber materials are not, however, restricted to natural rubber. They include a great variety of synthetic polymers of similar properties. An **elastomer** is a polymer which exhibits rubber elastic properties, i.e. a material which can be stretched to several times its original length without breaking and which, on release of the stress, immediately returns to its original length. That is to say, its deformation is reversible.

A very illuminating experiment is to subject a rubber band to about 100% strain by hanging a dead load on to it and then heat the rubber band. The elongation of the rubber band will suddenly decrease when it is heated. This may first seem anomalous, but after reading this chapter you will understand the reason.

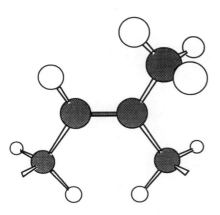

Figure 3.1 Repeating unit of *cis*-1,4-polyisoprene: carbons – shaded; oxygen – white; shaded bond indicates a double bond.

Metals or other highly crystalline materials exhibit Hookean elastic behaviour at strains typically less than 0.2%. Their elasticity is dominantly **energy-driven**. The displacement of the atoms in the lattice accompanying the stress causes an increase in internal energy (Fig. 3.2). The entropy remains approximately constant. If entropic effects are neglected, the elastic force causing the displacement of the atoms from their equilibrium states is equal to the slope in the plot of potential (U) against displacement (r) (Fig. 3.2). The potential exhibits the following approximate r dependence near the equilibrium point:

$$U = C(r - r_0)^2 \quad (3.1)$$

where C is a constant and r_0 is the equilibrium bond length. The elastic force is:

$$f = \frac{\partial U}{\partial r} = 2C(r - r_0) \quad (3.2)$$

The stress (σ) is:

$$\sigma = \frac{2C}{A}(r - r_0) = E\varepsilon \quad (3.3)$$

where $E = 2CR_0/A$, A is the cross-sectional area and ε is the strain. Equation (3.3) is **Hooke's law**.

Rubbers exhibit predominantly **entropy-driven** elasticity. This was discovered by Gough (1805) and later by Lord Kelvin (1857) and Joule (1859) through measurements of force and specimen length at different temperatures. They discovered the thermoelastic effects: (a) that a stretched rubber sample subjected to a constant uniaxial load contracts reversibly on heating; and (b) that a rubber sample gives out heat reversibly when stretched. These observations were consistent with the view that the entropy of the rubber decreased on stretching. The molecular picture of the entropic force originates from Meyer, von Susich and Valko (1932), Kuhn (1934)

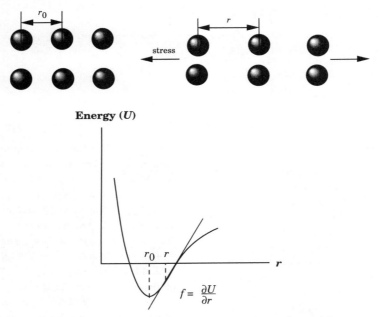

Figure 3.2 Energy-driven elasticity typical of crystalline solids.

and Guth and Mark (1934), who suggested that the covalently bonded polymer chains were oriented during extension. The theoretical development during the 1940s was due to James and Guth (1942), Wall (1942, 1943), and Flory and Rehner (1943), who suggested, in slightly different theoretical forms, that the elastic force was due to changes in the conformational entropy. This view that the long chain molecules are stretched out to statistically less favourable states still prevails (Fig. 3.3). The force acting on the rubber network is equal to the slope of the plot of free energy (G) against displacement. The instantaneous deformation occurring in rubbers is due to the high segmental mobility and thus rapid changes in chain conformation of the molecules. The energy barriers between different conformational states must

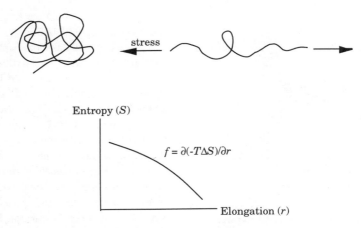

Figure 3.3 Entropy-driven elasticity of rubber materials.

therefore be small compared to the thermal energy (RT).

Analogies can be found between rubbers and gases. An increase in the chaotic state of the molecules (increase in entropy) occurs in both cases with increasing temperature. As a result, the rubber decreases in length (extension) with increasing temperature under a constant load. The same trend is indeed obtained in a compressed gas, the latter expanding (deformation = (volume)$^{-1}$) when the temperature is increased. The pressure of the gas is predominantly entropically driven and so is the stress in the rubber. The reversible character of the deformation is a consequence of the fact that rubbers are lightly crosslinked materials (Fig. 3.4). The crosslinks prevent the chains from adopting 'new' positions relative to their neighbouring chains in the unstressed state. The length of the chains between adjacent crosslinks is typically several hundred main chain atoms. The crosslinks can be permanent, i.e. the crosslinks are covalent bonds.

From a historical perspective, the accomplishment by Charles Goodyear in 1839 of a method to vulcanize natural rubber with sulphur was a crucial breakthrough. Sulphur links attached to the cis-1,4-polyisoprene molecules formed the network structure which is a prerequisite for obtaining elastic properties (Fig. 3.5).

Later development of vulcanization technology has involved peroxide crosslinking and thermoplastic elastomers. The latter consist of block copolymers with hard segments (physical crosslinks) and flexible segments (Fig. 3.6). The crosslink domains are either glassy amorphous or crystalline. These materials can be processed by conventional thermoplastic process-

Figure 3.4 Crosslinked rubber. The crosslinks are indicated by filled circles.

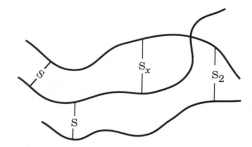

Figure 3.5 Sulphur bridges linking cis-1,4-polyisoprene.

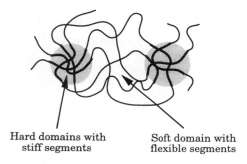

Hard domains with stiff segments   Soft domain with flexible segments

Figure 3.6 Structure of thermoplastic elastomers.

ing techniques at temperatures above the glass transition temperature or above the crystal melting point of the hard segment domains.

3.2 THERMO-ELASTIC BEHAVIOUR AND THERMODYNAMICS: ENERGETIC AND ENTROPIC ELASTIC FORCES

Figure 3.7 shows the classical data of Anthony, Caston and Guth from the 1940s. At small strains, typically less than $\lambda = L/L_0 < 1.1$, the stress at constant strain decreases with increasing temperature, whereas at λ values greater than 1.1, the stress increases with increasing temperature. This change from a negative to a positive temperature coefficient is referred to as **thermo-elastic inversion**. The reason for the negative coefficient at small strains is the positive thermal expansion and that the curves are obtained at constant length. An increase in temperature causes some thermal expansion (increase in L_0 and also a corresponding length extension in the perpendicular directions) and consequently a decrease in the true λ

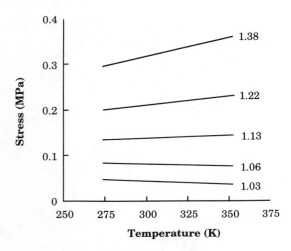

Figure 3.7 Stress at constant length (shown adjacent to each line is a strain value referring to this length, assuming constant L_0) as a function of temperature for natural rubber. Drawn after data from Anthony, Casta and Guth (1942).

at constant L. The effect would not appear if L_0 were measured at each temperature and if the curves were taken at constant λ (relating to L_0 at the actual temperature). The positive coefficient is typical of entropy-driven elasticity.

The reversible temperature increase which occurs when a rubber band is deformed can be sensed with your lips, for instance. It is simply due to the fact that the internal energy remains relatively unchanged on deformation, i.e. $dQ = -dW$ (when $dE = 0$), where Q is heat and W is work. If work is performed on the system, then heat is produced, leading to an increase in temperature. The temperature increase under adiabatic conditions can be substantial. Natural rubber stretched to $\lambda = 5$ reaches a temperature which is 5–10 K higher than that prior to deformation. When the external force is removed and the specimen returns to its original, unstrained state, an equivalent temperature decrease occurs.

The stress acting on a crosslinked rubber causes orientation of the chain segments which may lead to changes in the intramolecular-related internal energy. In addition, small changes in volume occur which also lead to an increased internal energy. The following thermodynamic treatment yields an expression differentiating between the entropic and energetic contributions to the elastic force.

According to the first and second laws of thermodynamics, the internal energy change (dE) in a system exchanging heat (dQ) and work (dW) reversibly is given by:

$$dE = T\,dS - p\,dV + f\,dL \quad (3.4)$$

where dS is the differential change in entropy, $p\,dV$ is the pressure–volume work and $f\,dL$ is the work done by deformation.

The Gibbs free energy (G) is defined as:

$$G = H - TS = E + pV - TS \quad (3.5)$$

where H is the enthalpy. Differentiating eq. (3.5) gives:

$$dG = dE + p\,dV + V\,dp - T\,dS - S\,dT \quad (3.6)$$

Insertion of eq. (3.4) in eq. (3.6) gives:

$$dG = f\,dL + V\,dp - S\,dT \quad (3.7)$$

The partial derivatives of G with respect to L and T are:

$$\left(\frac{\partial G}{\partial L}\right)_{p,T} = f \quad (3.8)$$

$$\left(\frac{\partial G}{\partial T}\right)_{L,p} = -S \quad (3.9)$$

G is a function of state, which means that the order of derivation is unimportant:

$$\left(\frac{\partial}{\partial T}\left(\frac{\partial G}{\partial L}\right)_{p,T}\right)_{p,L} = \left(\frac{\partial}{\partial L}\left(\frac{\partial G}{\partial T}\right)_{L,p}\right)_{p,T} \quad (3.10)$$

By combining eqs (3.8)–(3.10), the following expression is obtained:

$$\left(\frac{\partial f}{\partial T}\right)_{L,p} = -\left(\frac{\partial S}{\partial L}\right)_{p,T} \quad (3.11)$$

The partial derivative of G with respect to L at constant p and constant T (from eq. (3.5)) is:

$$\left(\frac{\partial G}{\partial L}\right)_{p,T} = \left(\frac{\partial H}{\partial L}\right)_{p,T} - T\left(\frac{\partial S}{\partial L}\right)_{p,T} \quad (3.12)$$

Combining eqs (3.11) and (3.12) gives:

$$f = \left(\frac{\partial H}{\partial L}\right)_{p,T} + T\left(\frac{\partial f}{\partial T}\right)_{p,L} \quad (3.13)$$

The derivative of H with respect to L at constant p and constant T (from eq. (3.5)) is:

$$\left(\frac{\partial H}{\partial L}\right)_{p,T} = \left(\frac{\partial E}{\partial L}\right)_{p,T} + p\left(\frac{\partial V}{\partial L}\right)_{p,T} \quad (3.14)$$

Experiments show that the volume is approximately constant during deformation, $(\partial V/\partial L)_{p,T} \approx 0$. Hence,

$$\left(\frac{\partial H}{\partial L}\right)_{p,T} = \left(\frac{\partial E}{\partial L}\right)_{p,T} \quad (3.15)$$

and

$$f = \left(\frac{\partial E}{\partial L}\right)_{p,T} + T\left(\frac{\partial f}{\partial T}\right)_{p,L} \quad (3.16)$$

The first term, $(\partial E/\partial L)_{p,T}$, is associated with the change in internal energy accompanying deformation at constant pressure and temperature. The other term originates from changes in entropy (degree of order) by deformation; note that $(\partial f/\partial T)_{L,p} = -(\partial S/\partial L)_{p,T}$. Figure 3.8 shows schematically the partition of the force into energetic and entropic parts.

It should be noted that the entropy- and energy-related parts of the elastic force are not only associated with chain orientation. An additional and important contribution originates from a change in volume:

$$\left(\frac{\partial E}{\partial L}\right)_{p,T} = \left(\frac{\partial E}{\partial L}\right)_{T,V} + \left(\frac{\partial E}{\partial V}\right)_{T,L}\left(\frac{\partial V}{\partial L}\right)_{p,T} \quad (3.17)$$

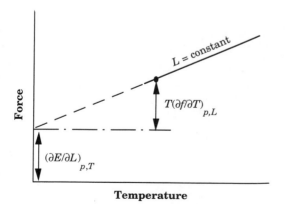

Figure 3.8 Energetic $(\partial E/\partial L)_{p,T}$ and entropic $T(\partial f/\partial T)_{p,L}$ components of the elastic force.

Typical of rubbers is that the volume remains approximately constant on deformation, i.e. $(\partial V/\partial L)_{p,T}$ is small. The change in internal energy which accompanies a change in volume, on the other hand, is substantial, i.e. $(\partial E/\partial V)_{T,L}$ is very large.

An analogous expression can be derived for constant volume conditions:

$$f = \left(\frac{\partial E}{\partial L}\right)_{V,T} + T\left(\frac{\partial f}{\partial T}\right)_{V,L} \quad (3.18)$$

This equation is difficult to verify experimentally. A hydrostatic pressure has to be adjusted in order to keep the volume constant to counteract changes in volume caused by the stress–strain work. Flory showed that

$$-\left(\frac{\partial S}{\partial L}\right)_{T,V} \approx \left(\frac{\partial f}{\partial T}\right)_{p,\lambda} \quad (3.19)$$

where $\lambda = L/L_0$, L_0 being the length for zero stress at temperature T. Combining eqs (3.18) and (3.19) gives:

$$f = \left(\frac{\partial E}{\partial L}\right)_{V,T} + T\left(\frac{\partial f}{\partial T}\right)_{p,\lambda} \quad (3.20)$$

Equation (3.20) is very useful: by measuring the force (f) as a function of temperature (L_0 has to be determined at each temperature) at constant pressure and elongation (λ) the change in internal energy and entropy at constant volume can be obtained. The latter refer to the changes in the quantities caused by **orientation** (directional preference of the chains) alone. Figure 3.9 presents data for natural rubber indicating that the entropic part constitutes 80–85% of the elastic force. According to eq. (3.18), the energetic force component (f_e) under constant volume conditions is given by:

$$\frac{f_e}{f} = 1 - \frac{T}{f}\left(\frac{\partial f}{\partial T}\right)_{V,L} \quad (3.21)$$

Under constant pressure conditions the following equation holds:

$$\frac{f_e}{f} = 1 - \frac{T}{f}\left(\frac{\partial f}{\partial T}\right)_{p,L} + \frac{\beta T}{\lambda^3 - 1} \quad (3.22)$$

where β is the thermal expansion coefficient of the

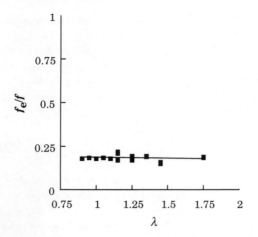

Figure 3.9 The fraction of the elastic force with energetic origin as a function of strain for natural rubber at room temperature. Drawn after data from Wolf and Allen (1975).

rubber. Flory showed that

$$\frac{f_e}{f} = T\left(\frac{d(\ln\langle r^2\rangle_0)}{dT}\right) \quad (3.23)$$

These equations can be understood in terms of the fact that a deformation of the network causes a change in conformational energy which is an intramolecular effect. Table 3.1 shows results from experiments on a number of polymers, some of them diluted with swelling agents. Both negative and positive values of f_e/f have been found.

Polyethylene shows a negative value (−0.42), which is expected since the extended all-trans conformation is of lower energy than the random coil conformation. Stretching molten and crosslinked PE causes a decrease in both entropy and energy.

Other polymers such as natural rubber and poly(dimethyl siloxane) exhibit positive values, i.e. the extended conformation is of higher energy than the unstrained structure. The preferred conformation of poly(dimethyl siloxane) is all-trans, but this gives the chain a non-extended form, in this case due to the difference in bond angles along the chain, i.e. for O–Si–O and Si–O–Si. Swelling of the polymers in various solvents causes no systematic change in f_e/f, demonstrating the intramolecular nature of the energetic force component (Table 3.1).

3.3 THE STATISTICAL MECHANICAL THEORY OF RUBBER ELASTICITY

The early molecular-based statistical mechanics theory was developed by Wall (1942) and Flory and Rehner (1943), with the simple assumption that chain segments of the network deform independently and on a microscopic scale in the same way as the whole sample (affine deformation). The crosslinks are assumed to be fixed in space at positions exactly defined by the specimen deformation ratio. James and Guth (1943) allowed in their 'phantom network model' a certain free motion (fluctuation) of the crosslinks about their average affine deformation positions. These two theories are in a sense 'limiting cases', with the affine network model giving an upper

Table 3.1 Energetic stress ratio of a few polymers

Polymer	Diluent	v_2[a]	f_e/f
Polyethylene	none	1.00	−0.42
Polyethylene	n-$C_{30}H_{62}$	0.50	−0.64
Polyethylene	n-$C_{32}H_{66}$	0.30	−0.50
Natural rubber	none	1.00	0.17
Natural rubber	n-$C_{16}H_{34}$	0.34–0.98	0.18
Natural rubber	decalin	0.20	0.14
Poly(dimethyl siloxane)	none	1.00	0.25
Trans (1,4-polyisoprene)	none	1.00	−0.10
Trans (1,4-polyisoprene)	decalin	0.18	−0.20

Source: Mark (1984).
[a] Volume fraction of polymer in network.

bound modulus and the phantom network model theory the lower bound.

Figure 3.10 shows schematically the difference between the affine network model and the phantom network model. The affine deformation model assumes that the junction points (i.e. the crosslinks) have a specified fixed position defined by the specimen deformation ratio (L/L_0, where L is the length of the specimen after loading and L_0 is the length of the unstressed specimen). The chains between the junction points are, however, free to take any of the great many possible conformations. The junction points of the phantom network are allowed to fluctuate about their mean values (shown in Fig. 3.10 by the points marked with an A) and the chains between the crosslinks to take any of the great many possible conformations.

The starting point here is the affine network model which is founded on the following assumptions:

- The chain segments between crosslinks can be represented by Gaussian statistics of phantom (unperturbed) chains.
- The network consists of N-chains per unit volume. The entropy of the network is the sum of the entropies of the individual chains.
- All different conformational states have the same energy.
- The deformation on the molecular level is the same as that on a macroscopic level, i.e. deformation is **affine**.
- The unstressed network is **isotropic**.
- The volume remains constant during deformation.

The distribution of the end-to-end vectors is as follows (section 2.5):

$$P(x, y, z)dx\,dy\,dz = \left(\frac{3}{2\pi\langle r^2\rangle_0}\right)^{3/2}$$
$$\times \exp\left[-\frac{3(x^2 + y^2 + z^2)}{2\langle r^2\rangle_0}\right]dx\,dy\,dz \quad (3.24)$$

where $\langle r^2\rangle_0$ is the average end-to-end distance of the phantom chains. Boltzmann's entropy relationship ($S = k \ln P$) is useful here:

$$S = k \ln(P(x, y, z)dx\,dy\,dz)$$
$$= k\left(\frac{3}{2}\ln\left(\frac{3}{2\pi\langle r^2\rangle_0}\right)\right.$$
$$\left. -\left(\frac{3(x^2 + y^2 + z^2)}{2\langle r^2\rangle_0}\right) + \ln(dx\,dy\,dz)\right) \quad (3.25)$$

which after simplification becomes

$$S = C - k\frac{3r^2}{2\langle r^2\rangle_0} \quad (3.26)$$

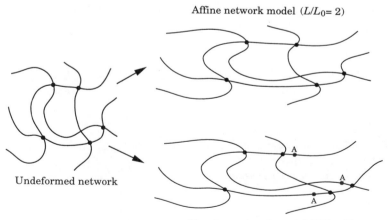

Figure 3.10 Schematical representation of the deformation of a network according to the affine network model (upper) and the phantom network model (lower). The points marked with an A indicate the position of the crosslinks assuming affine deformation.

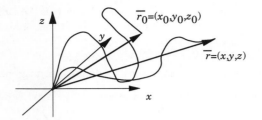

Figure 3.11 Deformation of a random chain from unstressed state $\bar{r}_0$ to stressed state $\bar{r}$.

where C is a constant. The unstressed state is characterized by the end-to-end vector $\bar{r}_0 = (x_0, y_0, z_0)$, shown in Fig. 3.11. The end-to-end vector $\bar{r} = (x, y, z)$ corresponding to the stressed state is related to $\bar{r}_0$ through the deformation matrix $(\lambda_1, \lambda_2, \lambda_3)$ according to:

$$x = \lambda_1 x_0 \quad y = \lambda_2 y_0 \quad z = \lambda_3 z_0$$

The entropies of the chain before (S_0) and after (S) the stress have been applied are:

$$S_0 = C - k\left(\frac{3(x_0^2 + y_0^2 + z_0^2)}{2\langle r^2\rangle_0}\right) \quad (3.27)$$

$$S = C - k\left(\frac{3(\lambda_1^2 x_0^2 + \lambda_2^2 y_0^2 + \lambda_3^2 z_0^2)}{2\langle r^2\rangle_0}\right) \quad (3.28)$$

and the difference in entropy between the two states is:

$$\Delta S = S - S_0$$
$$= -3k\left(\frac{(\lambda_1^2 - 1)x_0^2 + (\lambda_2^2 - 1)y_0^2 + (\lambda_3^2 - 1)z_0^2}{2\langle r^2\rangle_0}\right) \quad (3.29)$$

The change in entropy for the network (ΔS_N) is the sum of the contributions of all chain segments of the network. It is here assumed that deformation is affine, i.e. matrix (λ_i) is the same for all chain segments:

$$\Delta S_N = \sum_1^n \Delta S = -3k\left(\frac{(\lambda_1^2 - 1)\sum_1^n x_0^2 + (\lambda_2^2 - 1)\sum_1^n y_0^2 + (\lambda_3^2 - 1)\sum_1^n z_0^2}{2\langle r^2\rangle_0}\right) \quad (3.30)$$

It is also assumed that the original system is isotropic, i.e.:

$$\sum_1^n x_0^2 = \sum_1^n y_0^2 = \sum_1^n z_0^2$$

$$\sum_1^n x_0^2 + \sum_1^n y_0^2 + \sum_1^n z_0^2 = \sum_1^n r_0^2$$

$$\sum_1^n x_0^2 = \sum_1^n y_0^2 = \sum_1^n z_0^2 = \frac{1}{3}\sum_1^n r_0^2 = \frac{n\langle r^2\rangle_0}{3}$$

$$(3.31)$$

where n is the number of Gaussian chain segments in the system. Insertion of eq. (3.31) into eq. (3.30) gives:

$$\Delta S_N = -\tfrac{1}{2}nk(\lambda_1^2 + \lambda_2^2 + \lambda_3^2 - 3)$$
$$\Delta G = -T\Delta S_N = \tfrac{1}{2}nkT(\lambda_1^2 + \lambda_2^2 + \lambda_3^2 - 3)$$

$$(3.32)$$

Equation (3.32) is general and is not restricted to any particular state of stress. Let us derive a stress–strain equation for a rubber specimen subjected to a constant uniaxial stress. The deformation along the stress is denoted λ. It may also be assumed that the transverse deformations are equal; $\lambda_2 = \lambda_3$. The assumption that the volume remains constant during deformation can be formulated as follows:

$$\lambda_1 \lambda_2 \lambda_3 = 1 \quad (3.33)$$

Insertion of $\lambda_1 = \lambda$ and $\lambda_2 = \lambda_3 = \lambda_t$ in eq. (3.33) gives:

$$\lambda \lambda_t^2 = 1$$

$$\lambda_t = \frac{1}{\sqrt{\lambda}}$$

Hence, the following deformation matrix is obtained: $(\lambda, 1/\sqrt{\lambda}, 1/\sqrt{\lambda})$. The force ($f$) is obtained by using eq. (3.32) and $\lambda = L/L_0$:

$$f = \left(\frac{\partial(\Delta G)}{\partial L}\right)_{T,V} = \left(\frac{\partial(\Delta G)}{\partial \lambda}\right)_{T,V}\left(\frac{\partial \lambda}{\partial L}\right)_{T,V}$$

$$f = \frac{\partial}{\partial \lambda}\left(\frac{1}{2}nkT\left(\lambda^2 + \frac{2}{\lambda} - 3\right)\right)\frac{\partial}{\partial L}\left(\frac{L}{L_0}\right)$$

$$= \frac{nkT}{L_0}\left(\lambda - \frac{1}{\lambda^2}\right)$$

$$f = \sigma\left(\frac{A_0}{\lambda}\right) = \frac{NRT}{L_0}\left(\lambda - \frac{1}{\lambda^2}\right) \quad (3.34)$$

where σ is the real stress, A_0 is the original cross-sectional area, L_0 is the original length of the sample parallel to the stress, N is the number of moles of Gaussian chain segments and R is the gas constant. After simplification, the following stress–strain equation is obtained:

$$\sigma = \frac{NRT}{V_0}\left(\lambda^2 - \frac{1}{\lambda}\right) \quad (3.35)$$

where V_0 is the volume of the system. Equation (3.35) contains two system-size-dependent quantities which are removed by considering that:

$$\frac{N}{V_0} = \left(\frac{N\bar{M}_c}{V_0}\right)\left(\frac{1}{\bar{M}_c}\right) = \left(\frac{m_0}{V_0}\right)\left(\frac{1}{\bar{M}_c}\right) = \frac{\rho}{\bar{M}_c}$$

$\bar{M}_c$ is the number average molar mass of the Gaussian chain segments and ρ is the density. The true stress–extension equation becomes:

$$\sigma = \frac{\rho RT}{\bar{M}_c}\left(\lambda^2 - \frac{1}{\lambda}\right) \quad (3.36)$$

The 'modulus', $\rho RT/\bar{M}_c$, increases in a linear manner with increasing temperature. This is typical of entropy-elastic materials. The other important aspect of eq. (3.36) is that the modulus is inversely proportional to $\bar{M}_c$. Rubbers with a high crosslinking density, i.e. low $\bar{M}_c$, behave stiffly. Figure 3.12 shows the thermoelastic behaviour of the ideal entropy-elastic rubber material. All lines meet at the origin at 0 K. Any deviation from the origin, positive or negative, indicates an energetic contribution to the elastic force.

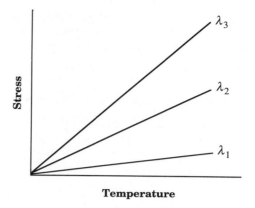

Figure 3.12 Stress–temperature relationships (constant λ) for an ideal entropy-elastic material.

Equation (3.36) may also be written:

$$\sigma = N_e RT\left(\lambda^2 - \frac{1}{\lambda}\right) \quad (3.37)$$

where N_e is the molar number of chain segments per unit volume: $N_e = N/V_0 = \rho/\bar{M}_c$.

The number of crosslinks (v) is proportional to the number of chain segments and inversely proportional to the functionality of the crosslink (ψ, which denotes the number of chains emanating from the junction point):

$$v = \frac{2N_e}{\psi}$$

which may be inserted into eq. (3.37) to give:

$$\sigma = \frac{\psi v RT}{2}\left(\lambda^2 - \frac{1}{\lambda}\right) \quad (3.38)$$

The phantom network model of James and Guth predicts a different free energy-deformation expression (cf. eq. (3.32)):

$$\Delta G = \left(1 - \frac{2}{\psi}\right)\frac{nkT}{2}(\lambda_1^2 + \lambda_2^2 + \lambda_3^2 - 3) \quad (3.39)$$

which, when applied to the uniaxial constant stress case, yields the following stress–strain equation:

$$\sigma = \left(1 - \frac{2}{\psi}\right)\frac{\psi v RT}{2}\left(\lambda^2 - \frac{1}{\lambda}\right) \quad (3.40)$$

48 The rubber elastic state

For a network with a crosslink functionality (ψ) of 4, the phantom network model predicts a modulus which is 1/2 of the modulus predicted by the affine network model.

3.4 SWELLING OF RUBBERS IN SOLVENTS

Rubbers may swell considerably in good solvents. The swelling leads to isotropic extension of the network (Fig. 3.13).

The deformation matrix becomes:

$$\lambda_1 \lambda_2 \lambda_3 = \frac{V}{V_0} = \lambda_v$$
$$\lambda_1 = \lambda_2 = \lambda_3 = \lambda \qquad (3.41)$$
$$\lambda = \sqrt[3]{\lambda_v}$$

The volume expansion factor $\lambda_v = 1/v_2$, where v_2 is the volume fraction of polymer component in the swollen gel, is inserted into eq. (3.41):

$$\lambda = v_2^{-1/3} \qquad (3.42)$$

which leads to an increase in free energy for the network expansion (ΔG_{el}) (eq. 3.32):

$$\Delta G_{el} = \frac{3\rho RT}{2\bar{M}_c}(v_2^{-2/3} - 1) \qquad (3.43)$$

The free energy for the network expansion (eq. (3.43)) can be converted to the molar free energy ($\Delta G_{1,el}$) of dilution by substituting $1/v_2 = 1 + n_1 V_1$, where n_1 is the mole fraction of the solvent and V_1 is the molar volume of the solvent, and by taking the partial derivative with respect to n_1:

$$\Delta G_{1,el} = \frac{\rho RT}{\bar{M}_c} V_1 v_2^{1/3} \qquad (3.44)$$

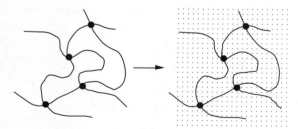

Figure 3.13 Isotropic swelling of a network polymer.

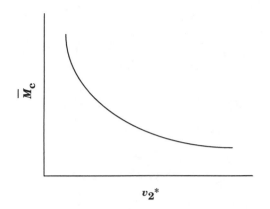

Figure 3.14 Schematic demonstration of the relationship between equilibrium swelling (v_2^*) and the average molar mass of the chain segments of the network ($\bar{M}_c$).

A detailed presentation of the thermodynamics of polymer solutions is given in Chapter 4; here only the final Flory–Huggins equation (eq. (3.45)) is presented. The decrease in free energy comes from the mixing enthalpy and entropy and the molar free energy of mixing ($\Delta G_{1,m}$) becomes:

$$\Delta G_{1,m} = RT\left(\ln(1-v_2) + \left(1 - \frac{1}{x}\right)v_2 + \chi v_2^2\right)$$
$$\approx RT(\ln(1-v_2) + v_2 + \chi v_2^2) \qquad (3.45)$$

where χ is the Flory–Huggins interaction parameter. Equilibrium is obtained when $\Delta G_1 = \Delta G_{1,el} + \Delta G_{1,m} = 0$:

$$\ln(1-v_2^*) + v_2^* + \chi v_2^{*2} + \frac{\rho V_1}{\bar{M}_c} v_2^{*1/3} = 0 \qquad (3.46)$$

The equilibrium degree of swelling, represented by v_2^*, is described by eq. (3.46) and the type of relationship between v_2^* and $\bar{M}_c$ is shown in Fig. 3.14.

3.5 DEVIATIONS FROM CLASSICAL STATISTICAL THEORIES

In the classical theory, it is assumed that the network is infinite, i.e. that no loose chain ends exist. Loose chain ends transfer the stress less efficiently than the other parts of the chain and it may be assumed that

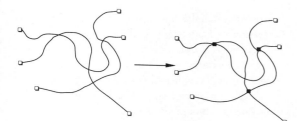

Figure 3.15 Crosslinking of a polymer with finite molar mass indicating the formation of loose chain ends.

they do not contribute to the elastic force. Figure 3.15 illustrates the fact that the number of loose chain ends present in a crosslinked polymer is the same as the number of chain ends in the polymer prior to crosslinking. Hence, the concentration of loose chain ends is solely determined by the molar mass of the uncrosslinked polymer.

Loose chains which do not contribute to the elastic force reduce the number of load-carrying chain segments (N_e) to:

$$N_e = 2v_e = 2v_0\left(1 - \frac{N_p}{v_0}\right) \quad (3.47)$$

where v_0 is the number of crosslinks of which only v_e are effective, and N_p is the number of original molecules prior to crosslinking in a polymer of molar mass M. It is here assumed that the functionality of the links is 4. In a unit volume of uncrosslinked rubber:

$$N_p = \frac{\rho N_A}{M} \quad (3.48)$$

where N_A is the Avogadro number. In a unit volume of crosslinked rubber, the number of chain segments (effective and non-effective) is:

$$2v_0 = \frac{\rho N_A}{\bar{M}_c} \quad (3.49)$$

where $\bar{M}_c$ is the average molar mass of the chain segments between the crosslinks. The number of effective chains can be derived by combining eqs (3.47)–(3.49):

$$N_e = 2v_0\left(1 - \frac{2\bar{M}_c}{M}\right) \quad (3.50)$$

which can be used to modify the stress–strain relationship of a uniaxially stressed rubber:

$$\sigma = \frac{\rho RT}{\bar{M}_c}\left(1 - \frac{2\bar{M}_c}{M}\right)\left(\lambda^2 - \frac{1}{\lambda}\right) \quad (3.51)$$

Other types of network defect also exist: physical crosslinks and closed loops (Fig. 3.16). Physical crosslinks may be permanent with a locked-in conformation (Fig. 3.16(a)) of temporary by entanglement. The presence of the latter type leads to visco-elastic behaviour, i.e. to creep and stress relaxation. Intramolecular crosslinks decrease the interconnectivity and reduce the number of load-carrying chains.

Figure 3.17 demonstrates that the simple statistical mechanical theory involving only a single materials constant correctly describes the properties of a real,

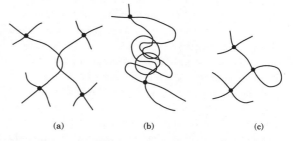

Figure 3.16 'Chain defects' in networks: (a) permanent physical crosslink; (b) temporary physical crosslink; (c) intramolecular crosslink.

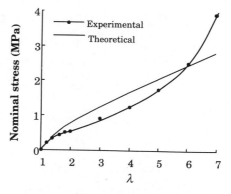

Figure 3.17 Nominal stress (force per unit unstrained area) of crosslinked NR ($\rho RT/\bar{M}_c = 0.39$ MPa) as a function of the extension ratio $\lambda = L/L_0$. Drawn after data presented in Treloar (1975).

unfilled rubber to a first approximation. The positive curvature of the experimental data appearing at high extension ratios, above $\lambda = 4$, has two possible causes: first, the segment vectors do not follow Gaussian statistics when the rubber is highly stretched, i.e. the probability of having the vector at any x and y values ($P(x)$ and $P(y)$) is a function of the z value of the particular bond vector; and second, crystallization may occur in the rubber. The deviation appearing at more moderate extension values is not presently understood.

The classical statistical theory assumes that the end-to-end chain vectors ($\bar{r}$) obey Gaussian statistics. Figure 3.18 illustrates the concepts 'Gaussian' and 'non-Gaussian' in two dimensions. Gaussian statistics in this case means that the probability distribution function $P(y)$ is not affected by the x value, i.e. $P(y) = f(y)$, whereas in the non-Gaussian case $P(y) = f(y, x)$. Gaussian statistics is a good approximation only if $(\bar{r}^2)^{1/2}$ is considerably smaller than the contour length (nl) of the chain. Experiments have shown that the marked upturn in the curve of nominal stress (f) against extension (λ) is due to a transition from Gaussian to non-Gaussian behaviour. The non-Gaussian statistical treatment takes into account the finite extensibility of the chain, and thus leads to a more realistic form of distribution function which is valid over the whole range of r values up to the fully extended length. Another consequence of this reasoning is that the Gaussian approximation becomes increasingly inadequate with increasing crosslink density. For very short chains, e.g. $n \leq 5$, the mean chain extension even in the unstrained state already exceeds that for which the Gaussian approximation is valid. For networks of such short chains, a non-Gaussian treatment is essential for the accurate representation of the behaviour even at the smallest strains.

We may return to Fig. 3.18 and note that non-Gaussian statistics predicts fewer conformations in the extended state (high x value) than the Gaussian statistics. Qualitatively this means that the entropy change and the elastic force are predicted to be higher according to the non-Gaussian theory than according to the classical Gaussian treatment.

The full derivation of the nominal stress (f)–strain relationship for a chain obeying non-Gaussian statistics is complicated. The final equation as derived by Kuhn is:

$$f = \left(\frac{kT}{l}\right)L^{-1}\left(\frac{r}{nl}\right) = \frac{kT}{l}\left[3\left(\frac{r}{nl}\right) + \frac{9}{5}\left(\frac{r}{nl}\right)^3 + \frac{297}{175}\left(\frac{r}{nl}\right)^5 + \frac{1539}{875}\left(\frac{r}{nl}\right)^7 + \cdots\right] \quad (3.52)$$

where $L^{-1}(r/nl)$ is the inverse Langevin function of (r/nl), which is expanded in series form in eq. (3.52). The Langevin function is defined as:

$$L(x) = \coth x - \frac{1}{x}$$

A chain obeying Gaussian statistics adapts to the following equation:

$$f = \frac{kT}{l}\left(3\left(\frac{r}{nl}\right)\right) \quad (3.53)$$

It may be noted that eqs (3.52) and (3.53) are identical for chains with $r \ll nl$. Figure 3.19 compares the stress–strain curves derived under the two conditions (eqs (3.50) and (3.51)). The strong upturn in stress experimentally verified can only be obtained by the non-Gaussian equation. X-ray diffraction on natural rubber has shown that the initial upturn in the load–extension curve is a genuine non-Gaussian effect, unrelated to crystallization. At higher extension

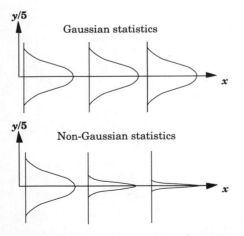

Figure 3.18 Illustration of Gaussian and non-Gaussian statistics.

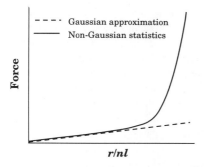

Figure 3.19 Force against extension (r/nl) relation for a random chain. Schematic curves.

values, crystallization was, however, observed in crystallizable polymers, e.g. in natural rubber.

It is also important to note that a whole family of non-Gaussian statistical treatments is reported in the literature. Treloar (1975, section 3.10) presents an excellent review on this topic.

3.6 SMALL-ANGLE NEUTRON SCATTERING DATA

Small-angle neutron scattering (SANS) of labelled (deuterated) amorphous samples and rubber samples detects the size of the coiled molecules and the response of individual molecules to macroscopic deformation and swelling. It has been shown that uncrosslinked bulk amorphous polymers consist of molecules with dimensions similar to those of theta solvents in accordance with the Flory theorem (Chapter 2). Fernandez et al. (1984) showed that chemical crosslinking does not appreciably change the dimensions of the molecules. Data on various deformed network polymers indicate that the individual chain segments deform much less than the affine network model predicts and that most of the data are in accordance with the phantom network model. However, definite SANS data that will tell which of the affine network model and the phantom network model is correct are still not available.

3.7 THE THEORY OF MOONEY AND RIVLIN

Mooney (1940, 1948) and Rivlin (1948) derived eq. (3.54) on the basis that the rubber was incompressible and isotropic in its unstrained state and that the material behaves as a Hookean solid in simple shear. This equation is valid for a uniaxially stressed specimen. The coefficients C_1 and C_2 are different for different materials and can in that sense be considered as material constants. The equation shows a striking resemblance to the equations derived from the statistical theory.

$$\sigma = 2\left(C_1 + \frac{C_2}{\lambda}\right)\left(\lambda^2 - \frac{1}{\lambda}\right) \quad (3.54)$$

It is interesting to compare eq. (3.54) with the expressions obtained from the statistical theories (Fig. 3.20). According to both the affine network model and the phantom network model of James and Guth, the reduced stress remains constant and independent of strain, which is not the case for the Mooney–Rivlin equation.

According to Flory, the coefficient C_2 is related to the looseness with which the crosslinks are embedded within the structure. This is supported by the fact that C_2 has been found to decrease with increasing solvent content in swelled rubbers. At a polymer content of $v_2 = 0.2$, C_2 approaches zero.

3.8 SUMMARY

The elasticity of rubbers is predominantly entropy-driven which leads to a number of spectacular phenomena. The stiffness increases with increasing temperature. Heat is reversibly generated as a consequence of an applied elastic force and stretching.

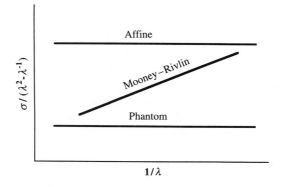

Figure 3.20 Reduced stress, $\sigma/(\lambda^2 - \lambda^{-1})$, as a function of the reciprocal strain.

The elastic force can be thought of as consisting of an entropic and an elastic part. The energetic contribution to the elastic force is generally small and is controlled by intramolecular factors, the energetic state of the extended conformation. Polymers such as polyethylene, with an extended conformation as their low-energy state, exhibit a negative energetic force contribution, whereas in other cases (e.g. natural rubber) the energetic elastic force is positive.

The affine network model assumes that the network consists of phantom Gaussian chains, that all network changes are entropical, that deformation is affine and that the volume remains constant during deformation, and yields the following expression for the free energy change accompanying the deformation ($\lambda_1, \lambda_2, \lambda_3$):

$$\Delta G = \tfrac{1}{2}NRT(\lambda_1^2 + \lambda_2^2 + \lambda_3^2 - 3) \quad (3.55)$$

For the case of a rubber specimen subjected to a constant uniaxial stress, the following true stress (σ)–strain (λ) expression can be derived:

$$\sigma = \frac{\rho RT}{\bar{M}_c}\left(\lambda^2 - \frac{1}{\lambda}\right) \quad (3.56)$$

where ρ is the density, R is the gas constant, T is the temperature (in kelvin), and $\bar{M}_c$ is the number average molar mass of the chains between the crosslinks. The modulus $\rho RT/\bar{M}_c$ is proportional to the absolute temperature and increases with increasing crosslink density, i.e. with decreasing $\bar{M}_c$.

A closely related theory was developed by James and Guth for the so-called phantom network, in which the positions of the junctions were allowed to fluctuate about mean positions prescribed by the affine deformation ratio. They derived an expression very similar to eq. (3.56):

$$\sigma = \frac{\rho RT}{2\bar{M}_c}\left(\lambda^2 - \frac{1}{\lambda}\right) \quad (3.57)$$

The transection of the chain segment in the phantom network leads to a fall by a factor of 2 in the apparent modulus compared with that predicted by the affine network model.

None of these theories is, however, adequate to describe the stress–strain behaviour at large strains, a typical upper strain limit for the validity of the Gaussian theories being $\lambda \approx 4$ for natural rubber of normal crosslink density. Expressions derived from non-Gaussian statistics of chain segments yield the right type of upturn in stress–strain behaviour at higher strain levels. Loose chain ends, temporary and permanent chain entanglements, intramolecular cross-linking, mechanical- and oxidation-induced network degradation are complications not directly addressed by the classical statistical mechanics theory.

Swelling equilibrium in a network polymer can be predicted by the Flory–Rehner equation, derived on the basis of the affine network model and the Flory–Huggins expression for polymer solutions.

The theory of Mooney and Rivlin, which is based on continuum mechanics, yields the following expression for a uniaxially stressed rubber:

$$\sigma = 2\left(C_1 + \frac{C_2}{\lambda}\right)\left(\lambda^2 - \frac{1}{\lambda}\right) \quad (3.58)$$

where C_1 and C_2 are constants. Equation (3.58) shows a close resemblance to the equations derived from statistical mechanics (eqs (3.56) and (3.57)).

3.9 EXERCISES

3.1. The rubber in a blown-up balloon is stretched in a biaxial fashion. Derive the force–strain relationship under the assumption that the rubber follows the Gaussian statistical theory of rubber elasticity.

3.2. Derive the relationship between the internal pressure (p) and the degree of expansion of the balloon ($\alpha = D/D_0$). Assume that the ideal gas law ($pV = nRT$) is valid.

3.3. At what α value is maximum internal pressure attained?

3.4. Suppose the balloon has a small nose. Is it possible to get the nose to expand to the same degree as the rest of the balloon?

3.5. Many rubber materials exhibit a time dependence in their mechanical properties (see Fig. 3.21). Make a list of possible reasons.

3.6. Polyethylene can be crosslinked by decomposition of organic peroxides, hydrolysis of vinyl-silane grafted polyethylene or by high-energy (β or γ) irradiation. Design a suitable experiment to determine the crosslink density and present the relevant equations.

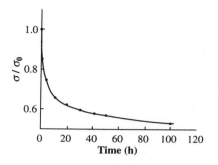

Figure 3.21 Results from measurements of continuous stress relaxation of nitrile rubber (low network density). Data from Björk (1988).

3.7. Calculate from the Kuhn model equation the modulus at room temperature of natural rubber ($\rho \approx 970$ kg m^{-3}) crosslinked with n molar fraction of organic peroxide. Assume that each peroxide molecule results in one crosslink.

3.8. Calculate the temperature increase occurring in natural rubber with $\bar{M}_c = 5000$ g mol^{-1} when it is stretched to $\lambda = 5$ at room temperature. Use the following data: $\rho \approx 970$ kg m^{-3}, $c_p = 1900$ J kg^{-1} K^{-1}.

3.10 REFERENCES

Anthony, P. C., Caston, R. H. and Guth, E. (1942) *J. Phys. Chem.* **46**, 826.

Björk, F. (1988) PhD thesis on Dynamic stress relaxation of rubber materials, Department of Polymer Technology, Royal Institute of Technology, Stockholm, Sweden.

Fernandez, A. M., Widmaier, J. M., Sperling, L. H. and Wignall, G. D. (1984) *Polymer* **25**, 1718.

Flory, P. J. (1976) *Proc. Roy. Soc. Lond. A.* **351**, 351.
Flory, P. J. and Rehner, J. (1943) *J. Chem. Phys.* **11**, 512.
Gough, J. (1805) *Proc. Lit. and Phil. Soc., Manchester, 2nd Ser.* **1**, 288.
Guth, E. and Mark, H. (1934) *Monats. Chem.* **65**, 93.
James, H. M. and Guth, E. (1942) *Ind. Eng. Chem.* **34**, 1365.
James, H. M. and Guth, E. (1943) *J. Chem. Phys.* **11**, 455.
Joule, J. P. (1859) *Trans. Roy. Soc. Lond. A.* **149**, 91.
Lord Kelvin (1857) *Quarterly J. Math.* **1**, 57.
Kuhn, W. (1934) *Kolloid Zeitschrift* **68**, 2.
Mark, J. E. (1984) The rubber elastic state, in *Physical Properties of Polymers* (J. E. Mark, ed.). American Chemical Society Washington, DC.
Meyer, K. H., von Susich, G. and Valko, E. (1932) *Kolloid Zeitschrift* **59**, 208.
Mooney, M. (1940) *J. Appl. Phys.* **11**, 582.
Mooney, M. (1948) *J. Appl. Phys.* **19**, 434.
Rivlin, R. S. (1948) *Phil. Trans. Roy. Soc. London Ser. A* **241**, 379.
Treloar, L. R. G. (1975) *The Physics of Rubber Elasticity*, 3rd edn. Clarendon Press, Oxford.
Wall, F. T. (1942) *J. Chem. Phys.* **10**, 132.
Wall, F. T. (1943) *J. Chem. Phys.* **11**, 527.
Wolf, F. P. and Allen, G. (1975) *Polymer* **16**, 209.

3.11 SUGGESTED FURTHER READING

Aklonis, J. J. and MacKnight, W. J. (1983) Chemical stress relaxation, in *Introduction to Polymer Viscoelasticity*, 2nd edn. Wiley, New York.

Boyd, R. H. and Phillips, P. J. (1993) *The Science of Polymer Molecules*. Cambridge University Press, Cambridge.

Mark. J. E. (1993) The rubber elastic state, in *Physical Properties of Polymers*, 2nd edn. (J. E. Mark, ed.). American Chemical Society, Washington, DC.

Mark, J. E. and Erman, B. (eds) (1992) *Elastomeric Polymer Networks*. Prentice-Hall, Englewood Cliffs, NJ.

Tobolsky, A. V. (1960) Chemical stress relaxation, in *Properties and Structure of Polymers*. Wiley, New York.

POLYMER SOLUTIONS 4

4.1 INTRODUCTION

A **solution** is any phase containing more than one component. It may be a gas, liquid or a solid. The thermodynamics of polymers in solutions is one of the major topics in the science and technology of polymers. The usefulness of a polymer in a specified environment may be limited by its physical and chemical stability. Mild, local swelling in a stressed polymer may lead to a very pronounced decrease in fracture toughness, a phenomenon referred to as environmental stress cracking. The swelling by solvents of higher solvent power is more extensive and may lead to softening and ultimately to dissolution. Solutions of polymers are used in several important applications, e.g. adhesives and coatings. The classical analyses of polymers are conducted on dilute solutions, e.g. size exclusion chromatography, osmometry, viscometry and light scattering. Much of the knowledge about polymers has been obtained by work conducted on solutions.

From a thermodynamic point of view, a condition for miscibility is given by the following expression:

$$\Delta G_{mix} = \Delta H_{mix} - T\Delta S_{mix} < 0$$

where ΔG_{mix} is the free energy of mixing, ΔH_{mix} is the enthalpy of mixing and ΔS_{mix} is the entropy of mixing. It can be stated that $\Delta G_{mix} < 0$ is a necessary but not sufficient condition for miscibility. If the graph of $\Delta G_{mix}(x_1)$, where x_1 is the molar fraction of component 1, is concave with no inflection point, miscibility is complete over the entire composition range. If the equation shows two or more inflection points, then miscibility is limited to the compositions 'outside' the two compositions with common tangent, the so-called binodal points. Blends of intermediate composition, phase-separate into two solutions.

This chapter deals only with fully amorphous polymers. Crystalline polymers constitute a more complex case, since the separation of the polymer molecules in the solvent matrix needs to be preceded by melting of the crystals. This subject is treated in Chapter 8.

We shall in section 4.2 deal with regular solutions of small-molecule substances. The construction of phase diagrams from the derived equations is demonstrated. The Flory–Huggins mean-field theory derived for mixtures of polymers and small-molecule solvents is dealt with in section 4.3. It turns out that the simple Flory–Huggins theory is inadequate in many cases. The scaling laws for dilute and semi-dilute solutions are briefly presented. The inadequacy of the Flory–Huggins approach has led to the development of the equation-of-state theories which is the fourth topic (section 4.6) Polymer–polymer mixtures are particularly complex and they are dealt with in section 4.7.

4.2 REGULAR SOLUTIONS

The first attempt to treat theoretically the change in enthalpy and entropy when two liquids are mixed was made in 1910 by von Laar. The molecules are placed in a regular lattice and the mixing enthalpy (always positive) is calculated from nearest-neighbour interaction. Volume is assumed to be invariant. The idea of **regular solution** was introduced independently by Hildebrand and Wood (1932) and Scatchard (1931) and can be viewed as an improvement on van Laar's theory. Similar theories were independently proposed by Wood and Scatchard in the early 1930s. The regular solution theory is a useful theory for the description of mixtures of non-polar small-molecule liquids with positive or zero mixing enthalpy.

The volume of the regular solution is equal to the sum of the volumes of its components, i.e. there is no change in volume on mixing the components. It is convenient to think about regular solution in terms of a regular lattice with positions that can be occupied by either of the components. The entropy and enthalpy changes on mixing are calculated separately.

Let us now consider a binary mixture. The mixing entropy is determined by the number of possible ways (P) of arranging the mixture of two low molar mass components, denoted 1 and 2, it being assumed that each molecule occupies only one lattice position, as shown in Fig. 4.1. P is given by:

$$P = \frac{n!}{n_1! \, n_2!} \quad (4.1)$$

where n is the total number of lattice positions present, n_1 is the number of molecules of type 1 and n_2 is the number of molecules of type 2. Boltzmann's entropy law ($S = k \ln P$) may then be applied to obtain:

$$\Delta S_{mix} = k(\ln n! - \ln n_1! - \ln n_2!)$$

and, using Stirling's approximation ($\ln x! \approx x \ln x - x$), the following equation is obtained:

$$\Delta S_{mix} = k(n \ln n - n - n_1 \ln n_1 + n_1 - n_2 \ln n_2 + n_2)$$

which can be simplified, since $n = n_1 + n_2$:

$$\Delta S_{mix} = k(n \ln n - n_1 \ln n_1 - n_2 \ln n_2)$$
$$= k((n_1 + n_2)\ln n - n_1 \ln n_1 - n_2 \ln n_2)$$
$$= k(n_1(\ln n - \ln n_1) + n_2(\ln n - \ln n_2))$$
$$= -k(n_1 \ln x_1 + n_2 \ln x_2) \quad (4.2)$$

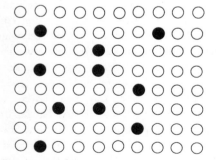

Figure 4.1 Lattice of a binary mixture of two low molar mass components.

where x_1 and x_2 are the molar fractions of components 1 and 2, respectively. Equation (4.2) can be rewritten using quantities in molar terms:

$$\frac{\Delta S_{mix}}{N} = -R(x_1 \ln x_1 + x_2 \ln x_2) \quad (4.3)$$

where N is the number of moles of the two components, i.e. $N = N_1 + N_2$.

The enthalpy of mixing (ΔH_{mix}) can be calculated from the interaction energies of the contacting atoms (1–1, 2–2 and 1–2 contacts):

$$\frac{\Delta H_{mix}}{N} = (\sqrt{\Delta E_1} - \sqrt{\Delta E_2})^2 x_1 x_2 = B x_1 x_2 \quad (4.4)$$

where ΔE_1 and ΔE_2 are the energies of vaporization of components 1 and 2. Equation (4.4) is strictly valid only for solutions consisting of components of equal size in the absence of specific interaction such as hydrogen bonding.

The free energy of mixing (ΔG_{mix}) for a regular solution is obtained by inserting the entropic and enthalpic components into the equation $\Delta G_{mix} = \Delta H_{mix} - T \Delta S_{mix}$:

$$\frac{\Delta G_{mix}}{N} = B x_1 x_2 + RT(x_1 \ln x_1 + x_2 \ln x_2) \quad (4.5)$$

Figure 4.2 shows the effect of increasing the B value on the free energy–composition relationship. Curves a–c are concave, indicating full miscibility at all compositions. Curves d and e show two symmetrically placed minima, two symmetrically placed inflection points and a central maximum. Figure 4.3 shows the same features in more detail.

The inflection points, given by $(d^2(\Delta G_{mix}/N)/dx_1^2) = 0$, are the so-called **spinodal** points, which define the thermodynamic limits of metastability. Within this concentration region, i.e. between the concentrations associated with spinodal points, the second derivative of ΔG_{mix} with respect to x_1 is negative, resulting in an unstable system with respect to any fluctuation in composition and temperature. **Spinodal decomposition** refers to the process that occurs inside the spinodal region by which a homogeneous blend phase separates. Concentration gradients develop driven by the free energy. Diffusion is 'negative' in the sense that the net flow is in the direction of increasing

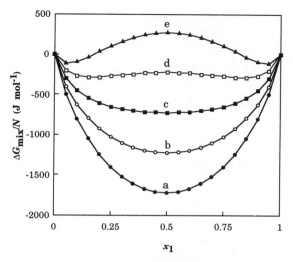

Figure 4.2 The free energy of mixing at 300 K according to eq. (4.5) for the following B values: (a) $B = 0$; (b) $B = 2000$ J mol^{-1}; (c) $B = 4000$ J mol^{-1}; (d) $B = 6000$ J mol^{-1}; (e) $B = 8000$ J mol^{-1}.

concentration. Spinodal decomposition has no nucleation free energy barrier. A binary system within the spinodal region forms initially a fine bicontinuous morphology (both components form continuous phases). The morphology gradually coarsens with time and at the later stages one of the components forms a discontinuous phase.

The so-called **binodal** points have a common tangent in the plot of ΔG_{mix} against x_1. The equilibrium (binodal) points fulfil the condition that the chemical potential $(\mu_i = (dG/dn_i)_{T,p,n_j})$ is equal for both components in both phases (Fig. 4.3):

$$\mu_1' - \mu_1^0 = \mu_1'' - \mu_1^0 \quad (4.6)$$

and

$$\mu_2' - \mu_2^0 = \mu_2'' - \mu_2^0 \quad (4.7)$$

where the primes and double primes denote the different phases.

The calculation of the chemical potentials of the components can be done by the intercept method which is graphically shown in Fig. 4.3. It relies on the fundamental thermodynamic equations:

$$\frac{\Delta G_{mix}}{N} = x_1(\mu_1 - \mu_1^0) + x_2(\mu_2 - \mu_2^0) \quad (4.8)$$

$$x_1 \, d\mu_1 + x_2 \, d\mu_2 = 0 \quad (4.9)$$

Equation (4.9) is known as the Gibbs–Duhem equation. The following two equations can be derived from eqs (4.8) and (4.9):

$$\mu_1 - \mu_1^0 = \frac{\Delta G_{mix}}{N} + x_2 \frac{d(\Delta G_{mix}/N)}{dx_1} \quad (4.10)$$

$$\mu_2 - \mu_2^0 = \frac{\Delta G_{mix}}{N} + x_1 \frac{d(\Delta G_{mix}/N)}{dx_2} \quad (4.11)$$

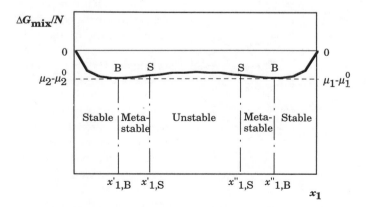

Figure 4.3 Free energy as a function of composition (x_1) according to the regular solution model showing binodal (B) and spinodal (S) concentrations. The chemical potentials of the two binodal points obtained by the intercept method are shown.

In these symmetrical cases (Figs 4.2 and 4.3), the binodal points are given by the two minima. The binodal curve is the equilibrium curve differentiating stable one-phase systems from heterogeneous systems. In the concentration range between the binodal and spinodal points, the free energy curvature is positive and the solution is metastable towards composition flutuations. Nucleation and growth are associated with the phase separation and only one of the components forms a continuous phase.

Let us now instead vary the temperature and keep the enthalpic term (B) constant (Fig. 4.4). At low temperatures where the entropy term is small, phase separation dominates, whereas at temperatures above 350 K miscibility occurs at all compositions.

The binodal and spinodal data of Fig. 4.4 are presented in the temperature–composition phase diagram in Fig. 4.5. This kind of phase diagram is typical of mixtures of small-molecule substances and also in many cases of mixtures of polymers and small-molecule substances. The spinodal and binodal curves meet in the so-called **upper critical solution temperature** (UCST). At temperatures greater than the UCST miscibility occurs at all compositions. The critical solution temperature is obtained by applying the following condition on the mixing free energy

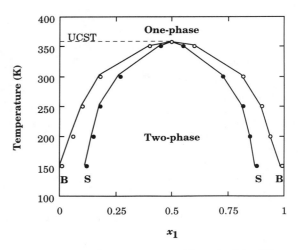

Figure 4.5 Phase diagram constructed from the data of Fig. 4.4 showing the binodal (B) and spinodal (S) curves. The upper critical solution temperature (UCST) is shown.

function:

$$\left(\frac{d^3(\Delta G_{mix}/N)}{dx_1^3}\right) = 0 \quad (4.12)$$

This condition may, in the search for the critical temperature, be combined with the condition for the spinodal, $(d^2(\Delta G_{mix}/N)/dx_1^2) = 0$.

4.3 THE FLORY–HUGGINS THEORY

The regular solution approximation is not valid for solutions containing polymers. However, it is no great step to the Flory–Huggins mean-field theory. This theory was independently introduced by Flory (1942) and Huggins (1942a, b, c). Both models are based on the idea of a lattice in which the components of the mixture are placed. Both models assume that the volume is unchanged during mixing. The mixing entropy is strongly influenced by the chain connectivity of the polymer component. The mixing enthalpy should for polymer–small-molecule mixtures have a form similar to that for regular solutions. The mixing entropy and the mixing enthalpy are first calculated separately in the Flory–Huggins treatment and then brought together in the free energy of mixing as in the case of the regular solution model. Figure 4.6 shows schematically the lattice model used

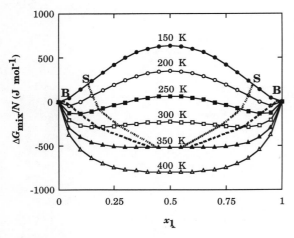

Figure 4.4 The free energy of mixing at different temperatures is shown in the diagram. B is set to 6000 J mol^{-1}. The binodal (B) and spinodal (S) points are indicated in the figure.

The Flory–Huggins theory 59

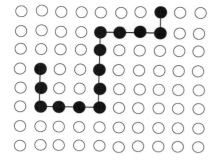

Figure 4.6 Lattice of a binary mixture of a polymer solute and a low molar mass solvent.

in the calculations. Each repeating unit of the polymer, here called a **segment**, occupies one position in the lattice and so does each solvent molecule.

The n lattice positions are partitioned between n_1 low molar mass solvent molecules and n_2 polymer solute molecules. Each polymer molecule occupies x lattice positions. The situation at which the analysis starts is when i polymer molecules have already been placed in the lattice. The number of vacant positions is then $n - xi$, which is equal to the number of different ways of placing the first segment of the $(i+1)$th molecule. The number of different ways of arranging the next segment is equal to the product of the coordination number (z) of the lattice and the fraction of remaining vacant position $(1 - f_i)$: i.e. $z(1 - f_i)$. The third segment (and all the rest) have one adjacent position occupied by the 'previous' segment, hence they can take any of $(z - 1)(1 - f_i)$ positions. The number of different ways of arranging the $(i + 1)$th molecule becomes:

$$v_{i+1} = (n - xi)z(z - 1)^{x-2}(1 - f_i)^{x-1} \quad (4.13)$$

The number of different ways to arrange all the polymer molecules is:

$$P_2 = \frac{v_1 v_2 v_3 v_4 \ldots v_{n_2}}{n_2!} \quad (4.14)$$

The reason for introducing the divisor $n_2!$ is that the 'different' polymer molecules cannot be distinguished.

The fraction of vacant positions is $1 - f_i = (n - xi)/n$ which, when inserted into eq. (4.13), yields:

$$v_{i+1} = (n - xi)^x \left(\frac{z - 1}{n}\right)^{x-1} \quad (4.15)$$

Only diluted solutions are considered here, i.e. $n \gg x$. Under these conditions, eq. (4.15) can be expressed approximately as:

$$v_{i+1} = \frac{(n - xi)!}{(n - x(i + 1))!} \left(\frac{z - 1}{n}\right)^{x-1} \quad (4.16)$$

Insertion of eq. (4.16) into eq. (4.14) gives:

$$\begin{aligned} P_2 &= \left(\frac{z-1}{n}\right)^{n_2(x-1)} \frac{1}{n_2!} \frac{(n-x)!(n-2x)!(n-3x)!\ldots(n-n_2x)!}{(n-2x)!(n-3x)!\ldots(n-(n_2+1)x)!} \\ &= \left(\frac{z-1}{n}\right)^{n_2(x-1)} \frac{1}{n_2!} \frac{(n-x)!}{(n-(n_2+1)x)!} \approx \left(\frac{z-1}{n}\right)^{n_2(x-1)} \frac{n!}{(n-n_2x)!n_2!} \end{aligned} \quad (4.17)$$

This approximation can only be made when $n \gg x$ (dilute solution). The entropy of the solution is calculated from the Boltzmann equation ($S = k \ln P_2$):

$$S = k\{\ln n! - \ln(n - xn_2)! - \ln n_2! + n_2(x - 1)\ln(z - 1) - n_2(x - 1)\ln n\}$$

which, after inserting Stirling's approximation ($\ln x! \approx x \ln x - x$), becomes:

$$\begin{aligned} S &= k\{n \ln n - n - (n - xn_2)\ln(n - xn_2) + n - xn_2 - n_2 \ln n_2 + n_2 + n_2(x - 1)\ln(z - 1) - n_2(x - 1)\ln n\} \\ &= k\{n \ln n - (n - xn_2)\ln(n - xn_2) - n_2 \ln n_2 - n_2(x - 1) + n_2(x - 1)\ln(z - 1) - n_2(x - 1)\ln n\} \end{aligned} \quad (4.18)$$

Equation (4.18) can be simplified by considering that $n = n_1 + n_2 x$:

$$S = k\{(n_1 + n_2 x)\ln(n_1 + n_2 x) - (n_1 + n_2 x - n_2 x)\ln(n_1 + n_2 x - n_2 x) \\ - n_2 \ln n_2 - n_2(x-1)\ln(n_1 + n_2 x) + n_2(x-1)(\ln(z-1) - 1)\}$$

$$= \left\{k(n_1 + n_2 x)\ln(n_1 + n_2 x) - n_1 \ln n_1 - n_2 \ln n_2 - n_2(x-1)\ln(n_1 + n_2 x) + n_2(x-1)\ln\left(\frac{z-1}{e}\right)\right\}$$

$$S = k\left\{(n_1 + n_2)\ln(n_1 + n_2 x) - n_1 \ln n_1 - n_2 \ln n_2 + n_2(x-1)\ln\left(\frac{z-1}{e}\right)\right\}$$

$$S = k\left\{-n_1 \ln\left(\frac{n_1}{n_1 + n_2 x}\right) - n_2 \ln\left(\frac{n_2}{n_1 + n_2 x}\right) + n_2(x-1)\ln\left(\frac{z-1}{e}\right)\right\} \quad (4.19)$$

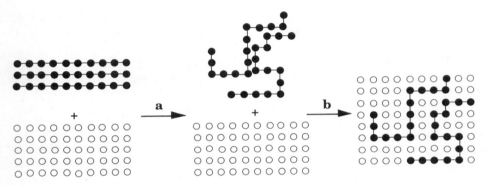

Figure 4.7 Schematic representation of steps a and b in the disordering and dissolution of a polymer.

It is fruitful to consider that this state has been reached by two different steps (Fig. 4.7):

(a) Disorientation of the polymer molecules, i.e. the formation of an amorphous polymer.
(b) Dissolution of the amorphous polymer in the solvent.

Our interest is focused on step b. The entropy increase obtained when the molecules are disorientated according to step a is subtracted from the total entropy increase (eq. (4.19)) to obtain the entropy of mixing (step b). The entropy (S_a) of the amorphous state prior to mixing the polymer with solvent is calculated by inserting $n_1 = 0$ into eq. (4.19), which gives:

$$S_a = kn_2 \ln x + kn_2(z-1)\ln\left(\frac{z-1}{e}\right) \quad (4.20)$$

Equation (4.20) is then subtracted from eq. (4.19) to obtain ΔS_{mix}:

$$\Delta S_{mix} = k\left\{-n_1 \ln\left(\frac{n_1}{n_1 + n_2 x}\right) - n_2 \ln\left(\frac{n_2}{n_1 + n_2 x}\right) \right. \\ \left. + n_2(x-1)\ln\left(\frac{z-1}{e}\right)\right\} - kn_2 \ln x \\ - kn_2(x-1)\ln\left(\frac{z-1}{e}\right)$$

$$\Delta S_{mix} = k\left\{-n_1 \ln\left(\frac{n_1}{n_1 + n_2 x}\right) \right. \\ \left. - n_2\left(\ln\left(\frac{n_2}{n_1 + n_2 x}\right) + \ln x\right)\right\}$$

$$= k\left\{-n_1 \ln\left(\frac{n_1}{n_1 + n_2 x}\right) - n_2 \ln\left(\frac{n_2 x}{n_1 + n_2 x}\right)\right\} \quad (4.21)$$

which, after considering that

$$v_1 = \frac{n_1}{n_1 + n_2 x} \quad \text{and} \quad v_2 = \frac{n_2 x}{n_1 + n_2 x}$$

where v_1 and v_2 are the volume fractions of solvent and polymer respectively, becomes:

$$\Delta S_{mix} = -k(n_1 \ln v_1 + n_2 \ln v_2) \quad (4.22)$$

which in molar terms converts to:

$$\frac{\Delta S_{mix}}{N} = -R\left(v_1 \ln v_1 + \frac{v_2}{x} \ln v_2\right) \quad (4.23)$$

where $N = N_1 + xN_2$.

The mixing enthalpy is calculated by considering the interaction energies between the solvent molecules and the solute segments. Let us use the ideas and notation already defined in the calculation of the mixing entropy. Consider a solute segment surrounded by $v_1 z$ neighbours of solvent molecules and $v_2 z$ neighbours of solute segments (Fig. 4.8; left-hand side). The interaction energy is $v_1 z w_{12} + v_2 z w_{22}$. The lattice consists of $n v_2$ solute segments each interacting in this manner. The total interaction energy (H_2) is thus.

$$H_2 = \tfrac{1}{2} n v_2 (v_1 z w_{12} + v_2 z w_{22}) \quad (4.24)$$

The factor $\tfrac{1}{2}$ enters this expression because each pair of interacting species is counted twice by this procedure. In the same way (Fig. 4.8, right-hand side), the interaction energy (H_1) for the solvent molecules is obtained as follows:

$$H_1 = \tfrac{1}{2} n v_1 (v_1 z w_{11} + v_2 z w_{12}) \quad (4.25)$$

Let us then consider the change in interaction energy accompanying the mixing process ($1 + 2 \to$ mixture). The 11 and 22 interaction energies

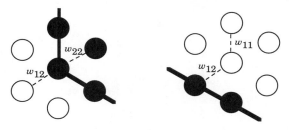

Figure 4.8 Schematic illustration of the interactions between solvent (unfilled) and solute=polymer (filled) molecules.

characteristic of the pure states, here denoted H_{01} and H_{02}, are thus:

$$H_{01} = \tfrac{1}{2} n v_1 z w_{11} \quad (4.26)$$

$$H_{02} = \tfrac{1}{2} n v_2 z w_{22} \quad (4.27)$$

Equations (4.24)–(4.27) are then combined to obtain the enthalpy of mixing (ΔH_{mix}):

$$\Delta H_{mix} = (H_1 + H_2) - (H_{01} + H_{02})$$
$$= \tfrac{1}{2} n z (v_2(v_2 w_{22} + v_1 w_{12})$$
$$+ v_1(v_1 w_{11} + v_2 w_{12}) - v_1 w_{11}$$
$$- v_2 w_{22}) \quad (4.28)$$

which, after rearrangement and insertion of the relationship $v_1 + v_2 = 1$, becomes:

$$\Delta H_{mix} = \tfrac{1}{2} n z v_1 v_2 (2 w_{12} - w_{11} - w_{22}) \quad (4.29a)$$

We then define

$$\Delta w_{12} = w_{12} - \tfrac{1}{2}(w_{11} + w_{22}) \quad (4.29b)$$

which is inserted into eq. (4.29) to give:

$$\Delta H_{mix} = n z v_1 v_2 \Delta w_{12} = n_1 z v_2 \Delta w_{12} \quad (4.30)$$

When the 12 attractions are stronger than the 11 and 22 attractions, both Δw_{12} and ΔH_{mix} are negative, and the mixing process is exothermal. Note that strong interaction forces mean high negative w values. If the 11 and 22 attractions are stronger than the 12 interaction, Δw_{12} and ΔH are positive and the mixing is endothermal. The intermediate case, when ΔH_{mix} and Δw are zero, is called athermal mixing. The different attraction forces are the same here.

Equation (4.30) can be written in the following form (using molar notation):

$$\Delta H_{mix} = RT \chi_{12} N_1 v_2 \quad (4.31)$$

where $\chi_{12} = z \Delta w_{12}/RT$ is called the **interaction parameter**. Equation (4.31) can be rewritten in the form:

$$\frac{\Delta H_{mix}}{N} = RT \chi_{12} v_1 v_2 \quad (4.32)$$

where $N = N_1 + N_2 x$. A combination of eqs (4.23) and (4.32) gives the change in free energy on mixing:

$$\frac{\Delta G_{mix}}{N} = RT\left(v_1 \ln v_1 + \frac{v_2}{x} \ln v_2 + \chi_{12} v_1 v_2\right) \quad (4.33)$$

The first two terms contain the entropic contribution which arises from the different placements of the solute molecules in the matrix of solvent molecules, but it neglects the possible entropy contribution originating from specific interactions between neighbouring solvent and solute molecules. Equation (4.33) states that these interactions only influence the enthalpy.

The chemical potentials (μ_i) and the activities (a_i) of the solvent and the solute can be calculated by application of eqs (4.10) and (4.11) to the Flory–Huggins equation (eq. (4.33)):

$$\frac{\mu_1 - \mu_1^0}{RT} = \ln a_1$$

$$= \ln v_1 + \left(1 - \frac{1}{x}\right)v_2 + \chi_{12} v_2^2 \quad (4.34)$$

$$\frac{\mu_2 - \mu_2^0}{RT} = \ln a_2$$

$$= \ln v_2 + (1 - x)v_1 + x\chi_{12} v_1^2 \quad (4.35)$$

Phase (spinodal) separation occurs when:

$$\frac{d^2(\Delta G_{mix}/N)}{dx_1^2} = 0 \Rightarrow \frac{d\mu_1}{dx_1} = 0 \quad (4.36)$$

and the critical temperature, where the spinodal and binodal curves meet, is associated with the condition:

$$\frac{d^3(\Delta G_{mix}/N)}{dx_1^3} = 0 \Rightarrow \frac{d^2\mu_1}{dx_1^2} = 0 \quad (4.37)$$

which after application to eq. (4.34) yields:

$$-\frac{1}{v_1} + \left(1 - \frac{1}{x}\right) + 2\chi_{12} v_2 = 0 \quad (4.38)$$

$$-\frac{1}{v_1^2} + 2\chi_{12} = 0 \quad (4.39)$$

By combining eqs (4.38) and (4.39), the following critical values for phase separation are obtained:

$$v_{2,c} = \frac{1}{1 + \sqrt{x}} \quad (4.40)$$

$$\chi_{12,c} = \frac{1}{2} + \frac{1}{2x} + \frac{1}{\sqrt{x}} \quad (4.41)$$

Note that the critical value for the interaction parameter is $\frac{1}{2}$ for a polymer of infinite molar mass. The critical temperature associated with phase separation is given by:

$$T_c = \frac{B}{R\chi_{12,c}} \quad (4.42)$$

which for a high molar mass polymer takes the value $T_c = 2B/R$. The critical concentration for regular binary solutions of components of equal size is always $v_1 = v_2 = 0.5$. However, the Flory–Huggins theory predicts a pronounced asymmetry, a shift towards very low values of v_2. For example, v_c takes the approximate value 0.10 for $x = 100$. At temperatures below T_c, both the solvent-rich and polymer-rich phases consist predominantly of solvent. This is according to experience. The critical temperature T_c is also considerably higher (at the same B value) for the polymer–small-molecule mixtures than for the regular solutions:

$$T_c = \frac{2B}{R} \quad \text{(for a low molar mass component-polymer } (M = \infty) \text{ mixture)}$$

$$T_c = \frac{B}{2R} \quad \text{(for a regular solution)}$$

This considerable difference can be traced to the entropy term originating from the polymer component; $(v_2/x)\ln v_2$ (eq. (4.33)) is for large x values considerably smaller than the corresponding term for the regular solution, $x_2 \ln x_2$ (eq. (4.5)).

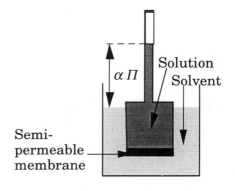

Figure 4.9 Simple experimental set-up for osmosis.

The measurement of osmotic pressure constitutes one of the main techniques for studying polymer solutions. Osmosis is the passage of a pure solvent into a solution separated from it by a semi-permeable membrane, i.e. a membrane permeable only to the solvent molecules and not to the polymer molecules (Fig. 4.9). The osmotic pressure Π is the pressure that must be applied to the solution to stop the flow of solvent molecules through the membrane. The osmotic pressure is related to the solution activity and chemical potential as follows:

$$-\Pi V_1 = RT \ln a_1 = \mu_1^0 - \mu_1 \quad (4.43)$$

where V_1 is the molar volume of the solvent. The chemical potential is, as was shown in eq. (4.34), given by:

$$\mu_1 - \mu_1^0 = RT\left(\ln v_1 + \left(1 - \frac{1}{x}\right)v_2 + \chi_{12}v_2^2\right) \quad (4.44)$$

Combining eqs (4.43) and (4.44):

$$\ln\left(\frac{a_1}{v_1}\right) - \left(1 - \frac{1}{x}\right)v_2 = \chi_{12}v_2^2 \quad (4.45)$$

The interaction parameter χ_{12} can be obtained as the slope coefficient in a plot of $\ln(a_1/v_1) - (1 - 1/x)v_2$ against v_2^2. Deviations from linearity are relatively common, i.e. $\chi_{12} = f(v_2)$. Furthermore, it is also found that the contact energy, $\chi_{12}RT$, is temperature-dependent. This indicates a serious flaw in the initial version of the Flory–Huggins theory. Flory noticed this and modified the theory within the framework of the lattice model. His reasoning was that interactions also lead to reorientation of the solute molecules, which as a result of this interaction differs from the pure component. Thus, it was postulated that there is a contribution to the entropy of mixing which comes from neighbour interaction. The magnitude of this entropy term should be proportional to the number of paired contacts by analogy with the enthalpy contribution. Equation (4.33) is thus correct according to this view, but the third term should not be considered to be purely enthalpic as was originally said. It should be regarded as a free energy term, i.e.:

$$\Delta w_{12} = \Delta w_{h12} - T\Delta w_{s12} \quad (4.46)$$

where Δw_{h12} and Δw_{s12} are the enthalpic and entropic contributions. The interaction parameter is the sum of entropy and enthalpy terms according to the equation:

$$\chi_{12} = \frac{z(\Delta w_{h12} - T\Delta w_{s12})}{RT} = \frac{z\Delta w_{h12}}{RT} - \frac{z\Delta w_{s12}}{R}$$

$$= \chi_{h12} + \chi_{s12} \quad (4.47)$$

The entropy and enthalpy change on mixing are, according to the modified Flory–Huggins theory, given by:

$$\frac{\Delta S_{mix}}{N} = -R\left(v_1 \ln v_1 + \frac{v_2}{x} \ln v_2 + v_1v_2\left(\frac{\partial(\chi_{12}T)}{\partial T}\right)\right) \quad (4.48)$$

$$\frac{\Delta H_{mix}}{N} = -RT^2\left(\frac{\partial \chi_{12}}{\partial T}\right)v_1v_2 \quad (4.49)$$

In cases where the contact energy is temperature-independent, the entropy of mixing is purely combinatorial, i.e. it is expressed by the first two terms in eq. (4.48) and the third term in the Flory–Huggins equation is only enthalpic. Osmometry data from Schulz and Doll (1952) treated according to eq. (4.45) showed that the entropy part of the interaction parameter was significantly larger than the enthalpic part for solutions of polymethylmethacrylate in various solvents, e.g. for toluene as solvent at 20°C: $\chi_{h12} = 0.03$ and $\chi_{s12} = 0.42$.

Figure 4.10 shows the phase diagram for polystyrene in cyclohexane as reported by Saeki et al.

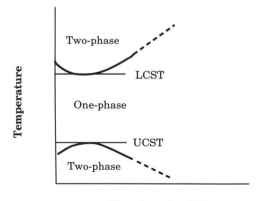

Figure 4.10 Schematic phase diagram for the system cyclohexane-polystyrene (PS). Drawn after data from Saeki et al. (1975).

(1975). The coexistence of upper and lower critical solution temperatures, abbreviated UCST and LCST, is remarkable. This complex phase diagram could not be described by the modified Flory–Huggins theory with a temperature-dependent contact energy and a temperature-independent $z\Delta w_{h12}$. However, it is possible to fit the Flory–Huggins equation to experimental data provided that the enthalpy part of the interaction parameter is given temperature dependence:

$$z\Delta w_{h12} = z\Delta w_{h12}^{\theta} + \int_{\theta}^{T} \Delta C_p \, dt \quad (4.50)$$

where θ is the so-called theta temperature, which is the UCST for a solution of a polymer of infinite molar mass and ΔC_p is the exchange heat capacity. The temperature dependence of the interaction parameter can be calculated from eq. (4.50) according to the scheme utilized in eqs (4.36)–(4.41):

$$\chi_{12} = \frac{1}{2} - \frac{z\Delta w_{h12}}{R\theta}\left(1 - \frac{\theta}{T}\right)$$
$$+ \frac{\Delta C_p}{R}\left(1 - \frac{\theta}{T} + \ln\left(\frac{\theta}{T}\right)\right) \quad (4.51)$$

The presence of both UCST and LCST can only be described by the Flory–Huggins equation in the case of a negative ΔC_p.

The concentration dependence of the osmotic pressure is another important test of the Flory–Huggins theory. We will first derive an expression relating osmotic pressure and polymer concentration and then define the second virial coefficient (A_2).

Combining eqs (4.43) and (4.44) gives:

$$-\Pi V_1 = RT\left(\ln(1-v_2) + \left(1 - \frac{1}{x}\right)v_2 + \chi_{12}v_2^2\right) \quad (4.52)$$

Expansion of $\ln(1-v_2)$ to the two terms $-v_2 - \frac{1}{2}v_2^2$ leads to the modification:

$$\frac{\Pi V_1}{RT} = v_2 + \frac{1}{2}v_2^2 - \left(1 - \frac{1}{x}\right)v_2 - \chi_{12}v_2^2 \quad (4.53)$$

which after simplification reads:

$$\frac{\Pi V_1}{RT} = \frac{v_2}{x} + (\tfrac{1}{2} - \chi_{12})v_2^2 \quad (4.54)$$

The concentration, c, of polymer in the solution (measured in kilograms of polymer per cubic metre of solution) is related to v_2 according to:

$$v_2 = cV_2^0 \quad (4.55)$$

where V_2^0 is the specific volume of the pure polymer. Insertion of eq. (4.55) into eq. (4.54) gives:

$$\frac{\Pi V_1}{RT} = \frac{cV_2^0}{x} + (\tfrac{1}{2} - \chi_{12})(V_2^0)^2 c^2 \quad (4.56)$$

Rearrangement and conversion of x into molar mass (M) give:

$$\frac{\Pi}{RTc} = \left(\frac{V_2^0 M_{rep}}{V_1}\right) \cdot \frac{1}{M} + (\tfrac{1}{2} - \chi_{12}) \cdot \frac{(V_2^0)^2}{V_1} c \quad (4.57)$$

where M_{rep} is the molar mass of the repeating unit of the polymer. The second virial coefficient (A_2) can be identified as the factor multiplied by concentration in the second term, i.e.:

$$A_2 = (\tfrac{1}{2} - \chi_{12}) \cdot \frac{(V_2^0)^2}{V_1} \quad (4.58)$$

When $A_2 = 0$, then $\chi_{12} = \tfrac{1}{2}$ and the solution behaves ideally. This means that the molar mass can in this case be determined on the basis of a single measurement of osmotic pressure. The extrapolation to infinite dilution ($c = 0$) is not necessary. The condition that $A_2 = 0$ marks the borderline between good and poor solvent conditions (Fig. 4.11). These conditions are referred to as **theta** (Θ)

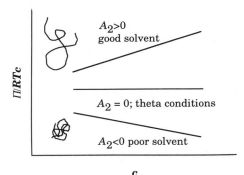

Figure 4.11 Osmotic pressure (Π/cRT) versus polymer concentration. The chain conformation is influenced by the 'goodness' of the solvent.

conditions. The theta state is, as has been discussed in Chapter 1, of particular importance. The polymer chains are unperturbed and the square of their average end-to-end distance $\langle r^2 \rangle_0$ is given by:

$$\langle r^2 \rangle_0 = Cnl^2 \qquad (4.59)$$

where C is a constant which depends on the segmental flexibility of the polymer, n is the number of segments and l is the segment length. The unperturbed state, i.e. the state of a polymer under theta conditions, is characterized by the absence of long-range interactions. The segments of a molecule under theta conditions are arranged in such a way as to indicate that they do not 'sense' the other segments of the same molecule. The molecule behaves as a 'ghost' or a 'phantom' and is sometimes also referred to as a phantom chain. Further details about the unperturbed chains and their conformational states are presented in Chapter 2.

Positive values of A_2 characterize good solvents. This is equivalent to negative or small positive values of χ_{12}. The latter parameter is proportional to both Δw and ΔH_{mix}, which means that these conditions correspond to an exothermic or a small endothermic enthalpy of mixing. Negative values of A_2 ($\chi_{12} > \frac{1}{2}$) are indicative of poor solvent conditions and a large endothermic enthalpy of mixing. This discussion relates ideally to polymers of infinite molar mass. Phase separation is thus predicted to occur at $\chi_{12} = \frac{1}{2}$. For low molecular mass polymers, on the other hand, this critical point is shifted somewhat towards higher χ_{12} values (eq. (4.41)).

The 'goodness' of a particular solvent–polymer system can be adjusted by changing the temperature. It has been shown by Flory that

$$\chi_{12} = \frac{1}{2} - \frac{z\Delta w_{h12}}{R\theta}\left(1 - \frac{\theta}{T}\right) \qquad (4.60)$$

where θ is known as the **theta temperature** or **Flory temperature**. When $T = \theta$, then χ_{12} is equal to $\frac{1}{2}$ and $A_2 = 0$.

The result of osmotic pressure measurements analysed according to eq. (4.57) showed, however, that A_2 is dependent on the molar mass of the polymer. This contradicts the Flory–Huggins theory; A_2 should be independent of molar mass (eq. (4.58)).

These problems arise from the mean-field assumption used to place the chain segments in the lattice. This picture is more correct in concentrated solutions where the polymer molecules interpenetrate and overlap. It is certainly not a good view of diluted solutions in which the polymer molecules are well separated. In the latter case it is obvious that the concentration of polymer segments is highly non-uniform in the solution.

4.4 CONCENTRATION REGIMES IN POLYMER SOLUTIONS

It is convenient to partition polymer solutions into three different cases according to their concentration. Dilute solutions involve only a minimum of interaction (overlap) between different polymer molecules. The Flory–Huggins theory does not represent this situation at all well due to its mean-field assumption. The semi-dilute case involves overlapping polymer molecules but still with a considerable separation of the segments of different molecules.

In this discussion we will now confine ourselves to good solvents with interaction parameter values smaller than 0.5. This simplifies the treatment since the enthalpy contribution becomes negligible. Let us first calculate the threshold concentration (c^*) for molecular overlap. It obviously depends on the molar mass (M) of the polymer, and the following relationship can be demonstrated:

$$c^* \propto M^{-4/5} \quad \text{(for a good solvent: } \langle r^2 \rangle^{1/2} \propto M^{3/5}\text{)} \qquad (4.61)$$

The dilute solution ($c \ll c^*$) with considerable separation of the coils can be treated as a perfect 'gas' and the osmotic pressure (Π) is simply proportional to the concentration of coils:

$$\Pi = \frac{c}{x} \cdot RT \qquad (4.62)$$

where x is the number of mers in the polymer. In a good solvent the coils repel each other and behave like hard spheres. A more accurate description of the concentration dependence, still for solutions in the diluted regime, is given by:

$$\frac{\Pi}{RT} = \frac{c}{x} + \text{const} \cdot \left(\frac{c}{x}\right)^2 R_s^3 \qquad (4.63)$$

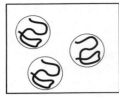  

Dilute solution $c \ll c^*$ Semi-dilute solution $c = c^*$ Con. solution $c > c^*$

Figure 4.12 Schematic descriptions of the state of polymer molecules in solutions with different polymer concentrations.

where R_s is the radius of the hard spheres. The second virial coefficient is thus related to the molar mass as follows:

$$A_2 \propto M^{-1/5}$$

The semi-dilute regime constitutes a sizeable concentration range: $v_2^* \ll v_2 \ll 1$, where v_2^* is the volume fraction corresponding to the concentration c^*. The semi-dilute solution case is in fact the simplest to describe. Molar mass effects are highly suppressed but still the solution is diluted. The solution activity and the osmotic pressure are independent of molar mass but dependent on the distance between the entanglement points in the loose network. The latter is controlled simply by the volume fraction of polymer (v_2), and the osmotic pressure is given by des Cloiseaux' law:

$$\frac{\Pi V_1}{RT} \propto (v_2)^{9/4} \quad (4.64)$$

or, expressed differently in concentration terms:

$$\frac{\Pi M}{cRT} \propto \left(\frac{c}{c^*}\right)^{5/4} \quad (4.65)$$

Empirical osmotic pressure data yield an exponent of 1.33 instead of the theoretical value of 1.25 (Fig. 4.13).

The osmotic pressure can also be directly related to the mesh size ξ of the network:

$$\frac{\Pi}{RT} \propto \xi^{-3} \quad (4.66)$$

Another interesting phenomenon in semi-dilute solutions is that the molecular coils decrease in size with increasing polymer concentration. The segments of a given chain becomes 'screened' by the nearby chain segments of other molecules and it loses its memory from its own segments and behaves more like unperturbed chain despite the goodness of the solvent. Theory predicts the following scaling law:

$$\langle s^2 \rangle \propto c^{-1/4} \quad (4.67)$$

Neutron scattering data give an exponent of -0.16 compared to the theoretical value of -0.25.

4.5 THE SOLUBILITY PARAMETER CONCEPT

In cases where the attraction forces are non-specific, e.g. when hydrogen bonds are absent, it can be assumed that w_{12} is the geometrical mean value of w_{11} and w_{22}:

$$w_{12} = -\sqrt{w_{11} w_{22}}$$

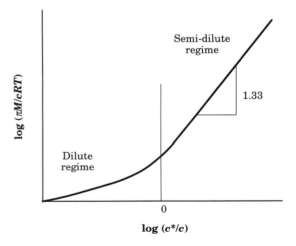

Figure 4.13 $\Pi M/cRT$ as a function of reduced polymer concentration c/c^*. Schematically drawn after data from Noda *et al.* (1981).

which, after insertion in eq. (4.29b), gives

$$\Delta w_{12} = -\left(\frac{w_{11} + w_{22}}{2} - \sqrt{w_{11} w_{22}}\right)$$

$$= \frac{(\sqrt{|w_{11}|} - \sqrt{|w_{22}|})^2}{2} \quad (4.68)$$

In this case, Δw_{12} and ΔH_{mix} are always greater than zero.

When specific interactions (hydrogen bonds) are involved eq. (4.68) is not applicable. The **cohesive energy density** (CED), is defined as the molar energy (ΔE_v) of vaporization per unit molar volume (V_1) which holds the molecules together, i.e. it is a measure of the **inter**molecular energy. The square root of the CED is called the **Hildebrand solubility parameter** or simply the **solubility parameter** and is denoted δ. When specific bonds are absent, the enthalpy of mixing becomes:

$$\Delta H_{mix} = z n_1 v_2 \Delta w_{12} = \tfrac{1}{2} z n_1 v_2 (\sqrt{|w_{11}|} - \sqrt{|w_{22}|})^2$$
$$\propto (\delta_1 - \delta_2)^2 \quad (4.69)$$

The CED is strictly defined as:

$$\delta^2 = \frac{\Delta E_v}{V_1} \quad (4.70)$$

The Hildebrand solubility parameter is readily obtained for liquids by inserting the appropriate experimentally determined quantities into eq. (4.70). Polymers degrade prior to vaporization and their solubility parameter values are determined indirectly by one of two essentially different techniques. In the first technique, the polymer is lightly crosslinked and then treated with a number of solvents with different solubility parameters. The best solvent, the one which swells the polymer the most, is then the one which has a solubility parameter which resembles the solubility parameter of the polymer (Fig. 4.14). The other technique involves the measurement of the intrinsic viscosity of solutions of the polymer in a number of solvents of different solubility parameters.

The intrinsic viscosity ($[\eta]$) is defined as follows:

$$[\eta] = \left(\frac{\ln \eta_{rel}}{c_2}\right)_{c_2 = 0} \quad (4.71)$$

where η_{rel} is the relative viscosity, which is equal to $\eta/\eta_s \approx t/t_s$, where t and t_s are the flow-through

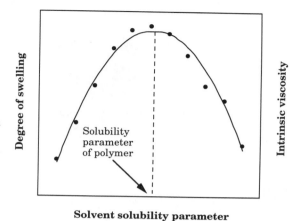

Figure 4.14 Illustration of methods for determining the solubility parameter of polymers.

times in the viscometer for the polymer solution and pure solvent, respectively, and c_2 is the concentration of polymer in the solution.

The solubility parameter of the polymer is obtained as the solubility parameter value corresponding to the maximum in the intrinsic viscosity–solubility parameter plot (Fig. 4.14).

The Hildebrand solubility parameter can be calculated theoretically using molecular group contributions on the basis of Small's formula:

$$\delta = \frac{\rho \sum G}{M} \quad (4.72)$$

where ρ is the density, $\sum G$ is the sum of the molar attraction constants of the groups of the repeating unit and M is the molar mass of the repeating unit.

Hansen (1967) recognized the fact that interactions are of different kinds, namely dispersive (non-polar), polar and hydrogen-bond interactions, and expressed the solubility parameter as:

$$\delta^2 = \delta_D^2 + \delta_P^2 + \delta_H^2 \quad (4.73)$$

where δ_D, δ_P and δ_H are the solubility parameter values relating to the dispersive, polar and hydrogen bond interactions, respectively. A low molar mass compound can thus be described by the three parameter values (δ_D, δ_P, δ_H). A polymer is described by four parameters: δ_D, δ_P, δ_H and R_{AO}, the latter being the radius of the 'solubility

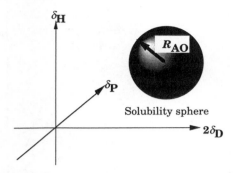

Figure 4.15 Solubility sphere in three-dimensional solubility parameter space.

sphere' in the $2\delta_D - \delta_P - \delta_H$ 'space' (Fig. 4.15). The necessity of doubling the δ_D axis was shown from Hansen's experimental data. Liquids within this sphere are solvents for the polymer and liquids appearing outside the sphere are non-solvents. This is expressed in the following expressions:

$$R_A^2 = (2\delta_{D,p} - 2\delta_{D,s})^2 + (\delta_{P,p} - \delta_{P,s})^2 + (\delta_{H,p} - \delta_{H,s})^2 \quad (4.74)$$

$$\frac{R_A}{R_{AO}} < 1 \quad \text{(solvent)}$$

$$\frac{R_A}{R_{AO}} > 1 \quad \text{(non-solvent)}$$

The indices 'p' and 's' refer to polymer and solvent, respectively.

4.6 EQUATION-OF-STATE THEORIES

Several flaws in the Flory–Huggins mean-field theory have been discussed in the previous section. The theory does not explicitly show any volume dependence of the free energy. The so-called **equation-of-state** theories have in common that they are founded on the expression $A = f(T, V)$ where A is the Helmholtz free energy. An equation of state, $p = f(T, V)$, can be formulated because $p = (\partial A / \partial V)_T$.

The common variables which are used in the different theories are the reduced variables $\tilde{p}$, $\tilde{v}$ and $\tilde{T}$ which are defined as follows:

$$\tilde{p} = \frac{p}{p^*} = \frac{pv^*}{z\varepsilon} \quad (4.75)$$

$$\tilde{v} = \frac{v}{v^*} \quad (4.76)$$

$$\tilde{T} = \frac{T}{T^*} = \frac{kT}{z\varepsilon} \quad (4.77)$$

where p, v and T are the pressure, specific volume and temperature, v^* is the hard-core specific volume, ε is the well depth in the interaction potential, z is the coordination number of the lattice, k is the Boltzmann constant, T^* is the hard-core temperature and p^* is the hard-core pressure.

The first equation-of-state theory developed for polymers was due to Flory, Orwoll and Vrij (1964) and Eichinger and Flory (1968), the so-called FOVE model. Each of the components is characterized by p^*, v^* and T^* obtained from data for the pure components. Two interaction terms, denoted X_{12} and Q_{12}, associated with the enthalpy and entropy of the mixture, are introduced. The equation of state is:

$$\frac{\tilde{p}\tilde{v}}{\tilde{T}} = \frac{\tilde{v}^{1/3}}{\tilde{v}^{1/3} - 1} - \frac{1}{\tilde{T}\tilde{v}} \quad (4.78)$$

which at normal pressure simplifies to:

$$\tilde{T} = \frac{\tilde{v}^{1/3} - 1}{\tilde{v}^{4/3}} \quad (4.79)$$

By introduction of the thermal expansion coefficient (α) and further rearrangement, the following expression is obtained:

$$\tilde{v}^{1/3} = \frac{3 + 4\alpha T}{3 + 3\alpha T} \quad (4.80)$$

The thermal pressure coefficient (γ) gives p^*:

$$\gamma = \left(\frac{\partial p}{\partial T}\right)_v = \frac{p^*}{T^*}\left(\frac{\partial \tilde{p}}{\partial \tilde{T}}\right)_{\tilde{v}} \quad (4.81)$$

The last derivative is obtained by differentiating eq. (4.78), which after insertion into eq. (4.81) gives:

$$p^* = \gamma T \tilde{v}^2 \quad (4.82)$$

So-called 'site fractions' (θ_1 and θ_2) are defined according to:

$$\theta_2 = \left(\frac{S_2}{S_1}\right) \frac{v_2}{(v_1 + (S_2/S_1)v_2)} \quad (4.83)$$

$$\theta_1 = 1 - \theta_2 \quad (4.84)$$

where S_i is the surface area per unit volume ratio for component i. The hard-core pressure of the mixture is given by:

$$p^* = v_1 p_1^* + v_2 p_2^* - v_1 \theta_2 X_{12} \quad (4.85)$$

and the hard-core temperature of the mixture by:

$$\frac{p^*}{T^*} = \frac{v_1 p_1^*}{T_1^*} + \frac{v_2 p_2^*}{T_2^*} \quad (4.86)$$

The mixing enthalpy can be expressed in terms of the quantities defined above as follows:

$$\Delta H_{mix} = (m_1 v_{sp1}^* + m_2 v_{sp2}^*)\left(\frac{v_1 p_1^*}{\tilde{v}_1} + \frac{v_2 p_2^*}{\tilde{v}_2} - \frac{p^*}{\tilde{v}}\right) \quad (4.87)$$

where m_i is the mass of component i, and v_{spi}^* is the specific volume of component i. The hard-core pressure depends on X_{12}, the latter being an adjustable parameter to be determined by fitting of eq. (4.87) to experimental mixing enthalpy data. The enthalpic interaction parameter (χ_H) is obtained from the partial molar heat of mixing ($\Delta \bar{H}_1$):

$$\Delta \bar{H}_1 = p_1^* V_1^* \left[\left(\frac{1}{\tilde{v}_1} - \frac{1}{\tilde{v}}\right) + \frac{\alpha T}{\tilde{v}}\left(\frac{\tilde{T}_1 - \tilde{T}}{\tilde{T}}\right)\right]$$

$$+ \frac{V_1^* X_{12}}{\tilde{v}}(1 + \alpha T)\theta_2^2$$

$$= RT\chi_H v_2^2 \quad (4.88)$$

where V_1^* is the molecular hard-core volume of component 1.

The entropic interaction parameter (χ_S) is obtained from the partial molar excess entropy change of mixing ($\Delta \bar{S}_{1,exc}$) (not including the combinatorial part):

$$T\Delta\bar{S}_{1,exc} = -p_1 V_1^* \left[3\tilde{T}_1 \frac{\ln(\tilde{v}_1^{1/3} - 1)}{\tilde{v}^{1/3} - 1} - \alpha T \frac{(\tilde{T}_1 - \tilde{T})}{\tilde{T}\tilde{v}}\right]$$

$$+ V_1^* \theta_2^2 \left(\frac{\alpha T X_{12} + \tilde{T}\tilde{v}Q_{12}}{\tilde{v}}\right)$$

$$= RT\chi_S v_2^2 \quad (4.89)$$

The chemical potentials of the components associated with the binodal concentrations are given by:

$$\mu_1 - \mu_1^0 = RT\left(\ln v_1 + \left(1 - \frac{x_1}{x_2}\right)v_2\right)$$

$$+ p_1^* V_1^* \left[3T_1^* \frac{\ln(\tilde{v}_1^{1/3} - 1)}{(\tilde{v}^{1/3} - 1)}\right.$$

$$+ (\tilde{v}_1^{-1} - \tilde{v}^{-1}) + \tilde{p}_1(\tilde{v}^{-1} - \tilde{v}_1^{-1})\bigg]$$

$$+ V_1^* \theta_2^2 \frac{(X_{12} - T\tilde{v}Q_{12})}{\tilde{v}} \quad (4.90)$$

$$\mu_2 - \mu_2^0 = RT\left(\ln v_2 + \left(1 - \frac{x_2}{x_1}\right)v_1\right)$$

$$+ p_2^* V_2^* \left[3T_2^* \frac{\ln(\tilde{v}_2^{1/3} - 1)}{(\tilde{v}^{1/3} - 1)}\right.$$

$$+ (\tilde{v}_2^{-1} - \tilde{v}^{-1}) + \tilde{p}_2(\tilde{v}^{-1} - \tilde{v}_2^{-1})\bigg]$$

$$+ V_2^* \theta_1^2 \frac{(X_{12} - T\tilde{v}Q_{12})S_2}{S_1 \tilde{v}} \quad (4.91)$$

where X_1 and X_2 are the chain length of polymers 1 and 2, and V_2^* is the molecular hard-core volume of component 2.

The FOVE model can predict negative mixing enthalpy and mixing volume as well as both LCST and UCST. The reorientation of segments arising from specific interaction is, however, ignored. The theory uses an entropy contact parameter, Q_{12}, that counterbalances any effect that specific interaction has on the excess energy of the mixture.

Several other equation-of-state models have been proposed: The 'lattice-fluid theory' of Sanchez and Lacombe (1978), the 'gas-lattice model' proposed by Koningsveld (1987), the 'strong interaction model' proposed by Walker and Vause (1982), and the group contribution theory proposed by Holten-Anderson (1992), etc. These theories are reviewed by Miles and Rostami (1992) and Boyd and Phillips (1993). The lattice-fluid theory of Sanchez and Lacombe has similarities with the Flory–Huggins theory. It deals with a lattice, but with the difference from the Flory–Huggins model in that it allows vacancies in the lattice. The lattice is 'compressible'. This theory is capable of describing both UCST and LCST behaviour.

4.7 POLYMER–POLYMER BLENDS

4.7.1 ASSESSMENT OF MISCIBILITY OF POLYMER–POLYMER BLENDS

In strict thermodynamical terms, miscibility refers to a single-phase system at a molecular level. In practical terms, it simply means that the system appears to be homogeneous at a level assessed by the particular test performed. In the literature, apparently conflicting data are presented for a given polymer–polymer system. This is a consequence of the fact that the miscibility has been assessed with different experimental techniques.

Turbidity measurement is a classical method, based on the apparent turbidity of multi-phase systems with significant differences in refractive indices. (Im)miscibility can be judged down to a level of micrometres. To this group of methods belongs the light scattering technique. Scattering of radiation of short wavelengths, X-ray and thermal neutrons with $\lambda \approx 0.1$–0.3 nm, permits fine structures to be resolved. The small-angle techniques, small-angle X-ray scattering (SAXS) and small-angle neutron scattering (SANS), can make assessments down to 5 nm (SAXS) and 50 nm (SANS). The requirement for the analyses is that the electron densities (SAXS) and the atomic numbers (SANS) differ between the different polymers. Wide-angle X-ray scattering (WAXS) can resolve even finer detail, in this case down to true molecular levels (0.1–1 nm).

The glass transition temperature (T_g) of blends of fully amorphous polymers is frequently measured to assess miscibility. There are a number of experimental techniques: calorimetric (DSC/DTA), dilatometric, dynamic mechanical and dielectric. The resolution of the T_g methods has been the object of some discussion and the proposed resolution limit for domains should range from 2 to 15 nm. The presence of two T_g's in a binary system indicates phase separation on a level greater than this minimum domain size. Figure 4.16 shows schematically the recorded glass transition temperature(s) as a function of composition for case a (complete miscibility), case b (partial miscibility) and case c (complete immiscibility).

One prerequisite for the applicability of the T_g methods is that the difference in T_g (ΔT_g) must be

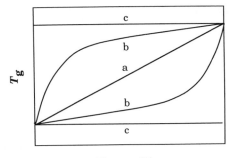

Figure 4.16 Glass transition temperatures as a function of composition in fully amorphous binary polymer–polymer blends. Cases a to c are described in the text.

sufficiently large. Thus, the T_g method should not be used in binary systems for which ΔT_g is smaller than 20 K. This is particularly true for systems with components showing broad glass transitions and also for systems with a distinct minor component present in a concentration of less than 20%. The width of the glass transition has also been used as an indicator of miscibility in 'borderline' cases. A narrow glass transition is typical of a miscible system, whereas immiscible systems exhibit a broad glass transition.

A number of equations have been proposed for relating the glass transition temperature of a miscible blend to the composition and the glass transition temperatures (T_{gi}) of its constituents. The Fox equation (1956) is one of the better known:

$$\frac{1}{T_g} = \sum_{i=1}^{n} \frac{w_i}{T_{gi}} \quad (4.92)$$

where w_i is the mass fraction of component i, and n is the number of components. Another equation was suggested by Utracki and Jukes (1984):

$$\frac{\ln T_g}{T_g} = \sum_{i=1}^{n} \frac{w_i \ln T_{gi}}{T_{gi}} \quad (4.93)$$

derived on the premise that $T_{gi}\Delta c_{pi}$ (Δc_{pi} is the change in specific heat associated with the glass transition of component i) was constant, which was in fact confirmed by these authors for most linear polymers.

It has been shown by Rodriguez-Parada and Percec (1986a, b) that certain miscible blends show a higher T_g than any of the constituents. This was explained as being due to a very strong intermolecular interaction between the components, in the form of electron-donor-acceptor complexes.

Spectroscopy in the form of nuclear magnetic resonance (NMR) spectroscopy and infrared (IR) spectroscopy is useful in the assessment of miscibility. The NMR method measures the mobility of specific groups which is influenced by miscibility. The parameters used in the assessment of miscibility are the half-lives of the spin relaxation times T_1, the spin-lattice, T_2, the spin-spin; and $T_{1\rho}$, the spin-lattice in the rotating frame. In the T_1 process the extra energy is transferred to the surrounding atoms, whereas T_2 involves transfer to the adjacent nuclei. The success of IR spectroscopy in the assessment of miscibility in certain polymer–polymer blends is due to the fact that hydrogen bonds are very strong and that they affect the molecular vibrations of nearby groups. Absorption bands which have been shown to be shifted in miscible polymer blends are the –OH band at 3500–3600 cm^{-1}, the C=O stretching bands at 1700–1750 cm^{-1}, and the –CH$_2$– symmetrical stretching band at 2886 cm^{-1}. Both NMR and IR are sensitive to very local structures, down to the molecular level.

Microscopy provides detailed information about miscibility and about phase morphology, i.e. the actual geometry of the phases. Optical microscopy resolves structures down to about 1 µm. The samples may need staining prior to examination. In other cases, where the refractive index mismatch is sufficiently large, direct examination can be made in the microscope using phase-contrast or interference-contrast optical microscopy.

Scanning electron microscopy (SEM) provides more detailed information on the morphology; domains down to a size of 10 nm can be resolved. Freeze fracturing and etching using solvents or degrading etchants are common techniques used to reveal the different phases.

Transmission electron microscopy (TEM) involves complex and tedious preparation of the samples, necessary to reveal the microphases. The work is well worthwhile, however, and the resolution of TEM is superior to optical microscopy and SEM. Domains of sub-nanometre size can be resolved by TEM. A classical example of a preparation technique is the

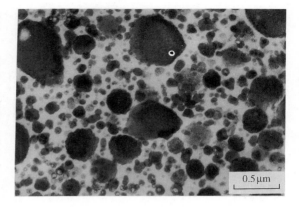

Figure 4.17 Transmission electron micrograph of ABS. Domains of poly(1,4-butadiene) appear dark. (B. Terselius, Dept of Polymer Technology, Royal Institute of Technology, Stockholm).

staining of unsaturated rubber polymers with OsO$_4$ which adds to the double bonds, giving this phase a very high density. Blends of, for example, polystyrene and polybutadiene can be stained in this way (Fig. 4.17). The rubber domains appear dark in the transmission electron micrographs. Staining with RuO$_4$, capable also of reacting with aromatic structures, has also been used, for instance in giving contrast to blends of polycarbonate and poly(butylene terephthalate) and polycarbonate and poly(styrene-co-maleic anhydride). More details about staining techniques and about optical and electron microscopy can be found in Chapter 11.

Figure 4.18 presents a summary of different techniques used for the assessment of polymer–polymer miscibility.

4.7.2 MISCIBILITY OF POLYMER–POLYMER BLENDS: PHASE DIAGRAMS AND MOLECULAR INTERPRETATION

The combinatorial entropy is very small in polymer–polymer blends. Miscibility in polymer–polymer blends can normally only be achieved when the heat of mixing is negative. In molecular terms, that is accomplished by specific interactions between different molecules. The most common type of specific interaction is the hydrogen bond. Examples of

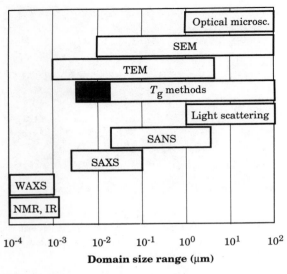

Figure 4.18 Size range covered by different experimental techniques for the assessment of miscibility.

indicates that the cause for miscibility is the similarity in the physical parameters of the components: $\alpha_1 \approx \alpha_2$, $\gamma_1 \approx \gamma_2$, and X_{12}, $Q_{12} \to 0$. Polyetheretherketone and polyetherimide, which are miscible at all compositions, belong to this class of blends. Infrared spectroscopy showed no sign of specific interaction for these blends. Polymethylmethacrylate and polyvinylacetate are other examples. Polyvinylfluoride and polymethylmethacrylate showed only a minor shift in the frequency of the carbonyl-stretch infrared band and since the system shows UCST behaviour, there is no reason to think about specific interactions in these blends.

Typical of polymer blends is that phase separation occurs when the temperature is increased to a certain lower critical solution temperature (LCST) (Fig. 4.19). The upper critical solution temperature (UCST) behaviour predominates in systems with small-molecule solvents (Fig. 4.19). The equation-of-state theories can relatively adequately describe the phase diagrams of polymer blends.

hydrogen-bonding mixtures are polymers with halogen atoms blended with polymers containing ester groups. Infrared spectroscopy confirms that weak hydrogen bonds were formed between chlorinated polyethylene and poly(ethylene-co-vinyl acetate). Electron–electron induction between phenyl groups and oxygen in carbonyls and ethers is another type of specific interaction which is operative in the miscible blends of poly(2,6-dimethyl-1,4-phenylene) oxide (PPO) and polystyrene.

Another group of miscible blends involves polymers of great similarity without the potential of specific interaction. Applying equation-of-state theory

There is a considerable technical interest in making stable and reproducible blends of immiscible polymers. Different strategies are possible to achieve finely dispersed systems and the following list may serve as examples:

- by introducing a third component, a block copolymer consisting of blocks of groups being miscible with the other two polymers of the blend. This kind of substance is often referred to as a compatibilizer.
- by promoting co-reactions between the polymers.
- by crosslinking one of the components.
- by modification of the polymers, introducing

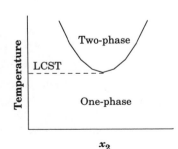 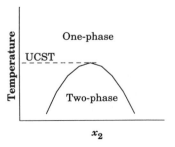

Figure 4.19 Schematic phase diagrams showing lower and upper critical solution temperatures, LCST and UCST.

groups with potential specific interaction, e.g. acid/base groups, hydrogen-bonding groups and charge-transfer complexes.

4.8 SUMMARY

The thermodynamics of polymers in solutions is one of the major topics in the science and technology of polymers. A necessary but not sufficient condition for miscibility is $\Delta G_{mix} < 0$ (ΔG_{mix} is the free energy of mixing). If the function $\Delta G_{mix}(x_1)$, where x_1 is the molar fraction of component 1, is concave with no inflection points, miscibility occurs at all compositions. The second case shows two or more inflection points, i.e. the second derivative of $\Delta G_{mix}(x_1)$ with respect to x_1 is equal to zero at these so-called 'spinodal' points. Complete miscibility is limited to compositions 'outside' the 'binodal' concentrations; this appears where the chemical potentials of the components are equal in both phases, i.e. the binodal points have a common tangent in the ΔG_{mix}–x_1 diagram. The binodal curve defines the equilibrium between stable one-phase systems and heterogeneous systems. Spinodal decomposition occurs in the concentration range between the spinodal points. This process has no free energy nucleation barrier.

The free energy of mixing for a binary mixture of low molar mass components in the absence of specific interaction between the components is adequately described by the regular solution model according to the following equation:

$$\frac{\Delta G_{mix}}{N} = Bx_1x_2 + RT(x_1 \ln x_1 + x_2 \ln x_2) \quad (4.94)$$

where $\Delta G_{mix}/N$ is the free energy of mixing per mole, x_i are molar fractions, B is excess enthalpy given by $(\sqrt{E_1} - \sqrt{E_2})^2$, in which E_i is the heat of vaporization of the components, R is the gas constant and T is temperature. The regular solution model predicts phase separation at low temperatures, i.e. an upper critical solution temperature (UCST).

The following expression was derived by Flory (1942) and Huggins (1942a, b, c,) for mixtures of a polymer and a low molar mass solvent:

$$\frac{\Delta G_{mix}}{N} = RT\left(v_1 \ln v_1 + \frac{v_2}{x} \ln v_2 + \chi_{12}v_1v_2\right) \quad (4.95)$$

where v_i are the volume fractions of the components, x is the number of segments of the polymer (component no. 2) and χ_{12} is the interaction parameter, which is equal to $\Delta w_{12}/RT$, Δw_{12} being the excess energy involved in nearest-neighbour interaction. The Flory–Huggins theory is based on a mean-field approximation assuming uniform concentration of the polymer segments in the solution.

In the original Flory–Huggins theory, the last term of eq. (4.95) was considered as purely enthalpic, but experimental data proved that $RT\chi_{12}$ shows temperature dependence. This led to a modification of the theory and the contact energy term ($RT\chi_{12}$) was considered as a free energy term with enthalpic and entropic parts. In fact, in many cases, the entropy part of the interaction parameter is far greater than the enthalpy part. The early versions of the Flory–Huggins theory were capable of describing only UCST behaviour. However, certain solutions showed phase separation at high temperatures, i.e. a lower critical solution temperature (LCST). In a further modification of the theory, by giving the enthalpic part of the interaction parameter a certain temperature dependence, it became possible also to describe LCST behaviour.

Solution activity data obtained by osmometry on dilute solutions showed that the second virial coefficient is dependent on molar mass, contradicting the Flory–Huggins theory. These problems arise from the mean-field assumption used to place the segments in the lattice. In dilute solutions, the polymer molecules are well separated and the concentration of segments is highly non-uniform. Several 'scaling laws' were therefore developed for dilute ($c < c^*$; c is the polymer concentration in the solution, c^* is the threshold concentration for molecular overlap) and semi-dilute ($c \geq c^*$) solutions. In a good solvent the threshold concentration is related to molar mass as follows:

$$c^* \propto M^{-4/5} \quad (4.96)$$

The osmotic pressure (Π) of dilute solutions is related to molar mass according to the following expression:

$$\frac{\Pi}{RT} \propto \frac{c}{x} + \text{const.}\left(\frac{c}{x}\right)^2 R_s^3 \quad (4.97)$$

where x is the number of mers in the polymer and R_s is the radius of the spherical molecular coil. The second virial coefficient is proportional to $M^{-1/5}$ in a good solvent. In the semi-dilute regime molar mass effects are suppressed and the osmotic pressure is only dependent on the polymer concentration:

$$\frac{\Pi}{RT} \propto (v_2)^{9/4} \propto \left(\frac{c}{c^*}\right)^{5/4} \quad (4.98)$$

Serious flaws in the Flory–Huggins mean-field theory – e.g. that it does not consider volume effects – led to the development of a whole family of theories called equation-of-state theories. These models can predict both LCST and UCST, and negative mixing enthalpy and mixing volume. Some of the equation-of-state theories are specifically developed to account for specific interaction.

The solubility parameter (δ) is defined as the square root of the molar energy (ΔE_v) of vaporization per unit molar volume (V_1) which holds the molecules together, i.e. a measure of the intermolecular energy. Miscibility prediction based on similarity in solubility parameter is widely used. Hansen recognized the fact that interactions are of different kinds: dispersive (non-polar), polar and hydrogen-bond interactions. Hansen expressed the solubility parameter as:

$$\delta^2 = \delta_D^2 + \delta_P^2 + \delta_H^2 \quad (4.99)$$

where δ_D, δ_P and δ_H are the solubility parameter values relating to the dispersive, polar and hydrogen-bond interactions, respectively. Miscibility could be predicted by measuring the difference in three-dimensional δ space, the only precaution being that the δ_D axis should be doubled.

Miscibility assessment of polymer blends can be made by several techniques: turbidity measurements, scattering methods using X-rays or light, measurement of the glass transition temperature, infrared spectroscopy, nuclear magnetic resonance spectroscopy, and optical and electron microscopy. These techniques show great differences in their resolution, some of them detecting miscibility down to true segmental levels whereas others only detect micrometre domains. In the literature, apparently conflicting data are presented for a given polymer–polymer system. This is a consequence of the fact that the miscibility has been assessed with different experimental techniques. Typical of polymer blends is that phase separation occurs when the temperature is increased to a certain LCST. The combinatorial entropy is very small in polymer–polymer blends. Miscibility in these blends can normally only be achieved when the heat of mixing is negative. In molecular terms, this is accomplished by specific interactions between different molecules. The most common type of specific interaction is the hydrogen bond. Another group of miscible polymer blends involves polymers of great similarity without the potential of specific interaction. Application of equation-of-state theory indicates that the cause for miscibility in these cases is the similarity in the physical parameters of the components.

4.9 EXERCISES

4.1. (a) Calculate the number of different ways of arranging 10 solute molecules in a lattice of 100 positions. Each solute molecule occupies one lattice position.
(b) Calculate the number of different ways of arranging an oligomer consisting of 10 repeating units in a lattice of 100 positions. Each repeating unit occupies one lattice position.
(c) What are the main differences between binary mixtures based on (i) low molar mass species (L) and polymer (P) and (ii) L/L?

4.2. Construct the binodal and spinodal curves in a $T–x_1$ diagram by using the regular solution model with $B = 8$ kJ mol^{-1}.

4.3. Derive an expression for the spinodal concentration based on the regular solution model.

4.4. Plot the critical parameters ($v_{2,c}$, $\chi_{12,c}$ and T_c) for a mixture of small-molecule solvent (1) and polymer (2) as a function of x (the number of segments in the polymer). Assume that $B = 2$ kJ mol^{-1}. Draw schematic phase diagrams ($T–v_2$) for polymers of different molar mass.

4.5. Plot the molar mass dependence of the theshold concentration (c^*) for molecular overlap in a polymer solution.

4.6. In what kind of conformational state are polymers in a: (i) good solvent, dilute solution; (ii) theta solvent, dilute solution; (iii) good solvent, semi-dilute solution; (iv) molten state?

4.7. It is well known that polyvinylchloride (PVC) is not miscible with its own monomer (VC). This gives the polymer a characteristic particle structure which remains even after melt-processing and leads to poor mechanical properties. The solubility parameters of PVC are: $\delta_D = 18.2$, $\delta_P = 7.5$, $\delta_H = 8.3$, $R_{AO} = 3.5$ (Barton 1983). The solubility parameter values of VC are: $\delta_D = 15.4$, $\delta_P = 8.1$, $\delta_H = 2.4$. Confirm that PVC is immiscible with VC on the basis of these data.

4.8. A miscible blend shows a higher glass transition temperature than the pure polymers. Suggest a possible explanation. Suggest also another experiment which will critically test your hypothesis.

4.10 REFERENCES

Barton, A. F. M. (1983) *Handbook of Solubility Parameters and Other Cohesion Parameters*. CRC Press, Boca Raton, FL.

Boyd, R. H. and Phillips, P. J. (1993) *The Science of Polymer Molecules*. Cambridge University Press, Cambridge.

Eichinger, B. E. and Flory, P. J. (1968) *Trans. Faraday Soc.* **64**, 2035, 2053, 2061, 2066.

Flory, P. J. (1942) *J. Chem. Phys.* **10**, 51.

Flory, P. J., Orwoll, R. A. and Vrij, A. (1964) *J. Am. Chem. Soc.* **86**, 3507, 3515.

Fox, T. G. (1956) *Bull. Am. Phys. Soc.* **1**, 123.

Hansen, C. M. (1967) Three Dimensional Solubility Parameter and Solvent Diffusion Coefficient. Importance in Surface Coating Formulation, Doctoral Dissertation, Danish Technical Press, Copenhagen.

Hildebrand, J. H. and Wood, S. E. (1932) *J. Chem. Phys.* **1**, 817.

Holten-Anderson, J., Fredenshind, A., Rasmussen, P. and Carvoli, G. (1992) (personal communication) in *Multicomponent Polymer Systems* (eds Miles, I. S. and Rostami, S.) Longman, UK.

Huggins, M. L. (1942a) *Ann. NY Acad. Sci.* **41**, 1.

Huggins, M. L. (1942b) *J. Phys. Chem.* **46**, 151.

Huggins, M. L. (1942c) *J. Am. Chem. Soc.* **64**, 1712.

Koningsveld, R., Kleintjens, L. A. and Leblans-Vinckl, A. (1987) *J. Phys. Chem.* **91**, 6423.

Miles, I. S. and Rostami, S. (eds) (1992) *Multicomponent Polymer Systems*. Longman, Harlow.

Noda, I., Kato, N., Kitano, T. and Nagasawa, M. (1981) *Macromolecules* **14**, 668.

Rodriguez-Parada, J. M. and Percec, V. (1986a) *Macromolecules* **19**, 55.

Rodriguez-Parada, J. M. and Percec, V. (1986b) *J. Polym. Sci., Polym. Chem. Ed.* **24**, 579.

Saeki, S., Kuwahara, S., Konno, S. and Kaneko, M. (1975) *Macromolecules* **6**, 246.

Sanchez, I. C. and Lacombe, R. H. (1978) *Macromolecules* **11**, 1145.

Scatchard, G. (1931) *Chem. Rev.* **8**, 321.

Schultz, G. V. and Doll, H. (1952) *Z. Elektrochem.* **56**, 248.

Utracki, L. A. and Jukes, J. A. (1984) *J. Vinyl Tech.* **6**, 85.

Von Laar, J. J. (1910) *Z. Phys. Chem.* **72**, 723.

Walker, J. S. and Vouse, C. A. (1982) in Eighth Symposium on Thermophysical Properties (J. V. Sengers, ed.) *Am. Soc. Mech. Eng.* **1**, 411.

4.11 SUGGESTED FURTHER READING

de Gennes, P. G. (1979) *Scaling Concepts in Polymer Physics*. Cornell University Press, Ithaca, NY, and London.

Flory, P. J. (1953) *Principles of Polymer Chemistry*. Cornell University Press, Ithaca, NY, and London.

Hiemenz, P. C. (1986) *Principles of Colloid and Surface Chemistry*. Marcel Dekker, New York and Basel.

Utracki, L. A. (1989) *Polymer Alloys and Blends*. Hanser, Munich.

van Krevelen, D. W. (1990) *Properties of Polymers*, 3rd edn. Elsevier, Amsterdam.

THE GLASSY AMORPHOUS STATE

5.1 INTRODUCTION TO AMORPHOUS POLYMERS

Let us consider a polymer which is in its molten state. What may happen when it cools down? The two possibilities are shown in Fig. 5.1. The polymer may either crystallize (route a) or cool down to its glassy, amorphous state (route b). The temperature at which the slope in the specific volume–temperature graph (route b) changes is referred to as the glass transition temperature, T_g. Rigid-rod polymers, i.e. polymers with very inflexible groups in the backbone chain or in side chains, may form liquid-crystalline states, an issue dealt with in Chapter 6.

What molecular factors determine whether a polymer will crystallize or not? The regularity of the polymer is the key factor: isotactic polypropylene crystallizes, whereas atactic polypropylene does not. Atactic polymers generally do not crystallize with two exceptions:

1. The X group in $(-CH_2-CHX-)_n$ is very small, allowing regular packing of the chains regardless of whether the different pendant groups are randomly placed. Poly(vinyl alcohol) with its small hydroxyl X groups is a good example of this kind of polymer.
2. The X group forms a long regular side chain. Side-chain crystallization may occur provided that the pendant groups are of sufficient length.

Random copolymers are incapable of crystallizing except when one of the constituents is at a significantly higher concentration than the other constituent. Linear low-density polyethylene with a crystallinity of about 50% contains 98.5 mol% of methylene units and 1.5 mol% of CHX units, where X is $-CH_2CH_3$ or a longer homologue. Polymers which are potentially crystallizable may be quenched to a glassy amorphous state. Polymers with large side groups having an inflexible backbone chain, are more readily quenched to a glassy, fully amorphous polymer than flexible polymers such as polyethylene. Figure 5.2 shows schematically the effects of polymer structure and cooling rate on the solidified polymer structure.

Cooling past the glass transition temperature is accompanied by a dramatic change in the mechanical properties. The elastic modulus increases by a factor of 1000 when the polymeric liquid is cooled below T_g and the modulus of the glassy polymers is relatively insensitive to changes in molar mass and repeating unit structure. The actual value of T_g is, however, very dependent on the repeating unit, the molecular architecture and the presence of low molar mass species, as is shown in section 5.2.

It is appropriate to point out that the T_g value recorded in any given experiment is dependent on the temperature scanning rate. This is further

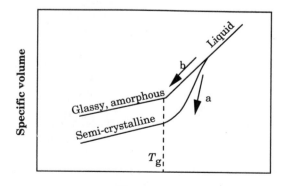

Figure 5.1 Cooling of a liquid following routes a (crystallization) or b (forming a glassy amorphous structure).

78 The glassy amorphous state

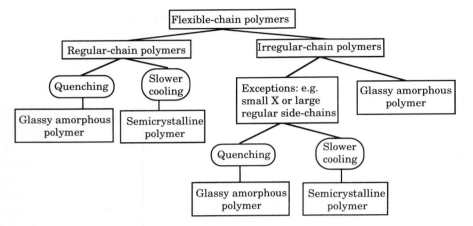

Figure 5.2 The effect of molecular and thermal factors on the structure of the solidified polymer.

discussed in section 5.3. A statement about a T_g value should always be accompanied by a description of the experiment. The remainder of this chapter presents a summary of the non-equilibrium nature of the glassy material, the current theories of the glass transition, the mobility of the molecules in their glassy state (subglass processes) and finally the structure of glassy, amorphous polymers.

5.2 THE GLASS TRANSITION TEMPERATURE

5.2.1 EFFECT OF REPEATING UNIT STRUCTURE ON THE GLASS TRANSITION TEMPERATURE

Let us first start with the class of polymers with the repeating unit $-CH_2-CHX-$. What are the effects of the X group on T_g? If X is a relatively inflexible group then T_g increases with increasing size of the group (Fig. 5.3). The pendant groups pose restrictions on the torsion about σ bonds in the backbone chain, i.e. the backbone chain becomes less flexible and this causes the increase in T_g.

Polymers with relatively flexible side chains, e.g. polyacrylates or polymethacrylates, exhibit the opposite behaviour with a very pronounced decrease in T_g with increasing length of the side chain (Fig. 5.4). The dominant effect of the side group is to increase the distance between adjacent backbone chains. Polymers with longer pendant groups may crystallize under certain conditions, which then may lead to an increase in T_g. The replacement of hydrogen with a methyl group at the α carbon causes a dramatic increase in T_g; the difference in T_g between polymethacrylates and polyacrylates is about 100°C.

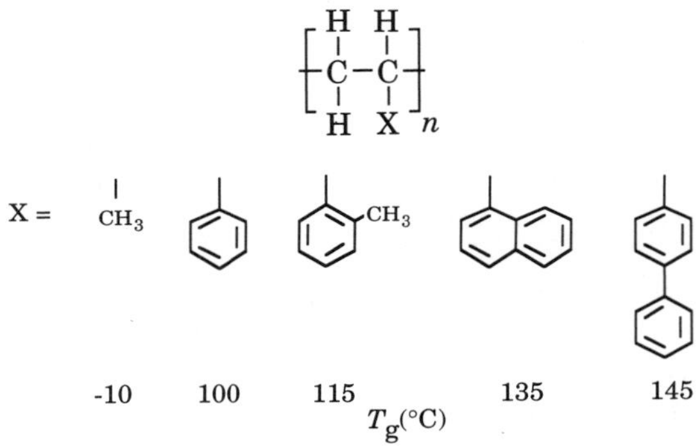

Figure 5.5 shows that T_g strongly increases with increasing chlorine content. The carbon–chlorine bond is polar. Thus, T_g increases with increasing polarity. The polarity of poly(vinylidene chloride) (PVDC) is relatively low because of the symmetry (Fig. 5.6). This observation can be generalized to the statement that T_g generally increases with increasing cohesive energy density (CED) as shown in the following equation:

$$T_g = \frac{2\delta^2}{mR} + C_1 \qquad (5.1)$$

Figure 5.3 Glass transition temperatures of different vinyl polymers showing the influence of the size of the pendant group. Data from Eisenberger (1984).

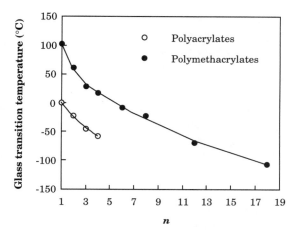

Figure 5.4 Glass transition temperature of polyacrylates and polymethacrylates as a function of number of carbons (n) in the oligo-methylene (R) group. Drawn after data from Rogers and Mandelkern (1957) (polymethacrylates) and Shetter (1963) (polyacrylates).

where δ^2 is the CED, m is a parameter which describes the internal mobility of the groups in a single chain, R is the gas constant and C_1 is a constant. The CED provides an integrated measure of the strength of the secondary bonds in a compound. Materials with strong secondary bonds show high CED values. Further details about the CED are given in Chapter 4.

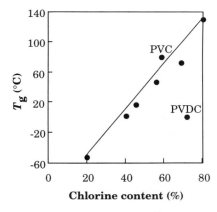

Figure 5.5 Effect of chlorine content on the glass transition temperature of chlorinated polyethylene. Data for polyvinylchloride (PVC) and polyvinylidene chloride (PVDC) are shown for comparison. Drawn after data from Schmieder and Wolf (1953).

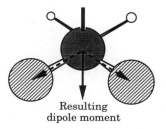

Figure 5.6 Repeating unit of polyvinylidene chloride and schematic representation of the resulting dipole moment.

Beaman (1953) and Boyer (1954) noticed in the early 1950s that both T_g and the melting point (T_m) of crystallizable polymers increased with decreasing chain flexibility and increasing CED. They established the following empirical correlations:

$$\frac{T_g}{T_m} = \frac{1}{2} \quad \text{(symmetrical molecules)} \quad (5.2)$$

$$\frac{T_g}{T_m} = \frac{2}{3} \quad \text{(asymmetrical molecules)} \quad (5.3)$$

According to more recent data, about 80% of the data for symmetrical and asymmetrical polymers are within the T_g/T_m limits 0.5–0.8.

5.2.2 THE CONCEPT OF FREE VOLUME

The poet Lucretius, who lived in the first century BC, wrote down the thoughts of Epicurus (341–271 BC) on empty space or the free volume:

> Therefore there is intangible space, voids, emptiness. But if there were none, things could not in any way move; for that which is the province of body, to prevent and to obstruct, would at all times be present to all things; therefore nothing would be able to move forward, since nothing could begin to give place.

These words were probably the first clear statement about the free volume and it is fascinating that the link between free volume and mobility was so clearly understood.

We may consider the existence of two types of volume in matter, namely occupied volume and free volume, the latter allowing mobility of the atoms (segments). Free volume cannot be directly measured

80 The glassy amorphous state

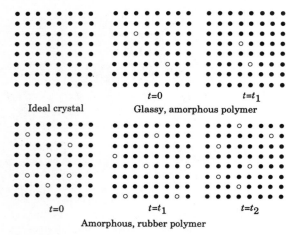

Figure 5.7 Illustration of the free volume concept. Occupied volume is marked by filled circles and free volume (hole) by open circles. Snapshots are taken at different times (t).

but must be deduced from other measurements, e.g. from the measurement of total volume. The free volume can be represented by mobile holes jumping around in the structure. The fractional free volume is the ratio of the free volume to the total volume. Figure 5.7 shows an ideal crystal which consists only of occupied volume, a glassy, amorphous polymer ($T < T_g$) with only a small fraction of slowly moving holes, and an amorphous, rubber polymer ($T < T_g$) with a higher concentration of rapidly moving holes.

The occupied volume is difficult to assess. It has been calculated from the van der Waals radii of the atoms. Others have used the crystalline volume at 0 K. A third approach is to take the difference between the total volume and the fluctuation volume. Doolittle (1951) used the extrapolated volume of liquids at 0 K as their occupied volume.

X-ray scattering and positronium annihilation have been used to estimate the size of empty volume ('hole') in different polymers. The volume of single holes in, for example, polycarbonate and polysulphone has been estimated at 0.02–0.07 nm³.

5.2.3 EFFECT OF MOLECULAR ARCHITECTURE ON THE GLASS TRANSITION TEMPERATURE

The molecular architecture, i.e. molar mass (M), degree of crosslinking, and chain branching also affect T_g. The variation in T_g due to variation in M for commercial polymers is insignificant and is almost always overridden by other factors. Associated with each chain end is a certain degree of extra mobility. We may assign a certain extra free volume to each chain end in terms of an extra fractional free volume (f_e). The concentration of chain ends is proportional to M^{-1}. The fractional free volume (f) at a specified temperature below T_g is:

$$f = f^o + \frac{C_2}{M} \quad (5.4)$$

where f^o is the fractional free volume of the polymer of infinite molar mass and C_2 is a constant proportional to f_e. The temperature dependence of the relative free volume can be expressed as follows:

$$f \approx (C_3 + C_4 T) + \frac{C_2}{M} \quad (5.5)$$

where C_3 and C_4 are constants. The free volume theory states that the material at the glass transition temperature takes a certain universal fractional free volume, denoted f_g:

$$f_g \approx C_3 + C_4 T_g + \frac{C_2}{M}$$
$$T_g \approx \frac{f_g - C_3}{C_4} - \frac{C_2}{C_4 M} = C_5 - \frac{C_6}{M} \quad (5.6)$$

where C_5 and C_6 are constants. This equation was first suggested by Fox (1956). Figure 5.8 shows data in adequate agreement with eq. (5.6). Other more complex models and equations were proposed later. Some of them consider chain entanglements as important elements. It has been recognized, particularly from rheological experiments, that there is a minimum molar mass below which chain entanglement does not occur.

Crosslinks reduce the available free volume, and hence T_g is expected to increase with increasing crosslink density (Fig. 5.9). The decrease in free volume should be proportional to the number of crosslinks (n_c):

$$f \approx f^o - n_c C_7 = f^o - \frac{C_8}{\overline{M}_c} \quad (5.7)$$

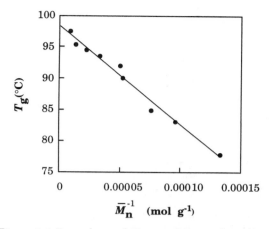

Figure 5.8 Dependence of T_g on molar mass for narrow fractions of atactic polystyrene. Data from Rietsch, Daveloose and Froelich (1976).

where C_7 and C_8 are constants, and $\bar{M}_c$ is the number average molar mass of the chains between the crosslinks. The following relationship is obtained by analogy with the molar mass dependence:

$$T_g \approx C_9 + \frac{C_{10}}{\bar{M}_c} \qquad (5.8)$$

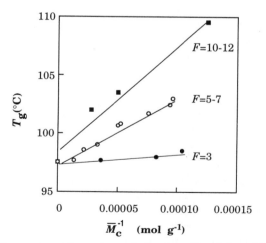

Figure 5.9 Effect of crosslinking on T_g of branched poly(styrene-co-divinyl benzene). The crosslink functionality (F) is shown adjacent to each regression line. Drawn after data from Rietsch, Daveloose and Froelich (1976).

where C_9 and C_{10} are constants. Figure 5.9 presents data for a copolymer of styrene and divinylbenzene, the latter giving rise to crosslinks.

The functionality of the crosslinks (F) was varied by allowing different amounts of divinylbenzene to polymerize (anionic mechanism) in the crosslink nodules. The crosslink functionality varied between 3 and 12. For a specified crosslink functionality, the data followed eq. (5.8) and the constant C_{10} increased with increasing F. Rietsch, Daveloose and Froelich (1976) showed that the increase in C_{10} at low F values (F < 6) was approximately proportional to F, but the increase in C_{10} with increasing F was smaller at higher F values. In a polymer which, prior to curing, has a molar mass of M, the following glass transition temperature equation is obtained:

$$T_g \approx C_4 - \frac{C_5}{M} + \frac{C_9}{\bar{M}_c} \qquad (5.9)$$

5.2.4 THE GLASS TRANSITION TEMPERATURE OF BLENDS AND COPOLYMERS

The glass transition temperature of a polymer blend is highly dependent on the morphology. An in-depth discussion on this subject is given in Chapter 4; fragments of it are repeated here. The most common case, shown in Fig. 5.10, is that the polymer components are immiscible, forming a two-phase system. Such a polymer blend, provided it is amorphous, exhibits two T_g's corresponding to each of the two different phases (Fig. 5.10).

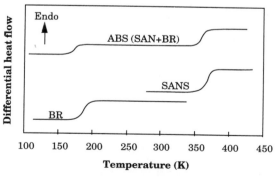

Figure 5.10 DSC traces of ABS and of its constituents poly(butadiene) (BR) and poly(styrene-co-acrylonitrile) (SAN). Schematic after data from Bair (1970).

Some polymer blends exhibit partial miscibility. They have a mutual, limited solubility indicated by a shift in the two T_g's accompanying a change in the phase composition of the blend. More uncommon is the type of miscibility indicated by the presence of only one T_g. Several equations relating T_g and composition in this case have been proposed. One of them is the Fox equation:

$$\frac{1}{T_g} = \frac{w_1}{T_{g1}} + \frac{w_2}{T_{g2}} \qquad (5.10)$$

where T_g is the glass transition temperature of the binary blend, T_{g1} and T_{g2} are the glass transition temperatures of polymers 1 and 2, and w_1 and w_2 are the mass fractions of polymers 1 and 2. Polystyrene and poly(phenylene oxide) are miscible over the whole composition range (Fig. 5.11).

Low molar mass liquids often have a strong T_g-depressive effect. They are commonly denoted **plasticizers**. Polyvinylchloride (PVC) is one of the most frequently plasticized polymers (Fig. 5.12). Polar and hygroscopic polymers such as polyamides absorb water with a plasticizing effect on the polymer. Plasticization causes a depression in T_g. It is an interesting fact that small amounts of plasticizer may result in a stiffer material at temperatures below the depressed T_g. This effect is referred to as **anti-plasticization** and is caused by strong intermolecular interaction between the polymer and the plasticizing species suppressing the subglass processes (see section 5.5.2).

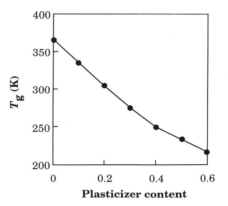

Figure 5.12 Plasticization of PVC: T_g as a function of di(ethylhexyl)-phthalate content. Drawn after data from Wolf (1951).

5.2.5 EFFECT OF PRESSURE ON THE GLASS TRANSITION TEMPERATURE

The hydrostatic pressure (p) affects T_g according to the following equation:

$$\left(\frac{dT_g}{dp}\right) = \frac{\Delta\beta}{\Delta\alpha} \qquad (5.11)$$

where $\Delta\beta$ and $\Delta\alpha$ are respectively the changes in compressibility and volume expansion coefficient associated with the glass transition. Typical values of this ratio for polymers are in the range of 0.2–0.4 K MPa^{-1}.

5.3 NON-EQUILIBRIUM FEATURES OF GLASSY POLYMERS AND PHYSICAL AGEING

5.3.1 GENERAL ASPECTS

The glass transition obeys second-order characteristics, i.e. volume and enthalpy are continuous functions through the transition temperature. However, their temperature derivatives, the thermal expansion coefficient and the specific heat, change in a discontinuous manner at the glass temperature.

The experiment illustrated in Fig. 5.13 and conducted by many researchers shows the non-

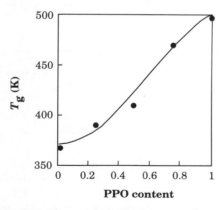

Figure 5.11 Glass transition temperature of compatible blends of polystyrene and polyphenylene oxide (PPO) as a function of PPO content. Drawn after data from Bair (1970).

Non-equilibrium features of glassy polymers and physical ageing   83

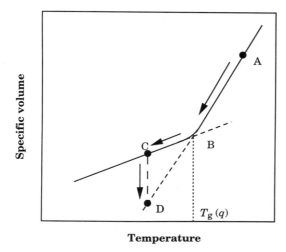

Figure 5.13 Illustration of the non-equilibrium nature of a glassy polymer.

equilibrium nature of a polymer that has been cooled at a constant rate (q) through the kinetic glass transition region. Volume may be continuously measured in a dilatometer. The sample is first heated to a temperature (point A in Fig. 5.13) well above the glass transition temperature (T_g). The sample is then cooled at constant rate (q). At point B, the volume decrease is retarded. A break in the curve occurs at the glass transition temperature ($T_g(q)$) which is interpreted as the kinetic glass transition. A few degrees below the breaking point, at C, the cooling is interrupted and the sample is then held at constant temperature. The specific volume of the material decreases under the isothermal conditions as a function of time, following the line CD. Equilibrium has not been attained at point C. It may be argued that equilibrium has not been reached in any of the points between B and C, i.e. the recorded glass transition has kinetic features. The process transferring the system from C towards D is denoted 'physical ageing' or simply in this case 'isothermal volume recovery'.

The term 'recovery' is often used instead of 'relaxation' to indicate that the process leads to the establishment (recovery) of equilibrium. We may replace specific volume by enthalpy and obtain a curve similar to that shown in Fig. 5.13. The curvature on both sides of the glass transition is slightly more developed in the enthalpy–temperature plot than in the volume–temperature plot. The approach of the non-equilibrium glass to the equilibrium state is accompanied by a decrease in enthalpy (isothermal enthalpy recovery), which can be detected *in situ* by high-resolution microcalorimetry.

Figure 5.14 shows the volumetric response to different cooling rates of a glass-forming polymer. The recorded T_g increases with increasing cooling rate; experimental work has shown that T_g is changed by approximately 3 K for a change by a factor of 10 in cooling rate. At higher cooling rates the time available to the system at each temperature is shorter than that at a slower cooling rate, and the curve begins to deviate from the equilibrium line at a higher temperature.

The volume recovery curves of amorphous atactic polystyrene previously equilibrated at different temperatures above and below a 'new' equilibration temperature (95.46°C) are shown in Fig. 5.15. The initially excessive and deficient volume states approach the same equilibrium volume in a nonlinear way. Also worthy of note is the asymmetrical character of the expansion and contraction curves, which is further demonstrated in Fig. 5.16. The contracting specimen is always closer to the equilibrium than the expanding specimen. The

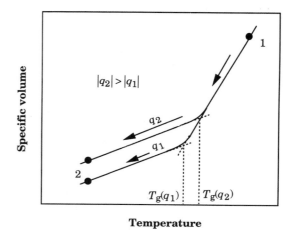

Figure 5.14 Schematic curves showing the cooling rate dependence of the specific volume of a glass-forming wholly amorphous polymer.

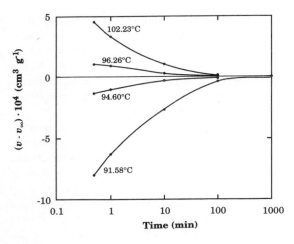

Figure 5.15 Volume recovery for atactic polystyrene at 1 atm and 95.46°C after prior equilibration at various temperatures as shown adjacent to each recovery curve. The equilibrium volume at 95.46°C is denoted v_∞. Drawn after data from Goldbach and Rehage (1967).

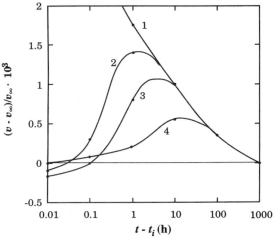

Figure 5.17 Volume recovery of polyvinyl acetate after quenching from 40°C to different temperatures T_1 at which the samples remained for different periods of time (t_1) before reheating to $T = 30°C$: (1) isotherm obtained by direct quenching from 40°C to 30°C; (2) $T_1 = 10°C$; $t_i = 160$ h; (3) $T_1 = 15°C$; $t_i = 140$ h; (4) $T_1 = 25°C$; $t_i = 90$ h. The equilibrium volume at 30°C is denoted v_∞. Drawn after data from Kovacs (1963).

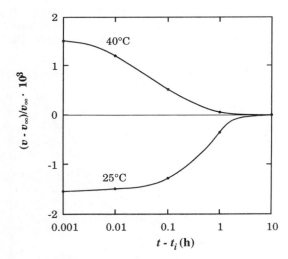

Figure 5.16 Isothermal volume recovery curves of glucose by contraction and expansion at $T = 30°C$ after quenching the samples to 25°C and 40°C, equilibration of the samples at those temperatures and reheating to 30°C. The equilibrium volume at 30°C is denoted v_∞. Drawn after data from Kovacs (1963).

asymmetry in approaching equilibrium by contraction and expansion is one of the most characteristic features of the structural recovery of glasses. The rate of recovery thus depends on the magnitude and the sign of the initial departure from the equilibrium state.

The so-called memory effect is illustrated by Fig. 5.17. The observed peak in the volume recovery curve appears only in samples that have not reached equilibrium at the lower temperature. These very complex relaxation effects are further discussed in section 5.3.2.

Changes in volume and enthalpy in any glassy polymer, i.e. physical ageing, are accompanied by significant changes in most engineering properties, e.g. tensile modulus, yield stress, fracture toughness, permeability, and electrical properties. An amorphous polymer which is cooled rapidly through the glass transition temperature region and then held at constant temperature shows a decrease in both specific volume and enthalpy as well as an increase in

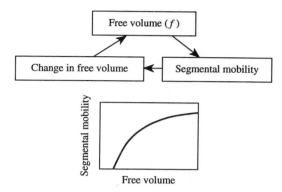

Figure 5.18 Free volume model for physical ageing.

tensile modulus and yield stress and a decrease in fracture toughness, impact strength and permeability. These important effects are thoroughly reviewed by Struik (1978).

The non-equilibrium nature of glassy polymers has been and still is a problem of significant interest for polymer scientists. A very simple qualitative model proposed by Struik is illustrated in Fig. 5.18. The closed loop contains the important ingredients of the problem: free volume, segmental mobility and volume change. In the initial state, the free volume and the segmental mobility of the polymer glass are significant, resulting in a rapid decrease in the free volume, etc.

5.3.2 KINETIC THEORIES

The important experimental findings, presented in section 5.3.1, can be summarized as follows:

1. The volume recovery is nonlinear with respect to the magnitude of the departure from equilibrium (Figs 5.15 and 5.16).
2. The rate of volume recovery depends not only on the magnitude of the departure from equilibrium but also on the sign of the departure: contraction is more rapid than expansion. The approach to equilibrium is thus significantly asymmetric (Figs 5.15 and 5.16).
3. Memory effects are observed after two or more changes in temperature (Fig. 5.17).

Findings 1 and 2 have been successfully described by models involving single structure-dependent retardation mechanisms, but these simple models fail to describe the memory effects. The latter can only be described by more complex models involving contributions from two or more independent retardation mechanisms.

Early theories of Tool (1946) and Davies and Jones (1953) proposed a one-parameter model for the volume (eq. (5.12)) and enthalpy (eq. (5.13)) recovery:

$$-\left(\frac{d\delta_v}{dt}\right) = q\Delta\alpha + \frac{\delta_v}{\tau_v} \quad (5.12)$$

$$-\left(\frac{d\delta_H}{dt}\right) = q\Delta c_p + \frac{\delta_H}{\tau_H} \quad (5.13)$$

where $\delta_v = (V - V_\infty)/V_\infty$, V is the actual volume, V_∞ is the equilibrium volume, q is the rate of temperature change, $\Delta\alpha$ is the change in volume expansion coefficient at T_g, τ_v is the isobaric volume retardation time, $\delta_H = H - H_\infty$, H is the actual enthalpy, H_∞ is the equilibrium enthalpy, Δc_p is the change in specific heat at T_g and τ_H is the isobaric enthalpy retardation time.

Under isothermal conditions, eq. (5.12) becomes:

$$-\frac{d\delta_v}{dt} = \frac{\delta_v}{\tau_v}$$

$$\delta_v = \delta_{v0} \exp\left(-\frac{t}{\tau_v}\right) \quad (5.14)$$

where δ_{v0} characterizes the volumetric state at time $t = 0$. Equation (5.14) cannot, however, be fitted to experimental isothermal volume recovery data. To overcome this shortcoming of the one-parameter model, Tool (1946) assumed that τ_v was not constant but dependent on δ_v. Doolittle's viscosity equation was used to relate τ_v to τ_{vg} or rather the fractional free volume (f) at temperature T:

$$\tau_v = \tau_{vg}\left(\frac{B}{f} - \frac{B}{f_g}\right) \quad (5.15)$$

where τ_v is the relaxation time at temperature T, τ_{vg} is the relaxation time at equilibrium at T_g, B is a constant and f_g is the equilibrium free volume at T_g. If eq. (5.15) is inserted into eq. (5.12), the following expression is obtained:

$$-\left(\frac{d\delta_v}{dt}\right) = q\Delta\alpha + \frac{\delta_v \exp\left(\frac{B}{f_g}\right)}{\tau_{vg} \exp\left(\frac{B}{f}\right)} \quad (5.16)$$

A similar equation can be derived for the enthalpy. It turned out that this modification to give the retardation time a free volume dependence, was successful in describing 'one-step' isothermal recovery but unsuccessful in describing memory effects. Kovacs, Aklonis, Hutchinson and Ramos (1979) attributed the latter to the contributions of at least two independent relaxation mechanisms involving two or more retardation times. These authors proposed a multiparameter approach, the so-called KAHR (Kovacs–Aklonis–Hutchinson–Ramos) model. The recovery process is divided into N subprocesses, which in the case of volumetric recovery may be expressed as:

$$-\frac{d\delta_i}{dt} = q\Delta\alpha_i + \frac{\delta_i}{\tau_i} \quad 1 \leq i \leq N$$

$$\Delta\alpha_i = g_i \Delta\alpha_i \quad (5.17)$$

$$\sum_{i=1}^{N} g_i = 1$$

Note that the subscript v is omitted for reasons of simplicity. The series of equations (5.17), N in number, are interrelated through the following equation:

$$\delta = \sum_{i=1}^{N} \delta_i \quad (5.18)$$

The temperature and structural dependences of the retardation times were expressed as follows:

$$\tau_i(T, \delta) = \tau_{i,r} \exp[-\theta(T - T_r)] \exp[-(1-x)\theta\delta/\Delta\alpha] \quad (5.19)$$

where the subscript r refers to the fixed reference temperature, x is a partitioning parameter (taking values between 0 and 1) characterizing the pure contribution of temperature to the retardation times and θ is given by the expression

$$\theta = \Delta h^*/(RT_g^2)$$

where Δh^* is the activation energy for the retardation times. The temperature shift function a_T, which characterizes the temperature dependence of τ_i at constant δ, is given by:

$$a_T = \frac{\tau_i(T, \delta)}{\tau_i(T_r, \delta)} = \exp[-\theta(T - T_r)] \quad (5.20)$$

The structural shift function a_δ is equal to:

$$a_\delta = \frac{\tau_i(T, \delta)}{\tau_i(T, 0)} = \exp[-(1-x)\theta\delta/\Delta\alpha] \quad (5.21)$$

When eqs (5.19)–(5.21) are combined, the following expression is obtained:

$$\tau_i = \tau_{i,r} a_T a_\delta \quad (5.22)$$

The retardation time spectrum ($G(\tau_{i,r})$) is obtained from the parameters g_i and $\tau_{i,r}$. It is assumed that g_i is independent of T and δ, whereas τ_i (at temperature T and structural parameter δ) is given by eq. (5.22). A change in temperature from T_r to T will thus shift the spectrum along the log τ axis by $\log(a_T a_\delta)$. The time-dependent variation in δ can be obtained for any thermal prehistory of the type

$$T(t) = T_0 + \int_{t=0}^{t} q \, dt \quad (5.23)$$

where T_0 is the initial temperature at which $\delta_i = 0$ (for all i), by inserting appropriate values for the material parameters (θ and x, which can be obtained by independent measurement), and by choosing a retardation time spectrum ($G(\tau_{i,r})$) at the reference temperature. The time-dependent variation in δ for a given thermal history (eq. (5.23)) is obtained by solving numerically the series of differential equations and related equations (5.17)–(5.19). The KAHR model is capable of predicting the experimental data obtained at different cooling rates (Fig. 5.14), the isothermal recovery after an initial constant-rate cooling, after heating a sample to a given constant temperature from a temperature at which it was in equilibrium ($\delta = 0$; Figs 5.15 and 5.16), and after heating a sample to a given constant temperature from a temperature at which it had not reached equilibrium ($\delta > 0$; Fig. 5.16). The KAHR model can also describe the asymmetrical recovery data shown in Fig. 5.17. The memory effect which is shown in Fig. 5.17 cannot be explained by the one-parameter models because they predict that $d\delta/dt$ would be 0 at the start when $\delta = 0$. The memory effect can only be explained by the multiplicity of the retardation times.

5.4 THEORIES FOR THE GLASS TRANSITION

A complete theoretical understanding of the glass transition phenomenon is not yet available. The current theories can be divided into three groups: free volume theories, kinetic theories and thermodynamic theories.

The **free volume** theories state that the glass transition is characterized by an iso-free volume state, i.e. they consider that the glass temperature is the temperature at which the polymers have a certain universal free volume. The starting point of the theory is that the internal mobility of the system expressed as viscosity is related to the fractional free volume. This empirical relationship is referred to as the Doolittle equation. It is a consequence of the universal William–Landel–Ferry (WLF) equation and the Doolittle equation that the glass transition is indeed an iso-free volume state. The WLF equation, expressed in general terms, is:

$$\log a_T = \frac{C_1(T - T_r)}{C_2 + T - T_r} \quad (5.24)$$

where $a_T = \eta_T/\eta_{T_s} = \tau_T/\tau_{T_s}$ (η represents the viscosity and τ represents a characteristic segmental relaxation time at temperatures T and T_r (reference temperature)), C_1 and C_2 are constants. When T_r is set equal to T_g, C_1 and C_2 in eq. (5.24) are almost universal constants for a wide range of polymers: $C_1 = -17.44$ and $C_2 = 51.6$ K. Thus,

$$\log a_T = \frac{-17.44(T - T_g)}{51.6 + T - T_g} \quad (5.25)$$

The WLF equation was originally based purely on empirical observations. It can, however, be derived from free volume theory starting from the empirical Doolittle viscosity equation:

$$\eta = A \exp\left(\frac{B}{f}\right) \quad (5.26)$$

where A and B are constants and f is the fractional free volume. The temperature shift factor (a_T) becomes:

$$a_T = \frac{\exp\left(\frac{B}{f}\right)}{\exp\left(\frac{B}{f_r}\right)} = \exp\left(B\left(\frac{1}{f} - \frac{1}{f_r}\right)\right) \quad (5.27)$$

where f is the fractional free volume at T and f_r is the fractional free volume at the reference temperature T_r. If the reference temperature is set to the glass transition temperature (T_g) and if it is assumed that $f = f_g$ at this temperature, the following equation is obtained:

$$a_T = \exp\left(B\left(\frac{1}{f} - \frac{1}{f_g}\right)\right) \quad (5.28)$$

The fractional free volume at temperature T can be expressed as:

$$f = f_g + \alpha_f(T - T_g) \quad (5.29)$$

where α_f is the coefficient of expansion of the fractional free volume. Insertion of eq. (5.29) into eq. (5.28) gives:

$$\begin{aligned} a_T &= \exp\left(B\left(\frac{1}{f} - \frac{1}{f_g}\right)\right) \\ &= \exp\left(B\left(\frac{1}{f_g + \alpha_f(T - T_g)} - \frac{1}{f_g}\right)\right) \\ &= \exp\left(B\left(\frac{-\alpha_f(T - T_g)}{f_g(f_g + \alpha_f(T - T_g))}\right)\right) \\ &= \exp\left(-\frac{B}{f_g}\left(\frac{\alpha_f(T - T_g)}{f_g + \alpha_f(T - T_g)}\right)\right) \\ &= \exp\left(\frac{\left[-\frac{B}{f_g}\right](T - T_g)}{\left[\frac{f_g}{\alpha_f}\right] + T - T_g}\right) \quad (5.30) \end{aligned}$$

The following expression is obtained by taking the logarithm of eq. (5.30):

$$\log a_T \approx \frac{\left[\frac{-B}{2.303 f_g}\right](T - T_g)}{\left[\frac{f_g}{\alpha_f}\right] + T - T_g} \quad (5.31)$$

Equation (5.31) may now be compared with eq. (5.25), the WLF equation, and the universal constants C_1 and C_2 can be identified as:

$$-\frac{B}{2.303 f_g} = -17.44 \Rightarrow f_g = 0.025 \quad (B = 1)$$

$$\frac{f_g}{\alpha_f} = 51.6\,\mathrm{K} \Rightarrow \alpha_f = \frac{f_g}{51.6} = \frac{0.025}{51.6} = 4.8 \times 10^{-4}\,\mathrm{K}^{-1}$$

The outcome of the fact that the WLF equation exhibits universal constants is that the fractional free volume at the glass transition temperature and the thermal expansion coefficient of the free volume also have universal values. Figure 5.19 shows schematically the temperature dependence of the free volume. Other approaches which can be categorized as free volume theories have been proposed. We refer to the further reading listed at the end of this chapter for more detailed information.

The **kinetic** theories are dealt with in section 5.3.2. According to these theories there is no true thermodynamic glass transition. The kinetic theories predict that the glass transition is a purely kinetic phenomenon and that it appears when the response time for the system to reach equilibrium is of the same size as the time-scale of the experiment. The theory predicts that a lowering of the cooling rate will lead to a decrease in the kinetic glass transition temperature.

The **thermodynamic** theory was formulated by Gibbs and DiMarzio (1958) who argued that, although the observed glass transition is a kinetic phenomenon, the underlying true transition can possess equilibrium properties. The thermodynamic theory attempted to explain the Kauzmann paradox, which can be stated as follows. If the equilibrium properties of a material, entropy (S) and volume (V), are extrapolated through the glass transition, the values of S and V for the glass will be lower than for the corresponding crystals. The equilibrium theory resolves the problem by predicting a thermodynamic glass transition reached when the conformational entropy takes the value zero (Fig. 5.20).

The Gibbs–DiMarzio theory is based on a lattice model of a similar type to that in the Flory–Huggins theory for polymer solutions. DiMarzio argued that the use of a lattice model to study polymers, as opposed to simple liquids, is more promising, because in polymers it is possible to form glasses from systems with no underlying crystalline phase. Atactic polymers cannot, generally speaking, form a crystalline phase.

The Gibbs–DiMarzio theory is based on a lattice of coordination number z filled with polymer molecules (n_x) each with a degree of polymerization x, and holes (n_0). The **intra**molecular energy is given by $xf\Delta\varepsilon$, where f is the number of bonds in the high-energy state (state 2) and $\Delta\varepsilon = \varepsilon_2 - \varepsilon_1$ is the

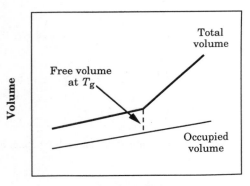

Figure 5.19 Schematic volume–temperature diagram of amorphous polymer showing the temperature dependence of the free volume.

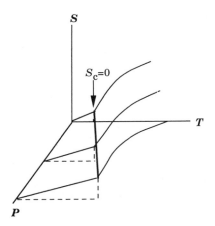

Figure 5.20 Entropy (S) as a function of temperature (T) and pressure (P) according to the Gibbs–DiMarzio theory. The second-order transition (denoted T_2 by the authors) is marked with a thick line. Drawn after DiMarzio (1981).

energy difference between high- and low-energy conformational states. The **inter**molecular energy is proportional to the number of holes (n_0) and the non-bonded interaction energy ΔE_h. The partition function is calculated by the same method as that adopted by Flory and Huggins. The partition function displays a second-order transition in the Ehrenfest sense (Fig. 5.20). The Ehrenfest equation for a second-order transition is

$$\frac{dp}{dT} = \frac{\Delta \alpha}{\Delta \beta}$$

where p is the pressure, T is the transition temperature, $\Delta \alpha$ is the volumetric thermal expansivity associated with the transition and $\Delta \beta$ is the change in compressibility associated with the transition. The hypothesis proposed by Gibbs and DiMarzio is that glass formation (denoted T_2) is associated with the condition $S_c = 0$, where S_c is the conformational entropy. As a glass-forming system is cooled down, the number of possible arrangements (i.e. the conformational entropy) of the molecules decreases gradually with decreasing temperature. This is due to a decrease in the number of vacancies (holes), a decrease in the permutation of holes and chain segments, and the gradual approach of the chains towards populating the low-energy state (state 1). Finally the condition $S_c = 0$ is fulfilled at T_2. The Gibbs-DiMarzio theory also allows the glassy state as a metastable phase of an energy greater than that of the low-energy crystalline state for crystallizable polymers.

Gibbs and DiMarzio were making comparisons with experimental data, evidently not T_2 but rather the kinetic T_g, and they were able to fit their equations to the experimental data. According to their theory, the molar mass dependence of T_g is given by:

where x is twice the degree of polymerization, v_0 is the volume fraction of holes and k is Boltzmann's constant. Equation (5.32) was fitted to experimental data for polyvinylchloride and the goodness of fit was greater than for eq. (5.6). The theory is capable of describing a whole range of experimentally established phenomena such as the molar mass dependence of crosslinked polymers of T_g, the change in specific heat associated with the glass transition (typical values for amorphous polymers are 0.3–0.6 J g^{-1} K^{-1}) and the compositional dependence of copolymers and polymer blends of T_g.

To make the story even more complicated, it turns out that the WLF equation may also be derived from both the kinetic and thermodynamic theories. Interesting and confusing!

5.5 MECHANICAL BEHAVIOUR OF GLASSY, AMORPHOUS POLYMERS

5.5.1 PHENOMENOLOGICAL MODELS

The deformation of polymers depends very strongly upon the state of order and molecular mobility. Figure 5.21 illustrates the mechanical behaviour of fully amorphous polymers. The presentation in this section is limited to the behaviour accompanying small-strain deformations, of the order of 1% or less. The strain (ε) is defined as $\varepsilon = \Delta L/L_0$, where ΔL is the stress-induced increase of specimen length and L_0 is the initial specimen length. Glassy amorphous polymers tend to deform in an approximately elastic manner and so also do crosslinked amorphous polymers above their glass transition temperature (Chapter 3). Uncrosslinked amorphous polymers deform viscously at temperatures above the glass transition temperature. At temperatures between the glass transition temperature and the temperature of the subglass relaxation (section 5.5.2), amorphous polymers are anelastic. Anelastic

$$\left(\frac{x}{x-3}\right)\left(\frac{\ln v_0}{1-v_0}\right) + \left(\frac{1+v_0}{1-v_0}\right) \ln\left[\frac{(x-1)(1-v_0)}{2xv_0} + 1\right] + \frac{\ln[3(x+1)]}{x}$$

$$= \frac{-\frac{2\Delta\varepsilon}{kT_g} \cdot \exp\left(-\frac{\Delta\varepsilon}{kT_g}\right)}{\left(1 + 2\exp\left(-\frac{\Delta\varepsilon}{kT_g}\right)\right)} - \ln\left[1 + 2\exp\left(-\frac{\Delta\varepsilon}{kT_g}\right)\right] \quad (5.32)$$

90 The glassy amorphous state

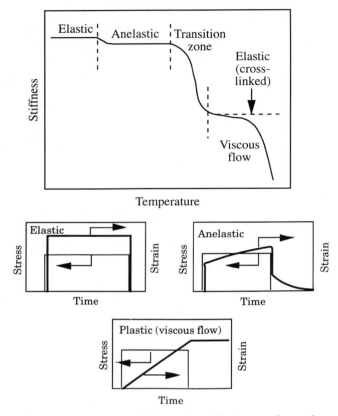

Figure 5.21 Schematic representation of the mechanical behaviour of amorphous polymers.

polymers creep upon loading but recover completely on removal of stress provided that the maximum strain is kept at small values, typically less than 1%.

The creep compliance $J(t) = \varepsilon(t)/\sigma$, where $\varepsilon(t)$ is the time-dependent strain and σ is the constant stress is, at very small strains, less than 1%, approximately independent of stress. The material is said to be **linear** anelastic or linear visco-elastic. Materials of this category follow the so-called Boltzmann superposition principle which can be expressed as follows:

$$\varepsilon(t) = \sum_{i=1}^{N} J(t - t_i)\Delta\sigma_i \qquad (5.33)$$

where $\varepsilon(t)$ is the strain at time t, J is the creep compliance, and t_i is the time for the application of a new incremental load $\Delta\sigma_i$. The strain response to a certain incremental applied stress is independent of previous and future applied stresses. The strain can in this way be calculated for any stress history provided that $J(t)$ is known. In this chapter we shall consider only the mechanical behaviour at such small strains that the material is approximately linear anelastic.

The lack of a satisfactory molecular theory for the deformation of polymers has led to the development of mechanical analogues and phenomenological models that represent the material. The task is to find combinations of elastic and viscous elements that reproduce the material behaviour. There is more than one combination that will reproduce the same behaviour. Some combinations are more convenient for a given kind of test and less convenient for another. However, a proper combination of elements should in principle be able to represent all the various tests. A condensed summary of the stress–strain behaviour of the various models is given below.

Mechanical behaviour of glassy, amorphous polymers

The ultimate (fracture) properties of a wholly amorphous polymer are strongly dependent on temperature. At low temperatures, in the elastic region, the fracture is predominantly brittle and the fracture toughness is low. A considerable increase in the fracture toughness accompanies the onset of the subglass process when approaching the anelastic region.

Polymers in the rubber plateau region have a low stiffness and a considerable fracture strain. Network polymers in this temperature region are elastic rubbers with high extensibility (Chapter 3).

Maxwell model

The Maxwell model consists of an elastic spring and a viscous element (dashpot) coupled in series. The stress–strain behaviour of this model in creep (constant stress) and relaxation (constant strain) is shown in Fig. 5.22 and its analytical expressions are as follows:

$$\varepsilon = \frac{\sigma_0}{E} + \frac{\sigma_0}{\eta} \cdot t \quad \text{(constant stress} = \sigma_0\text{)} \tag{5.34}$$

$$\sigma = \sigma_0 \exp\left(-\frac{E \cdot t}{\eta}\right) = \sigma_0 \exp\left(-\frac{t}{\tau}\right)$$
$$\text{(constant strain)} \tag{5.35}$$

where $\tau = \eta/E$ is the relaxation time. The dimension of τ is seconds, but it is only to be considered as a time constant for the model. In a stress relaxation experiment (constant ε), it is equal to the real time needed to lower the stress from its original value σ_0 to σ_0/e.

Voigt–Kelvin model

The Voigt–Kelvin model consists of an elastic element and a viscous element coupled in parallel (Fig. 5.23). The following constitutive equations are obtained by solving the differential equation:

$$\varepsilon = \frac{\sigma_0}{E}\left(1 - \exp\left(-\frac{E \cdot t}{\eta}\right)\right)$$
$$= \frac{\sigma_0}{E}\left(1 - \exp\left(-\frac{t}{\tau}\right)\right) \quad \text{(constant stress} = \sigma_0\text{)} \tag{5.36}$$

where τ is the retardation time. The retardation time (in seconds) is a system response time.

$$\sigma = \varepsilon_0 E \quad \text{(constant strain} = \varepsilon_0\text{)} \tag{5.37}$$

Burger's model

The Burger's model is obtained by combining the Maxwell and Voigt–Kelvin elements in series (Fig. 5.24). The following equation holds under constant stress conditions:

$$\varepsilon = \frac{\sigma_0}{E_1} + \frac{\sigma_0}{\eta_1}t + \frac{\sigma_0}{E_2}\left(1 - \exp\left(-\frac{E_2}{\eta_2}t\right)\right)$$
$$= \frac{\sigma_0}{E_1} + \frac{\sigma_0}{\eta_1}t + \frac{\sigma_0}{E_2}\left(1 - \exp\left(-\frac{t}{\tau_2}\right)\right) \tag{5.38}$$

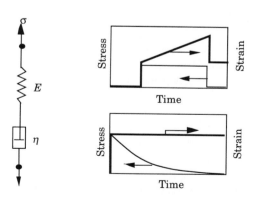

Figure 5.22 Maxwell element.

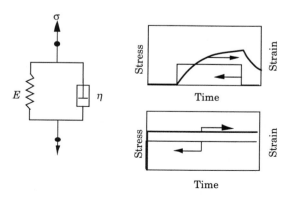

Figure 5.23 Voigt/Kelvin element.

92 The glassy amorphous state

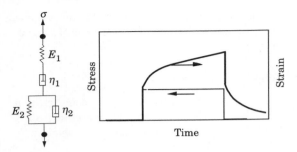

Figure 5.24 Burger's element.

The first term is the elastic response from spring 1, the second term is the viscous flow response from dashpot 1, and the final term is the response of the Voigt–Kelvin element.

Dynamic mechanical behaviour of the Maxwell element and relaxation spectra

A sinusoidal strain ($\varepsilon^* = \varepsilon_0 \exp(i\omega t)$) gives a sinusoidal stress in the Maxwell element and the following frequency dependence (ω is the angular velocity) is obtained for the storage modulus (E') and the loss modulus (E''):

$$E' = E\left(\frac{\omega^2 \tau^2}{1 + \omega^2 \tau^2}\right) \quad (5.39)$$

$$E'' = E\left(\frac{\omega \tau}{1 + \omega^2 \tau^2}\right) \quad (5.40)$$

Figure 5.25 shows the storage modulus and the loss modulus as functions of the angular velocity (ω).

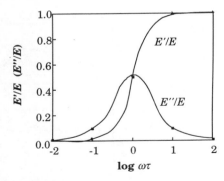

Figure 5.25 Storage and loss modulus as a function of frequency (angular velocity ω) for a single Maxwell element.

The loss modulus passes through a maximum at $\omega = 1/\tau$.

The stress–strain curves of anelastic polymers cannot be described by these simple expressions. A more complex combination of elements is required. Stress relaxation cannot be described by a single Maxwell element alone. A number of Maxwell elements coupled in parallel are needed. The resulting relaxation modulus becomes the sum of the responses of the individual elements:

$$E(t) = \sum_{i=1}^{n} E_i \exp\left(-\frac{t}{\tau_i}\right) \quad (5.41)$$

or, in continuous terms:

$$E(t) = E_{eq} + \int_{-\infty}^{+\infty} H(\tau) \exp\left(-\frac{t}{\tau}\right) d\ln\tau \quad (5.42)$$

where E_{eq} is the relaxed modulus (time $\to \infty$) and $H(\tau)$ is the relaxation time spectrum. The components of the dynamic modulus become:

$$E'(\omega) = E_{eq} + \int_{-\infty}^{+\infty} H(\tau) \frac{\omega^2 \tau^2}{1 + \omega^2 \tau^2} d\ln\tau \quad (5.43)$$

$$E''(\omega) = \int_{-\infty}^{+\infty} H(\tau) \frac{\omega \tau}{1 + \omega^2 \tau^2} d\ln\tau \quad (5.44)$$

Temperature dependence of relaxation time spectrum

The temperature dependence of the relaxation processes is expressed by one of two equations: the Arrhenius equation (eq. (5.45)) or the WLF equation (eq. (5.49)). The Arrhenius equation is:

$$\tau = \tau_0 \exp\left(+\frac{\Delta E}{RT}\right) \quad (5.45)$$

where ΔE is the activation energy, R is the gas constant and τ_0 is the pre-exponential factor (relaxation time). The Arrhenius equation can be fitted to data from isothermal plots of the loss modulus against frequency, recalling that the loss peak appears at the angular frequency:

$$\omega = \frac{1}{\tau} \Rightarrow f = \frac{1}{2\pi\tau} \quad (5.46)$$

where f is the frequency (in hertz). A so-called Arrhenius diagram is obtained by recording the frequency (f_{max}) associated with the loss maximum at different temperatures and by adapting these data to the equation:

$$\frac{1}{2\pi f} = \frac{1}{2\pi f_0} \exp[+(\Delta E/RT)]$$

$$\Rightarrow f = f_0 \exp[-(\Delta E/RT)] \quad (5.47)$$

It should be noted that the maximum of the loss peak corresponds to the **central** relaxation time. All real polymers exhibit a distribution in relaxation times, a relaxation time spectrum.

Taking the frequency at the loss maximum (f_{max}) for each temperature and plotting $\log f_{max}$ against $1/T$, the activation energy of the loss process is obtained as follows:

$$\Delta E = -2.303 R \left(\frac{d \log f_{max}}{d\left(\frac{1}{T}\right)} \right) \quad (5.48)$$

The subglass relaxation processes dealt with in section 5.5.2 obey Arrhenius temperature dependence.

The temperature dependence of the glass–rubber transition follows the Vogel–Fulcher equation, which is essentially a generalization of the WLF equation:

$$\tau = \tau_0 \exp[+C/(T-T_0)] \quad (5.49)$$

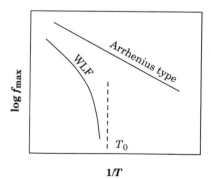

Figure 5.26 Illustration of different temperature dependences of relaxation processes.

Figure 5.26 illustrates eqs (5.45) and (5.49). The WLF behaviour appears curved in the Arrhenius plot and the curve approaches a singularity at temperature T_0. It should also be noted that the term 'activation energy' does not apply to relaxation processes showing WLF temperature dependence. In a narrow temperature region the curve may be approximated by a straight line and in that sense the activation energy may be used as a shift factor. The validity range of the WLF equation is from T_g to $T_g + 100$ K.

5.5.2 SELECTED EXPERIMENTAL DATA AND MOLECULAR INTERPRETATION

Figures 5.27 and 5.28 present mechanical data showing the presence of a series of different so-called relaxation processes. It is customary to call the isochronal high-temperature process, i.e. the glass transition, α, and to give processes appearing at lower temperatures the Greek letters β, γ, δ, ... as they appear in order of descending temperature (Fig. 5.28). If we are considering a system at constant temperature, as in Fig. 5.27, the processes appear in the reverse order chronologically, i.e. in the order ... δ, γ, β, α.

Amorphous polymers always show a glass transition process (α) and also one or more so-called subglass process(es), referred to as β, γ, δ, etc. In a

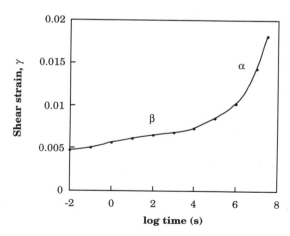

Figure 5.27 Shear strain as a function of time (log scale) for polymethyl methacrylate (PMMA) at a constant shear stress of 7.3 MPa at 30°C. Drawn after data from Lethersich (1950).

constant-stress experiment, they appear as steps in the strain (creep compliance)–log time plot (Fig. 5.27). It is difficult to obtain data covering nine orders of magnitude in time as shown in Fig. 5.27. However, by taking data over a more limited time period at **different** temperatures, it is possible to construct a so-called 'master curve' valid for a certain selected temperature (e.g. 30°C as in Fig. 5.27) by predominantly horizontal shifting of the creep curves in the creep compliance–log t diagram. It is here assumed that the retardation time distribution is shifted by a factor in accordance with eqs (5.44) and (5.48). From data obtained at different temperatures, data for an Arrhenius diagram could be obtained by pairing log t_{step} and $1/T$. The quantity t_{step} refers to the time associated with the inflection point in the step in the creep compliance–log time curve.

Figure 5.28 shows dynamic mechanical data (sinusoidally varying stress and strain) expressed in the loss modulus as a function of temperature at constant frequency. The peak maxima expressed in the pair (frequency and temperature) for each individual relaxation process can again be conveniently represented in an Arrhenius diagram.

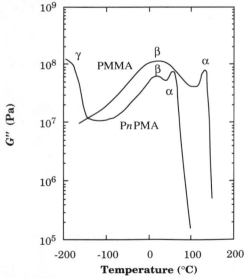

Figure 5.28 Temperature dependence of loss modulus (G'') as a function of temperature at 1 Hz for polymethyl methacrylate (PMMA) and poly(n-propyl methacrylate) (PnPMA). Drawn after data from Heijboer (1965).

Let us now return to the subglass processes. Their existence proves that the glassy polymer does have some limited segmental mobility which indeed is consonant with the observations of volume recovery (physical ageing) and, in general, the anelasticity. If we consider the polymer at such a low temperature or for such a short time that the subglass process(es) is (are) not triggered, the material deforms elastically and shows no measurable physical ageing. Struik (1978) showed that physical ageing only appeared at temperatures between T_β (the 'freezing-point' of the subglass process) and T_g. The change in modulus associated with the subglass process(es) is only small, of the order of 10% of the low-temperature, elastic value.

The molecular interpretation of the subglass relaxations has been the subject of considerable interest during the last 40 years. By varying the repeating unit structure and by studying the associated relaxation processes, it has been possible to make a group assignment of the relaxation processes. That is not to say that the actual mechanisms have been resolved. The relaxation processes can be categorized as side-chain or main-chain. Subglass processes appear both in polymers with pendant groups such as PMMA and in linear polymers such as polyethylene or poly(ethylene terephthalate). In the latter case, the subglass process must involve motions in the backbone chain.

The molecular interpretation of the relaxation processes of the methacrylates in general, and for the two shown in Fig. 5.28 in particular, is as follows.

The α process is clearly the glass transition. It is present in all polymethacrylates. It obeys WLF temperature dependence. The high-temperature subglass process β is present in all polymethacrylates. It shows both mechanical and dielectric activity and it is assigned to rotation of the side group. The β process shows Arrhenius temperature dependence.

The low-temperature subglass process (γ) is not present in PMMA and PEMA. It appears in PnPMA and longer alkyl homologues. It obeys Arrhenius temperature dependence and the activation energy is the same for all the higher polymethacrylates. It is assigned to motions in the flexible methylene sequence. It was concluded by Willbourn that a low-temperature process, essentially the same as the γ process in the higher polymethacrylates, with an activation energy of about 40 kJ mol^{-1} occurred in main-chain polymers with at least four methylene groups. This low-temperature process was attributed to restricted motion of the methylene sequence, so-called **crankshaft** motions. The same process appears in the amorphous phase of polyethylene and is in that case also denoted γ. There are two simple conformation rearrangements that are local and that leave the surrounding stems practically unchanged. The first, suggested by Boyer (1963), involves a change from ...TGT... to ...TG'T..., i.e. a three-bond motion. The second (Schatzki crankshaft, 1966) involves the TGTGT sequence, which remains unchanged and whose surrounding bonds change conformation and cause the mid-section to rotate as a crankshaft. The Schatzki crankshaft involves a considerable swept-out volume and can for that reason be excluded as a mechanism for the γ relaxation. The Boyer motion shows two energy barriers with an intermediate minimum. One of the options involves essentially an intramolecular activation energy and the swept-out volume is very small, whereas the other requires a significant swept-out volume and can for that reason be excluded. The mechanical activity, i.e. the strain associated with the conformational changes, is too small for these changes to be reasonable mechanisms for the γ process. Boyd and Breitling (1974) proposed an alternative explanation closely related to the three-bond motion. He called the mechanism a left-hand–right-kink inversion. It involves the following conformational change:

$$...\text{TTTTGTG'TTTT}... \rightarrow ...\text{TTTTG'TGTTTT}...$$

It has a small swept-out volume and requires only a modest activation energy. The stems are slightly displaced, which leads to a change in shape (strain). Hence, the process has mechanical activity. Model calculations showed that the suggested mechanism involves activation and activation entropy (almost) similar to the experimental, but the predicted relaxation strength was significantly lower than the experimental.

Polystyrene exhibits relatively complex relaxation behaviour. Apart from the glass transition (α), polystyrene exhibits four subglass relaxation processes, referred to as β, γ and δ in order of decreasing temperature. One view (McCammon, Saba and Work (1969); Sauer and Saba (1969)) is that the cryogenic δ process (55 K in PS at 10 kHz) is due to oscillatory motions of the phenyl groups, whereas others (Yano and Wada, 1971) believe that it arises from defects associated with the configuration of the polymer. The γ process appearing in PS at 180 K at 10 kHz has also been attributed to phenyl group oscillation or rotation. The high-temperature process denoted β occurs in PS between $T_g - 100$ K and T_g and is believed to be due to a rotation of the phenyl group with a main-chain cooperation.

5.6 STRUCTURE OF GLASSY, AMORPHOUS POLYMERS

Typical of glassy amorphous polymers is their transparency. They exhibit very low light scattering. X-ray diffraction reveals no Bragg reflections. This demonstrates that the ordered regions, if such exist, are smaller than 2–3 nm. The X-ray pattern consists of a broad so-called **amorphous halo** with a scattered intensity maximum corresponding to a d spacing of about 0.4 nm. This particular distance corresponds to the intermolecular distance. The discussion among scientists concerns the possibility that there is **some** degree of order in these polymers. The order is clearly only partial and limited to relatively small volumes. It is here instructive to boil spaghetti. Looking down on the spaghetti we see that small regions exist with a more or less parallel arrangement of the individual spaghetti cylinders. This order is, however, only local. Some researchers have modelled the packing and, based on density considerations, they have pointed out the necessity that the chains on a local scale must be closely parallel.

Experiments carried out by Stein and Hong (1976) showed that the birefringence did not change appreciably on swelling or extension and that the order along a single chain (axial order) did not change beyond a range of 0.5–1 nm, which is comparable with the range of ordering of low molar mass liquids. Other experiments, stress optical coefficient measurements, depolarized light scattering, magnetic birefringence, Raman scattering, Brillouin scattering,

NMR relaxation and small-angle X-ray scattering indicated that the intramolecular orientation is only slightly affected by the presence of other chains in the glassy amorphous state and that axial correlation is of the order of 1 nm or less.

Electron diffraction experiments indicate, according to some authors (Lovell, Mitchell and Windle, 1979), that there is some intermolecular order, i.e. parallelism of nearby chains in small volumes, 1–2 nm in characteristic length, of the sample.

The global dimensions of the chains in the bulk state were not revealed by experiment until the 1970s. The small-angle neutron scattering (SANS) technique was a revolution in this context. Using SANS, it was possible to determine the radius of gyration (s) of deuterated chains dissolved in a matrix of normal protonated chains. Atactic polystyrene is a well-documented fully amorphous polymer. It exhibits a value of $s^2/\bar{M}_w = 0.275$ Å mol$^{1/2}$ g$^{-1/2}$ in the glassy state, which is the very same value as is obtained by light scattering of solutions under theta conditions. The same observation was made in the case of amorphous polymethylmethacrylate. It is important to note that the radius of gyration in semicrystalline polymers is the same as that in its molten state prior to crystallization. The order introduced by crystallization evidently occurs on a more local basis without appreciably affecting the global dimensions of the chains.

We can conclude that the weak signs of order in the glassy amorphous state are both very local and incomplete. Such a vague structure is not yet easily described, but this is clearly presently an area of significant interest. Models have been presented featuring the random nature of the polymer chains both with separated chains and with entangled chains. A number of models with partial order have been presented, e.g. the Meander model by Pechold (1968). The reorientation of the polymer chains occurs here along certain well-defined lines giving the chain an overall meandering orientation.

5.7 SUMMARY

The reason why a polymer liquid is transformed into a glassy, amorphous state and not to a semicrystalline structure is two-fold: (a) the chain structure is irregular and the polymer cannot crystallize, even at the slowest possible cooling rate; (b) the polymer is cooled at a very high rate not permitting crystallization, even in a regular-chain polymer. The tendency for formation of a glassy, fully amorphous state increases with decreasing flexibility of the polymer.

The transformation of the liquid to a glass occurs at the glass transition temperature which is accompanied by a thousand-fold increase in the Young modulus. The glass transition is a second-order phase transformation, but not in the Ehrenfest sense. The experimental glass transition shows clear kinetic features.

Physical ageing occurs most prominently in a quenched, glassy polymer near the glass transition temperature. Physical ageing drives the structure towards the equilibrium at the prevailing temperature. The volume recovery is nonlinear with respect to the magnitude of the departure from equilibrium. The rate of volume recovery depends not only on the magnitude of the departure from equilibrium but also on the sign of the departure; contraction is more rapid than expansion. Memory effects are observed after two or more changes in temperature. The increase in density leads to changes in almost all engineering properties, e.g. creep compliance, impact strength and permeability. Physical ageing, i.e. volume and enthalpy recovery, can be described by the multi-parameter approach proposed by Kovacs, Aklonis, Hutchinson and Ramos, the so-called KAHR model.

The actual value of the glass transition temperature is affected by the intrinsic segmental flexibility of the polymer, the intermolecular interaction, the presence of plasticizing low molar mass species and the molecular architecture.

A complete theoretical understanding of the glass transition phenomenon is not yet available. The current theories can be divided into three main groups: free volume, kinetic and thermodynamic theories. The free volume theories state that the glass transition is characterized by an iso-free volume state. According to the kinetic theories, there is no thermodynamic glass transition; the phenomenon is purely kinetic. The thermodynamic theory, which was formulated by Gibbs and DiMarzio, states that, although the observed glass transition is a kinetic phenomenon, the underlying true transition possesses equilibrium properties according to the Ehrenfest equation. It is postulated by Gibbs and DiMarzio that the conformational entropy is zero at the thermodynamic glass transition T_2.

The segmental mobility is not zero at temperatures below the glass transition temperature. In fact, most polymers exhibit subglass relaxation processes involving localized segmental motions: rotations of side groups and restricted main-chain motions. The occurrence of these subglass processes is a prerequisite for physical ageing and also gives the polymer some useful ductility.

Glassy, amorphous polymers are typically optically clear. They show a liquid-like X-ray pattern. Discussions among scientists have dealt with possible short-range order. It may be concluded from currently available data that the possible order is vague and very local, it concerns regions of a few nanometres or less.

5.8 EXERCISES

5.1. Categorize the following polymers with regard to their low-temperature physical structure: polyethylene; isotactic polypropylene; poly(ethylene terephthalate); isotactic polystyrene; atactic polystyrene; atactic poly(vinyl alcohol); atactic poly(vinyl acetate); poly(ethylene-*stat*-propylene) with 50/50 molar composition, and the same polymer but with molar composition 98/2.

Some of the polymers can exist both in a fully glassy, amorphous state and in a semi-crystalline state. Say which and present explanations.

5.2. Describe an experiment by which physical ageing can be recorded.

5.3. Calculate the T_g of the polymer of infinite molar mass from the following data:

T_g (K)	182	278	354	361	362	369.5	375.5
M (g mol^{-1})	500	1000	4000	5000	6000	10 000	20 000

5.4. You have a polymer blend which shows one T_g but is opaque. Explain. Another polymer shows two T_g's but this material is transparent. Explain.

5.5. You have a polymer which is almost suitable for a given application. Propose practical methods to raise the T_g of this polymer.

5.6. Suppose that you have a polymer with a relatively low molar mass (M). There is consequently a significant difference (ΔT_g) in glass transition temperature between this polymer and a polymer of infinite molar mass (T_g^0). You have access to a peroxide which will crosslink. How much peroxide has to be added in order to increase the glass transition to the value T_g^0?

5.7. Present briefly the experimental techniques which can be used to differentiate between an amorphous and a semicrystalline polymer. What are the main results indicating full amorphism?

5.9 REFERENCES

Bair, H. E. (1970) *Polym. Eng. Sci.* **10**, 247.
Beaman, R. G. (1953) *J. Polym. Sci.* **9**, 472.
Boyd, R. H. and Breitling, S. M. (1974) *Macromolecules* **7**, 855.
Boyer, R. F. (1954) *J. Appl. Phys.* **25**, 825.
Boyer, R. F. (1963) *Rubber Chem. and Technol.* **36**, 1303.
Davies, R. O. and Jones, G. O. (1953) *Advances in Physics, Philosophical Magazine Supplement* **2**, 370.
DiMarzio, E. A. (1981) Equilibrium theory of glasses, in *Structure and Mobility in Molecular and Atomic Glasses* (J. M. O'Reilly and M. Goldstein, eds), Annals of the New York Academy of Sciences **371**). New York Academy of Sciences, New York.
Doolittle, A. K. (1951) *J. Appl. Phys.* **22**, 1471.
Eisenberger, A. (1984). The glassy state, in *Physical Properties of Polymers*, Am. Chem. Soc., Washington, DC.
Fox, T. G. (1956) *Bull. Am. Phys. Soc.* **1**, 123.
Gibbs, J. H. and DiMarzio, E. A. (1958) *J. Chem. Phys.* **28**, 373.
Goldbach, G. and Rehage, G. (1967) *Rheol. Acta* **6**, 30.
Heijboer, J. (1965) *Physics of Noncrystalline Solids*. North-Holland, Amsterdam.
Kovacs, A. J. (1963) *Fortschr. Hochpolym. Forsch.* **3**, 394.
Kovacs, A. J., Aklonis, J. J., Hutchinson, J. M. and Ramos, A. R. (1979) *J. Polym. Sci., Polym. Phys. Ed.* **17**, 1097.
Lethersich, W. (1950) *Brit. J. Appl. Phys.* **1**, 294.
Lovell, R., Mitchell, G. R. and Windle, A. H. (1979) *Faraday Discuss. Chem. Soc.* **68**, 46.
McCammon, R. D., Saba, R. G. and Work, R. N. (1969) *J. Polym. Sci., Part A2* **7**, 1271.
Pechold, W. (1968) *Kolloid Z. Z. Polym.* **228**, 1.
Rietsch, F., Daveloose, D. and Froelich, D. (1976) *Polymer* **17**, 859.
Rogers, S. S. and Mandelkern, L. (1957) *J. Phys. Chem.* **61**, 985.

Sauer, J. A. and Saba, R. G. (1969) *J. Macromol. Sci. Part A* **3**, 1217.
Schatzki, T. F. (1966) *J. Polym. Sci., Polym. Symp.* **14**, 139.
Schmieder, K. and Wolf, D. (1953) *Kolloid Z.* **134**, 149.
Shetter, J. A. (1963) *Polym. Lett.* **1**, 209.
Stein, R. S. and Hong, S. D. (1976) *J. Macromol. Sci., Part B* **12**, 125.
Struik, L. C. E. (1978) *Physical Ageing of Amorphous Polymers and Other Materials*. Elsevier, Amsterdam.
Tool, A. Q. (1946) *J. Am. Chem. Soc.* **29**, 240.
Wolf, D. (1951) *Kunststoffe* **41**, 89.
Yano, O. and Wada, Y. (1971) *J. Polym. Sci. Part A2* **9**, 669.

5.10 SUGGESTED FURTHER READING

Eisenberger, A. (1993) The glassy state and the glass transition, in *Physical Properties of Polymers*, 2nd edn. American Chemical Society, Washington, DC.
Ferry, J. D. (1980) *Viscoelastic Properties of Polymers*. Wiley, New York.
Keinath, S. E., Miller, R. L. and Rieke, J. K. (eds) (1987) *Order in the Amorphous State of Polymers*. Plenum, New York and London.
McCrum, N. G., Read, B. E. and Williams, G. (1967) *Anelastic and Dielectric Effects in Polymeric Solids*. Wiley, New York.

THE MOLTEN STATE

6.1 INTRODUCTION

A fluid phase is a liquid if the kinetic energies of the molecules and the potential energies of their interaction are comparable, so that the molecules can move 'viscously' relative to each other. A fluid phase is a gas if the kinetic energies greatly exceed the potential energies of their interaction. The translative kinetic energies of molecules in crystals are negligible. Conventional liquids possess only short-range order; long-range order is absent. Rheology, which is the first topic of this chapter (sections 6.2 and 6.3), is the mathematical discipline within which relationships between stress and strain in liquids are expressed.

The molten state of polymers is more dependent on the molar mass than any of their other physical states. Flexible-chain polymer molecules possess essentially random conformations in the molten state. The coiled molecules entangle in high molar mass polymers. These **chain entanglements** are very important for the rheological properties of the melt. The next part of this chapter (section 6.4) deals with the rheology of flexible-chain polymer melts. A discussion of the deformation mechanisms, including theoretical aspects, is also presented.

A relatively new class of polymers, the **liquid-crystalline polymers**, exhibits orientational order, i.e. alignment of molecules along a common director in the molten state. Liquid-crystalline polymers are used, after solidification, as strong and stiff engineering plastics and fibres. 'Functional' liquid-crystalline polymers with unique electrical and optical properties are currently under development. The fundamental physical and rheological aspects of liquid-crystalline polymers are the third subject of this chapter (section 6.5).

6.2 FUNDAMENTAL CONCEPTS IN RHEOLOGY

Rheology is the discipline which deals with the deformation of fluids. Stresses acting on a volume element of a liquid lead to a deformation of the element. The stress–strain relationships are referred to as constitutive equations. Such relationships are based on balance equations, which are statements of universal laws of the conservation of mass, momentum and energy. They can be expressed as follows:

$$\frac{\partial \rho}{\partial t} + \nabla \cdot (\rho \mathbf{v}) = 0 \quad \text{(mass)} \quad (6.1)$$

$$\rho \frac{D\mathbf{v}}{Dt} = -\nabla p + \nabla \cdot \boldsymbol{\sigma} + \rho \mathbf{g} \quad \text{(momentum)} \quad (6.2)$$

$$\rho C_v \cdot \frac{DT}{Dt} = -\nabla \cdot \mathbf{q} - T\left(\frac{\partial p}{\partial T}\right)_{\hat{V}} (\nabla \cdot \mathbf{v}) + \boldsymbol{\sigma} : \nabla \mathbf{v} + \dot{S}$$

$$\text{(energy)} \quad (6.3)$$

where ρ is the density, ∇ is the gradient operator, $\mathbf{v}$ is the velocity vector (three components), p is pressure, D/Dt is the material or substantial derivative, $\boldsymbol{\sigma}$ is the stress tensor, $\mathbf{g}$ is the body force vector, C_v is the specific heat at constant volume, T is the temperature, $\mathbf{q}$ is the heat flux vector, $\hat{V}$ is the specific volume and $\dot{S}$ is rate of heat generation from internal physical and chemical reactions.

Let us consider a fluid which is confined between two parallel plates (Fig. 6.1). The upper plate moves at a constant velocity $\mathbf{v}$ while the lower plate is at rest. The force needed to move the upper plate is denoted F and the contact area of the upper plate to

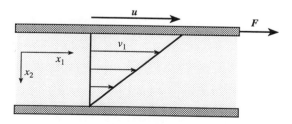

Figure 6.1 Steady simple shear flow.

the liquid is A. The conservation of momentum in this system can, according to eq. (6.2), be expressed as:

$$\frac{d\sigma_{21}}{dx_2} = 0 \qquad (6.4)$$

where σ_{21} is the shear stress. The shear stress is constant through the gap and is given by:

$$\sigma_{21} = \frac{F}{A} \qquad (6.5)$$

The velocities of the fluid elements are given by

$$\frac{dv_1}{dx_2} = \dot{\gamma} \qquad (6.6)$$

$$v_2 = 0 \qquad (6.7)$$

$$v_3 = 0 \qquad (6.8)$$

where $\dot{\gamma}$ is denoted the shear rate, which is the change in shear strain ($\gamma = dx_1/dx_2$) per unit time. The dimension of shear rate is s^{-1}. This flow case is referred to as simple shear flow and the state of stress may be described by the stress tensor:

$$\boldsymbol{\sigma} = \begin{Bmatrix} \sigma_{11} & \sigma_{12} & 0 \\ \sigma_{21} & \sigma_{22} & 0 \\ 0 & 0 & \sigma_{33} \end{Bmatrix} \qquad (6.9)$$

where the shear stresses $\sigma_{12} = \sigma_{21}$, and σ_{11}, σ_{22} and σ_{33} are normal stresses. The following three independent material parameters can be obtained from measurement of the components of stress tensor and the shear rate ($\dot{\gamma}$):

$$\eta = \frac{\sigma_{21}}{\dot{\gamma}} \qquad (6.10)$$

$$\psi_1 = \frac{\sigma_{11} - \sigma_{22}}{\dot{\gamma}^2} \qquad (6.11)$$

$$\psi_2 = \frac{\sigma_{22} - \sigma_{33}}{\dot{\gamma}^2} \qquad (6.12)$$

where η is the viscosity, ψ_1 is the primary normal stress coefficient and ψ_2 is the secondary normal stress coefficient.

Another deformation case is elongational flow, which is described by the following equation:

$$v_i = a_i x_i \quad (i = 1, 2, 3) \qquad (6.13)$$

where a_i are the strain rate coefficients. The following expression holds for an incompressible liquid:

$$\sum_{i=1}^{3} a_i = 0 \qquad (6.14)$$

Figure 6.2 shows a particularly simple case: uniaxial elongational flow, in which case the a_i coefficients take the values:

$$a_1 = \dot{\varepsilon} \qquad (6.15)$$

$$a_2 = -\frac{\dot{\varepsilon}}{2} \qquad (6.16)$$

$$a_3 = -\frac{\dot{\varepsilon}}{2} \qquad (6.17)$$

where $\dot{\varepsilon}$ is the elongation rate.

The normal stress difference, $\sigma_{11} - \sigma_{22}$, and the elongation rate ($\dot{\varepsilon}$) are here the only measurable quantities. The steady elongational viscosity ($\bar{\eta}$) is defined as

$$\bar{\eta} = \frac{\sigma_{11} - \sigma_{22}}{\dot{\varepsilon}} \qquad (6.18)$$

Biaxial elongational flow is defined by:

$$a_1 = a_2 = \dot{\varepsilon}_B \qquad (6.19)$$

$$a_3 = -2\dot{\varepsilon}_B \qquad (6.20)$$

where $\dot{\varepsilon}_B$ is the biaxial elongation rate. The biaxial elongational viscosity (η_B) is defined as:

$$\eta_B = \frac{\sigma_{11} - \sigma_{33}}{\dot{\varepsilon}_B} = \frac{\sigma_{22} - \sigma_{33}}{\dot{\varepsilon}_B} \qquad (6.21)$$

Dynamic mechanical data of polymer melts are also commonly measured. The melt is subjected to a sinusoidally varying shear strain (γ):

$$\gamma = \gamma_0 \cos \omega t \qquad (6.22)$$

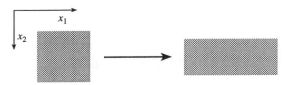

Figure 6.2 Simple elongational flow along the x_1 direction.

where γ_0 is the maximum shear strain, ω is the angular velocity (in rad s^{-1}) and t is time. The shear stress (σ) response will also be sinusoidal of the same frequency but shifted by a phase angle (δ) i.e.:

$$\sigma = \sigma_0 \cos(\omega t + \delta) \quad (6.23)$$

where σ_0 is the maximum stress. The shear stress and shear strain are commonly dealt with by using complex notation:

$$\gamma^* = \gamma_0(\cos \omega t + i \sin \omega t)$$
$$= \gamma_0 \exp(i\omega t) = \gamma' + i\gamma'' \quad (6.24)$$

$$\sigma^* = \sigma_0(\cos(\omega t + \delta) + i \sin(\omega t + \delta))$$
$$= \sigma_0 \exp[i(\omega t + \delta)] \quad (6.25)$$

The complex shear modulus is defined as:

$$G^* = \frac{\sigma^*}{\gamma^*} = \frac{\sigma_0}{\gamma_0}\cos\delta + i\frac{\sigma_0}{\gamma_0}\sin\delta = G' + iG'' \quad (6.26)$$

where $G'(\omega)$ is the storage modulus and $G''(\omega)$ is the loss modulus; G' is associated with the stored energy and G'' with the dissipation of energy as heat.

The complex viscosity is defined as:

$$\eta^* = \frac{\sigma^*}{\dot{\gamma}^*} = \frac{\sigma^*}{i\omega\gamma^*} = \frac{\sigma_0}{\gamma_0}\cdot\sin\delta\cdot\frac{1}{\omega} - i\frac{\sigma_0}{\gamma_0}\cos\delta\cdot\frac{1}{\omega}$$
$$= \frac{G''}{\omega} - \frac{iG'}{\omega} = \eta' - i\eta'' \quad (6.27)$$

where $\eta'(\omega)$ is the dynamic viscosity which is associated with the dissipation of energy; η'' is associated with the energy storage.

Let us now return to the simple shear case in order to illustrate the wealth of different possible types of rheological behaviour. The viscosity is given by eq. (6.9):

$$\eta = \frac{\sigma_{21}}{\dot{\gamma}}$$

For an ideal Newtonian liquid, the viscosity is a constant, independent of the shear rate ($\dot{\gamma}$). A pseudoplastic liquid exhibits a decreasing viscosity with increasing shear rate, where as a dilatant liquid shows an increasing viscosity with increasing shear rate (Fig. 6.3). Most polymer melts show pseudoplasticity. Wet beach sand is an example of a dilatant fluid.

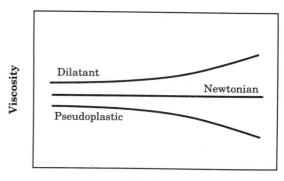

Figure 6.3 Viscosity as a function of shear rate for Newtonian, pseudoplastic and dilatant liquids.

Some fluids, so-called Bingham plastics, show a yield stress, σ_y (Fig. 6.4). No deformation occurs for shear stresses smaller than σ_y, whereas for shear stresses greater than σ_y these fluids show a plastic deformation characterized by a viscosity η. Examples of fluids with a yield stress are some molten polymers, oils, cement slurries and margarine.

Many polymer melts exhibit pseudoplastic behaviour. At low shear rates they show an almost linear shear stress–shear rate behaviour with a constant viscosity, the so-called zero-shear-rate viscosity, η_0 (Fig. 6.5). The normal stress differences approach the 'zero-shear rate' values $\psi_{1,0}$ and $\psi_{2,0}$.

At high shear rates, there is a linear decrease in viscosity with increasing shear rate in a log-log diagram (Fig. 6.5). The shear stress depends on the

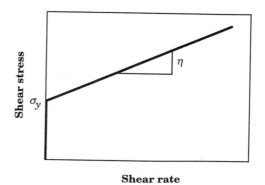

Figure 6.4 Shear stress as a function of shear rate for a Bingham plastic.

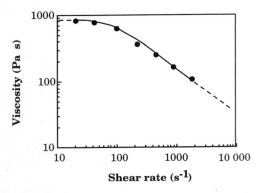

Figure 6.5 Apparent melt viscosity as a function of shear rate at 285°C for poly(butylene terephthalate). Drawn after data from Engberg *et al.* (1994b).

shear rate in this shear rate region according to the following power law expression, also known as Ostwald–de Waele equation:

$$\sigma_{21} = K\dot{\gamma}^n \quad (6.28)$$

where K and n are constants. The apparent viscosity ($\eta = \sigma_{21}/\dot{\gamma}$) is given by:

$$\eta = K\dot{\gamma}^{n-1} \quad (6.29)$$

Pseudoplastic liquids have n values smaller than unity, and the slope coefficient in the log-log diagram (Fig. 6.5) is negative.

Pseudoplasticity may be explained in simple, rational terms. At low shear rates, polymer molecules take conformations similar to those of the unperturbed chains. The great resistance to flow is due to the presence of numerous chain entanglements. The molecules are aligned along the shear direction when the melt is subjected to higher shear rates. This causes a reduction in the concentration of chain entanglements with reference to the unperturbed state. The viscosity thus decreases with increasing shear rate.

The time dependence of a non-Newtonian liquid is classified as either **thixotropic** or **rheopectic** (Fig. 6.6). The curves shown in Fig. 6.6 are obtained by increasing the shear rate at a constant rate, stopping at a certain shear rate and then decreasing the shear rate to the initial zero-shear-rate state. The rate at which the shear rate is changed affects the hysteresis loops, and for that reason it is difficult to give a general description of thixotropic and rheopectic liquids. Thixotropic liquids show thus at constant shear rate a decrease in the viscosity with increasing shear time, and the opposite behaviour is observed for the rheopectic liquids. High molar mass polymers, margarine, printing inks and paints are examples of thixotropic liquids. Gypsum suspensions and various soils are examples of rheopectic liquids.

Thixotropic liquids are unanimously pseudoplastic, i.e. the slope coefficient in the shear stress–shear rate diagram decreases with increasing shear rate. Pseudoplastic fluids are, however, not necessarily thixotropic. Dilatant liquids show generally rheopexy, but there are also exceptions to this rule. Rheopectic liquids are, on the other hand, always dilatant.

The most complex non-Newtonian liquids are the visco-elastic liquids, and to this group belong most high-polymer melts. Polymers show a rheological behaviour in between the ideal Newtonian liquid and the ideal Hookean solid. The response changes from solid-like (Hookean) at short shearing times to liquid-like at long shearing times. The history of loading for these visco-elastic materials, which are

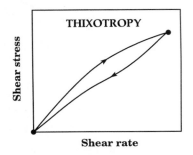
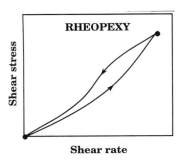

Figure 6.6 Hysteresis loops for time-dependent liquids.

being characterized by their relaxation shear modulus, $G(t) = \sigma_{21}/\gamma(t)$, is thus crucial. It is useful to define the following quantities:

$$\eta_0 = \int_0^\infty G(t)dt \qquad (6.30)$$

$$J_e^0 = \frac{1}{\eta_0^2} \int_0^\infty tG(t)dt \qquad (6.31)$$

$$\tau_0 = \frac{\int_0^\infty tG(t)dt}{\int_0^\infty G(t)dt} = \eta_0 J_e^0 \qquad (6.32)$$

where η_0 is the earlier defined zero-shear-rate viscosity, t is time and J_e^0 is the **steady-state recoverable shear compliance**. The latter is obtained from the total recoverable shear strain (γ_r) at steady state divided by the original applied shear stress (σ_{21}) in a constant-stress experiment:

$$J_e^0 = \lim_{\sigma_{21} \to 0; t \to \infty} \frac{\gamma_r}{\sigma_{21}}$$

The steady-state viscosity (η_0) is a measure of the energy dissipated during flow and J_e^0 determines the amount of elastically stored energy. The latter is manifested in several phenomena, e.g. die-swell, which is illustrated in Fig. 6.7. The convergent section causes a uniaxial elongational flow. The molecules are stretched out from their equilibrium coiled conformations. The chain entanglements act as temporary crosslinks and outside the die the release of the axial stress causes an axial shrinkage and a transverse expansion. The **relaxation time** (τ_0), the average relaxation time for $G(t)$, is a measure of the time required for final equilibration following a step strain.

The time dependence of the shear modulus of a typical molten polymer is shown in Fig. 6.8. The response at short times is almost Hookean, the molten polymer behaves like a glassy material. Any deformation is localized, and in the case of conformational changes only a few main-chain atoms

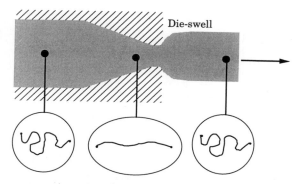

Figure 6.7 Illustration of die-swell also showing schematically the change in chain conformation of a chain between two entanglement points.

may be involved. The extensive deformation associated with the longest time-scales controls η_0 (eq. (6.30)).

Figure 6.9 shows the creep compliance (constant shear stress (σ_0) experiment) as a function of the loading time for a polymer melt. The creep compliance is defined as:

$$J(t) = \frac{\gamma(t)}{\sigma_0} \qquad (6.33)$$

where $\gamma(t)$ is the time-dependent shear strain. At short loading times, insufficient time is given to permit conformational rearrangement and the melt behaves like a glass. Conformational changes occur when

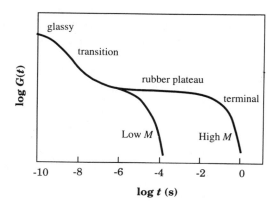

Figure 6.8 Schematic diagram showing shear relaxation modulus as a function of time for molten polymers of different molar mass (M).

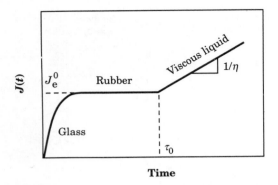

Figure 6.9 Creep compliance as a function of loading time for a viscoelastic polymer melt.

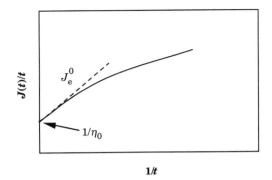

Figure 6.10 Creep compliance (J) as a function time (t). Calculation of zero-shear-rate viscosity and steady-state recoverable shear compliance.

reaching the plateau region, in which the entanglements give the material a rubber-like character. The compliance associated with the rubber plateau is simply the steady-state recoverable shear compliance. At loading times greater than the terminal relaxation time τ_0, the creep compliance increases linearly with time, with slope coefficient η^{-1}.

6.3 MEASUREMENT OF RHEOLOGICAL PROPERTIES OF MOLTEN POLYMERS

Most rheological measurements measure quantities associated with simple shear: shear viscosity, primary and secondary normal stress differences. There are several test geometries and deformation modes, e.g. parallel-plate simple shear, torsion between parallel plates, torsion between a cone and a plate, rotation between two coaxial cylinders (Couette flow), and axial flow through a capillary (Poiseuille flow). The viscosity can be obtained by simultaneous measurement of the angular velocity of the plate (cylinder, cone) and the torque. The measurements can be carried out at different shear rates under steady-state conditions. A transient experiment is another option from which both η_0 and J_e^0 can be obtained from creep data (constant stress) or stress relaxation experiment which is often measured after cessation of the steady-state flow (Fig. 6.10).

The normal stress differences, ψ_1 and ψ_2, could be obtained by measurement of, for example, the normal stress in a cone-plate geometry (x_1 is parallel to the direction of rotation, x_2 is parallel to the normal and x_3 is along the radius) experiment:

$$\sigma_{11} - \sigma_{22} = \frac{2F}{\pi R^2} \quad (6.34)$$

where F is the force trying to separate the plate and the cone, and R is the outer radius. The secondary normal stress difference can be obtained by also measuring the hydrostatic pressure (p) gradient along the radius:

$$-\frac{r \, \partial p}{\partial r} = \sigma_{11} + \sigma_{22} - 2\sigma_{33} \quad (6.35)$$

The hydrostatic pressure can be measured by having pressure transducers mounted on the surface (Fig. 6.11).

Elongation flow devices are less common than equipment recording shear flow. Measurement of elongational viscosity is complex to interpret and more difficult to perform than measurement of shear viscosity. It is necessary to maintain a constant elongation rate or axial stress while reaching steady-state conditions. It is not possible to control the elongation rate by pushing the melt through a converging section since that will lead to a combined shear flow and elongation flow. Typical tests involve stretching of filaments. It is by this technique only possible to perform measurement on high-viscosity melts, typically of a viscosity of 10 000 Pa s or greater, and only at low elongation rates, typically at 5 s^{-1} or less. High molar mass polyolefins, polystyrenes and

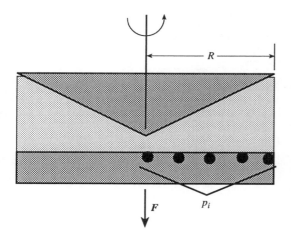

Figure 6.11 Cone-plate viscometer allowing measurement of the primary and secondary normal stress differences. Normal force (F) and pressure transducers recording hydrostatic pressures (p_i) at different radial positions are shown.

elastomers are typical polymers that can be studied by this technique.

Oscillatory shear experiments using, for example, cone-and-plate devices constitute the third main group of viscometric techniques. These techniques enable the complex dynamic viscosity (η^*) to be measured as a function of the angular velocity (ω). The fundamental equations are presented in section 6.2 (eqs (6.22)–(6.27)). Another arrangement is two rotating parallel excentric discs by which the melt is subjected to periodic sinusoidal deformation.

There are numerous other experimental methods that provide information about the rheological behaviour, including the molecular response to stresses and strains: swell and shrinkage tests, flow-birefringence measurements (Chapter 9) and flow-infrared-dichroism measurements (Chapter 9) to mention but a few.

6.4 FLEXIBLE-CHAIN POLYMERS

6.4.1 MOLAR MASS DEPENDENCE AND MOLECULAR INTERPRETATION

Small-angle neutron scattering has confirmed that the radius of gyration of polymer chains has the same molar mass dependence in molten flexible-chain polymers as under theta conditions. The reasoning behind this established fact is presented in Chapter 2. The time-scale for a certain change in the chain conformation depends strongly on the size of the group which is transformed (Fig. 6.12).

It should be noted that the different rearrangements of a polymer chain occur over many orders of magnitude in time. The rearrangement on a local scale is rapid and is primarily controlled by the nature of the repeating unit itself. The full equilibration of the polymer chain, on the other hand, requires a time which is many orders of magnitude longer. The longest times involved are strongly dependent on the 'global' molecular architecture – molar mass and chain branching. These slow processes, the terminal region relaxations, govern the flow properties. This is the reason why molar mass and molecular architecture, in general, are of such importance for the rheological properties of polymer melts.

Flow properties are very strongly dependent on molecular architecture, i.e. molar mass and chain branching. Figures 6.13 and 6.14 illustrate the effect of molar mass on the zero-shear-rate viscosity (η_0) and on the steady-state recoverable shear compliance.

Both η_0 and J_e^0 show an abrupt change in their molar mass coefficients at certain critical molar mass values, denoted M_c and M_c' (Figs 6.13 and 6.14). The critical molar mass (M_c) separating the two slopes in the plot of log η_0 against log M is not the same as the other characteristic molar mass M_c'. The low molar mass melts, with a molar mass lower than the critical M_c (M_c'), are characterized by few or no chain

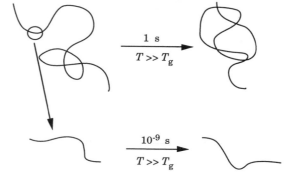

Figure 6.12 The time-dependent response of molten polymer chains – dependence of size of rearranging unit.

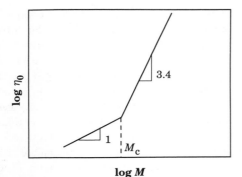

Figure 6.13 The logarithm of the zero-shear-rate viscosity plotted versus the logarithm of the molar mass. Schematic curve.

Table 6.1 Characteristic molar mass values (grams per mole) for linear polymers

Polymer	M_e	M_c	M'_c
Polyethylene	1 250	3 800	14 400
1,4-polybutadiene	1 700	5 000	11 900
cis-polyisoprene	6 300	10 000	28 000
Poly(vinyl acetate)	6 900	24 500	86 000
Poly(dimethyl siloxane)	8 100	24 400	56 000
Poly(α-methyl styrene)	13 500	28 000	104 000
Polystyrene	19 000	36 000	130 000

Source: Graessley (1984, p. 134).

entanglements and their resistance to flow is controlled by the 'monomeric frictional coefficient' times the length of the molecule (molar mass M). The high molar mass melts, on the other hand, are characterized by great many chain entanglements, which cause a restriction of the possible diffusion mechanism and a stronger molar mass dependence of the shear viscosity (η_0). The critical molar mass values may thus be associated with the minimum chain length needed to generate chain entanglements. The constant J_e^0 for the entangled melt systems means simply that the number of chain entanglements per volume unit of polymer is constant and not dependent on molar mass in this molar mass region. Table 6.1 presents a summary of data collected for different polymers.

The so-called 'entanglement molar mass' (M_e) is calculated from the plateau value of the shear modulus (G_e^0, Fig. 6.8) and, using classical rubber elasticity theory as described in Chapter 3, is given by:

$$M_e = \frac{\rho RT}{G_e^0} \quad (6.36)$$

where ρ is the density, R is the gas constant and T is the temperature. It is obvious that the critical molar mass values M_c and M'_c correlate with M_e. The terminal relaxation time τ_0 is approximately proportional to the product $\eta_0 J_e^0$ which means that for samples of 'subcritical' molar mass:

$$\eta_0 \propto M \quad \text{and} \quad J_e^0 \propto M \Rightarrow \tau_0 \propto M^2$$
$$\text{(for } M < M_c\text{)} \quad (6.37)$$

For samples of high molar mass, the molar mass dependence of τ_0 is given by:

$$\eta_0 \propto M^{3.4} \quad \text{and} \quad J_e^0 \approx \text{const.} \Rightarrow \tau_0 \propto M^{3.4}$$
$$\text{(for } M > M_c\text{)} \quad (6.38)$$

The flow behaviour in these two molar mass regions is thus governed by different flow mechanisms. In the low molar mass region the frictional forces on a very local scale control the flow, whereas in the high molar mass region chain entanglements play an important role. The discussion of the Rouse model and the reptation model is presented later in this section.

The temperature dependence of the shear viscosity can, according to the early work of Miller (1963), be adequately described by the WLF equation (see Chapter 4) for more details):

$$\eta_0 = A \exp\left(-\frac{B}{\alpha(T - T_0)}\right) \quad (6.39)$$

where A, B, α and T_0 are constants; T_0 is an adjustable variable that may take values considerably

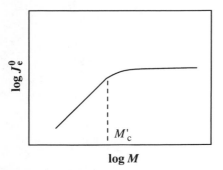

Figure 6.14 The logarithm of the steady-state recoverable shear compliance plotted versus the logarithm of molar mass. Schematic curve.

below the glass transition temperature. The ratio B/α is by some authors considered as a reduced activation energy ($\Delta E/R$, where ΔE is the activation energy and R is the gas constant). Equation (6.39) goes back on the classical Doolittle equation and the free volume concept which is described in detail in Chapter 4.

The pressure has a strong effect on the viscosity. This is not surprising because the free volume of the melt is one of the decisive factors of the shear viscosity. A typical value of the bulk modulus of a polymer melt is 10^9 Pa. Conventional melt processes, injection moulding and extrusion, operate at pressures of 10^6–10^7 Pa, at which the volume strain amounts to only 0.1–1%. However, high-pressure injection moulding, at pressures of 10^8 Pa, causes a volume strain of 10% which leads to a strong increase in the viscosity. It is established that $\log \eta$ is approximately proportional to both temperature and pressure. Under isoviscous conditions it is possible to obtain a conversion factor, $-(\Delta T/\Delta p)_\eta$, which expresses the drop in temperature which is equivalent to the imposed pressure increase in producing the same melt viscosity. Let us explain the meaning of this coefficient by the following example.

For low-density polyethylene the conversion factor is $-(\Delta T/\Delta p)_\eta = 5.3 \times 10^{-7}$°C Pa^{-1}. Assume that the pressure is 1000 atm (10^8 Pa) at 220°C. It is now possible to calculate the temperature (T) of the liquid at pressure 1 atm with the same viscosity as the liquid at 10^8 Pa and 220°C: $\Delta T = -(\Delta T/\Delta p)_\eta \cdot p = 5.3 \times 10^{-7} \times 10^8$ °C $= 53$°C; $T = 220 - 53 = 167$°C.

6.4.2 THE ROUSE MODEL

The Rouse (1953) model considers that the polymer chain consists of n completely flexible repeating units obeying Gaussian statistics moving in viscous surroundings (solution or melt). Three types of force act on each repeating unit: (i) a frictional force which is proportional to the relative velocity of the repeating unit with respect to the surrounding medium; (ii) a force originating from the adjacent repeating units of the same molecule; (iii) a random force due to Brownian motion. The Rouse model describes the effect of these forces on the flow dynamics. Excluded volume and long-range effects such as chain entanglements are not considered. The following expressions valid for an undiluted melt can be derived based on the model:

$$\tau_0 = \left(\frac{\zeta_0 N_A K_\theta}{\pi^2 M_{\text{rep}} RT}\right) M^2 \qquad (6.40)$$

$$\eta_0 = \left(\frac{\zeta_0 N_A K_\theta \rho}{6 M_{\text{rep}}}\right) M \qquad (6.41)$$

$$J_e^0 = \left(\frac{2}{5\rho RT}\right) M \qquad (6.42)$$

where ζ_0 is the monomeric friction coefficient, N_A is the Avogadro number, $K_\theta = s^2/M$ (s is the radius of gyration), M_{rep} is the molar mass of the repeating unit, R is the gas constant, T is the temperature, ρ is the density and M is the molar mass of the polymer.

The Rouse theory is clearly not applicable to polymer melts of a molar mass greater than M_c (M_c') for which chain entanglement plays an important role. This is obvious from a comparison of eqs (6.40)–(6.42) and experimental data (Figs 6.13 and 6.14) and from the basic assumptions made. However, for unentangled melts, i.e. melts of a molar mass less than M_c (M_c'), both the zero-shear-rate viscosity and recoverable shear compliance have the same molar mass dependence as was found experimentally (Figs 6.13 and 6.14). The Rouse model does not predict any shear-rate dependence of the shear viscosity, in contradiction to experimental data.

6.4.3 THE REPTATION MODEL

We may start by considering a coiled chain trapped in a network. This is a simpler case than the case of a thermoplastic melt since non-trivial entanglement effects are avoided. We may think about the system in only two dimensions (Fig. 6.15), with a coiled (C) chain and a great number of surrounding network chain segments which are obstacles, denoted O_i. The C chain is not allowed to cross any of the obstacles. It can, however, move in a worm-like fashion along its own axis. This motion was given the name **reptation** by its inventor Pierre Gilles de Gennes (1971). It is convenient to think that the C chain is trapped within a tube. This commonly used term 'tube' was introduced by Sam F. Edwards (1977). The

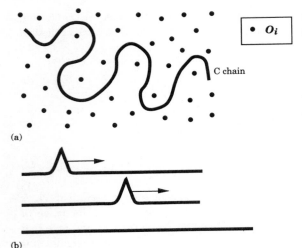

Figure 6.15 (a) Schematic description of the reptation process considering a single chain (C chain) surrounded by numerous obstacles (O_i). The C chain cannot move much laterally due to the obstacles. (b) Fundamental reptation process with the translative motion of a defect along the chain leading to a longitudinal shift of the reptating chain.

fundamental reptation process is illustrated in Fig. 6.15. The 'bump' (defect) moves along on the chain which results in a translative motion of the chain so that one of the chain ends 'comes out' of the tube and at the other chain end the tube 'disappears'.

Whether the above picture holds for polymer melts is still a matter for debate but there is much experimental evidence which suggests that reptation is the predominant mechanism for the dynamics of a chain in the highly entangled state.

The terminal relaxation time $\tau_0 = \tau_{ren}$ may be considered as the time needed to complete renewal of the molecular tube. The velocity (**v**) of the defects along the tube is given by:

$$\mathbf{v} = \mathbf{f} \cdot \mu_{tube} \quad (6.43)$$

where **f** is the force acting on the chain and μ_{tube} is the frictional coefficient. The frictional force ($\mathbf{v}/\mu_{tube}$) should be proportional to the chain length (molar mass M), i.e.:

$$\mu_{tube} = \frac{\mu_0}{M} \quad (6.44)$$

where μ_0 is a constant which is independent of molar mass. A similar equation can be derived for the tube diffusion coefficient (D_{tube}):

$$D_{tube} = \frac{D_0}{M} \quad (6.45)$$

where D_0 is a constant which is independent of molar mass. The time for the tube renewal process (τ_{ren}) can then be derived from the diffusion distance which is equal to the chain length (L):

$$\tau_{ren} = \frac{L^2}{D_{tube}} = \frac{L^2 M}{D_0} \quad (6.46)$$

The chain length is proportional to the molar mass, i.e.:

$$\tau_{ren} = \tau_{ren}^0 \cdot M^3 \quad (6.47)$$

where τ_{ren}^0 is a constant.

Masao Doi and Sam F. Edwards (1986) developed a theory on the basis of de Genne's reptation concept relating the mechanical properties of the concentrated polymer liquids and molar mass. They assumed that reptation was also the predominant mechanism for motion of entangled polymer chains in the absence of a permanent network. Using rubber elasticity theory, Doi and Edwards calculated the stress carried by individual chains in an ensemble of monodisperse entangled linear polymer chains after the application of a step strain. The subsequent relaxation of stress was then calculated under the assumption that reptation was the only mechanism for stress release. This led to an equation for the shear relaxation modulus, $G(t)$, in the terminal region. From $G(t)$, the following expressions for the plateau modulus, the zero-shear-rate viscosity and the steady-state recoverable compliance are obtained:

$$G_p \propto M^0 \quad (6.48)$$

$$\eta_0 \propto M^3 \quad (6.49)$$

$$J_e^0 \propto M^0 \quad (6.50)$$

Experimental data indicate that η_0 increases more strongly with M, $\eta_0 \propto M^{3.4}$ (Fig. 6.13) and the predicted viscosity values are in fact greater than the experimental. The predicted values for J_e^0 are lower than the experimental values. As pointed out by, for example, Graessley (1982), there is in real polymer melts a competition between reptation and other

Figure 6.16 Escape of a small portion of a chain from the tube.

relaxation mechanisms, and the viscosity values from the Doi–Edwards model considering only reptation should be considered as upper bound values. The contribution from the competing mechanisms should lead to an increase in relaxation and a lowering in the viscosity with reference to the upper bound value. De Gennes suggested that there is a release in the constrain in a chain originating from the reptation of the surrounding chains. This mechanism would be unique for thermoplastic melts and would not be operable in a polymer network.

The other competing mechanisms operate in both thermoplastic melts and networks. The walls of the reptation tube are not 'continuous' but contain holes through which the tube chains can bulge out. The tube is like a cage with bars, the surrounding chains, and the tube chain can find room between two bars to permit the escape of a portion of the chain (Fig. 6.16). This tube leakage process was in fact considered in the original work of de Gennes in the discussion of the reptation of star-shaped polymer molecules. Graessley presented in 1982 a reptation-based model considering the competing relaxation mechanisms. The model yields expressions similar to the Doi–Edwards model but with somewhat better agreement with experimental data.

The reptation model has also been successfully applied by Jud, Kausch and Williams (1981) to problems relating to crack healing of amorphous polymers.

6.5 LIQUID-CRYSTALLINE POLYMERS

6.5.1 HISTORICAL BACKGROUND AND BASIC CONCEPTS

Liquid-crystalline behaviour was first reported by the Austrian botanist, Friedrich Reinitzer, in a letter to the German physicist, Otto Lehmann, in 1888, in which he wrote about the observation of two first-order transitions in cholesteryl benzoate. At 418 K the solid crystal material became a hazy liquid, and at 452 K a clear transparent isotropic liquid was observed. Lehmann noticed that the hazy liquid was birefringent (anisotropic) and introduced the term **liquid crystal** for this category of spontaneously anisotropic liquids. The term 'liquid crystal' was introduced since the material above the lower transition temperature (418 K for cholesteryl benzoate) possessed intermediate order, i.e. it exhibited both the flow properties of a liquid and the anisotropy of a solid crystal. A synonym for 'liquid crystalline', which was introduced later, is **mesomorphic**, which is derived from the Greek 'mesos' meaning 'middle' and 'morphe' meaning 'form'.

The first important step in the polymer field was taken by the German chemist, D. Vorländer, who in 1923 initiated work which led to the synthesis of oligomers of *p*-phenylene groups linked by ester groups. He noted that the transition temperatures increased with increasing length of the oligomer/polymer. He also found that poly(*p*-benzamide) showed no melting prior to the chemical degradation occurring at high temperatures. Bawden and Pirie (1937) observed in the 1930s that solutions of the tobacco mosaic virus showed an anisotropic, liquid crystallinity. The early theoretical development was made by Onsager (1949), Ishihara (1951) and Flory (1956). Flory suggested, based on lattice theory, that rigid-rod polymers display an anisotropic liquid-crystalline phase at concentrations greater than a specific critical value, the latter being dependent on the length-to-width ratio ('aspect ratio') of the rigid-rod molecules. The extensive work of Robinson reported in 1956 that the phase behaviour of solutions of polyglutamates could be predicted by the Flory theory. This kind of anisotropic liquid-crystalline solution is referred to as **lyotropic** and, according to the theory of Flory, the critical concentration (v_2^*) for the formation of a liquid-crystalline phase is given by:

$$v_2^* = \frac{8}{x}\left(1 - \frac{2}{x}\right) \quad (6.51)$$

where x is the aspect ratio of the polymer.

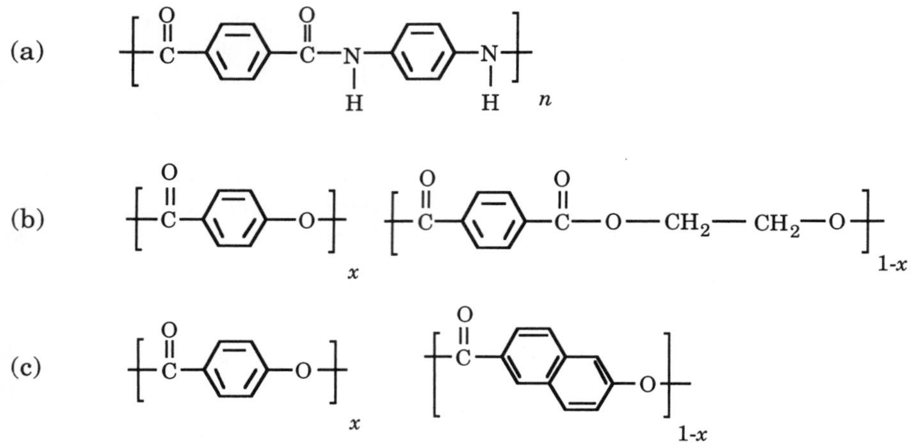

Figure 6.17 Repeating unit structures of (a) poly(p-phenylene terephthalamide) (Kevlar); (b) copolyester of p-hydroxybenzoic acid and ethylene terephthalate; (c) copolyester of p-hydroxybenzoic acid and 2,6-hydroxynaphthoic acid (Vectra).

The first commercially important liquid-crystalline polymer was Kevlar. Kwolek wrote in 1965 about anisotropic solutions of wholly aromatic polyamides in alkylamide and alkylurea solvents. This development led to Kevlar, i.e. ultra-oriented fibres of poly(p-phenylene terephthalamide) (Fig. 6.17). The solution of the polymer in concentrated sulphuric acid is nematic (the term 'nematic' will be explained in section 6.5.3) and fibres of high modulus and high strength can be spun from the solution.

In 1972, the first melt-processable (later categorized as **thermotropic**) liquid-crystalline polymer was reported by Cottis (1972). The term 'thermotropic' refers to the fact that the liquid-crystalline phase is stable within a certain temperature region. The polymer synthesized by Cottis was a copolyester based on p-hydroxybenzoic acid and biphenol terephthalate. This polymer is now available on the market under the name Xydar. In 1973, the first well-characterized thermotropic polymer, a copolyester of p-hydroxybenzoic acid and ethylene terephthalate (Fig. 6.17), was patented by Kuhfuss and Jackson (1976), when they reported the discovery of liquid-crystalline behaviour in this polymer. The availability of this polymer was of great importance to the scientific community but commercialization is yet to come. At the beginning of the 1980s, the Celanese Company developed a family of processable thermotropic liquid-crystalline polymers, later named Vectra. These polymers consist of p-hydroxybenzoic acid and p-hydroxynaphthoic acid, and possibly also other similar compounds (Fig. 6.17). The thermotropic polymers here referred to are all categorized as **main-chain** polymers. The rigid-rod moieties, commonly denoted mesogens, are located in the main chain. The current principal use of these materials is as high-performance engineering plastics. These polymers are readily oriented during melt-processing by elongational and/or shear flow and the stability of the oriented structure is very good, i.e. the relaxation time is very long.

Liquid-crystal displays based on low molar mass compounds have been on the market for many years. The use of liquid-crystalline polymers in electronics and optronics is predominantly focused on the so-called side-chain polymers with the mesogens located in the pendant groups (Fig. 6.18). The most important discovery in this area was made by Ringsdorf, Finkelmann and co-workers (1978). They found that a stable mesophase could only be formed if the mesogenic groups were uncoupled from the backbone chain by linkage via a flexible spacer group. During the last ten years there has been intense activity in the field, seeking materials with unique electrical and optical properties for applications in information technology.

The main-chain and side-chain polymers as well as the classical small-molecule liquid crystals consist essentially of one-dimensional molecules (rods). Low molar mass compounds of two-dimensional disc-shaped molecules were discovered by Chandrasekhar (1977) to possess liquid crystallinity. They were referred to as **discotic** liquid crystals. Later research has involved the synthesis and characterization of polymers with discotic moieties connected via flexible spacer groups.

6.5.2 CHEMICAL STRUCTURE OF LIQUID-CRYSTALLINE POLYMERS

Figure 6.18 shows schematically the essential features of the structure of main-chain and side-chain polymers. The **mesogenic** units give the polymers their anisotropic rod shape. The mesogenic group must be essentially linear and of high aspect ratio. The typical mesogenic group consists of at least two aromatic or cycloaliphatic rings connected in the *para* positions by a short rigid link which maintains the linear alignment of the rings. At first, it was believed that the mesogenic group had to be completely inflexible. It has, however, been shown more recently that the mesogenic group may contain a flexible sub-unit.

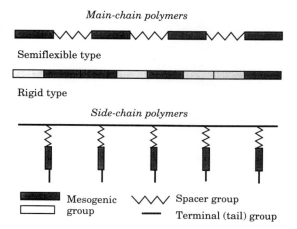

Figure 6.18 Main-chain and side-chain liquid crystalline polymers.

Figure 6.19 Schematic representation of mesogen.

Figure 6.19 displays the characteristic features of the mesogen.

Examples of cyclic units are 1,4-phenylene, 2,6-naphthalene and *trans*-1,4-cyclohexylene rings. The linking group connects the cyclic units on each side to give the chain a linear character. The preferred confirmation is planar (linear). A number of different groups fulfil these requirements: direct bond, ester, imino, azo, azoxy, and methylol groups (semiflexible).

The structure of the side-chain polymers is shown schematically in Fig. 6.18. Attached to each mesogenic group is a terminal group. Commonly reported examples of the latter are –OR, –R, –CN, –H, –NO$_2$, –NH$_2$, –Cl and –Br.

The flexible units (spacer groups) present in both main-chain and side-chain polymers are often oligomethylene groups. Oligosiloxane and oligo(ethylene oxide) groups are also used. The minimum number of –CH$_2$– groups required to decouple the mesogenic units from the backbone chain in side-chain polymers is 2–4. Some main-chain polymers, of which Vectra is possibly the most prominent example (Fig. 6.17), are fully aromatic and no flexible-chain group is present. The required lowering of the transition temperatures is accomplished with the random presence of lateral and longitudinal 'steps' from the 2,6-naphthalene units.

Discotic liquid-crystal compounds often consist of a central aromatic core with, for example, four aromatic rings and typically four or six substituents (e.g. alkoxy-, oligo-ester-groups) attached symmetrically to the aromatic core.

6.5.3 PHYSICAL STRUCTURE OF LIQUID-CRYSTALLINE POLYMERS

The liquid-crystalline state involves great many structures. A great number of different organizations between the isotropic, amorphus structure and the perfectly organized three-dimensional crystal may exist.

Let us consider a thermotropic polymer held at a high temperature. The melt will in this case possess complete optical transparency and is referred to as an isotropic liquid. The molecules lack both long-range orientational and positional order. The X-ray pattern is diffuse and liquid-like. When this polymer is cooled down, it will turn cloudy at a certain (clouding) temperature, which indicates the formation of a liquid-crystalline phase. The most probable liquid-crystalline phase to be formed directly from the isotropic melt is the **nematic**, which is characterized by long-range orientational order and positional disorder (Fig. 6.20). The term 'nematic' originates from the Greek word for 'threaded', reflecting the appearance of a nematic liquid-crystalline compound in an optical microscope without polarizers. The X-ray scattering pattern of nematic compounds is diffuse and liquid-like due to the absence of long-range positional order.

The order parameter (S), defined below, characterizes the state of orientation of the rigid-rod main chains or, in the case of side-chain polymers, of the pendant mesogenic groups in a sample or in a part of a sample, e.g. a liquid-crystalline 'domain':

$$S = \frac{3\langle \cos^2 \theta \rangle - 1}{2} \quad (6.52)$$

where θ is the angle between the rigid-rod chain (mesogenic group) and the director: $S = 1$ when all chains are parallel to the director; $S = 0$ when orientation is completely random. It is implicit in the use of a scalar quantity (S) to quantify orientation that the molecules have a cylindrical symmetry and that orientation is uniaxial. The order parameter is thus analogous with the Hermans orientation function, f (Chapter 9).

For a nematic liquid-crystal domain, selecting the average direction of the mesogens as the reference direction (director), this order parameter takes values between 0.3 and 0.8 (Fig. 6.21). The order parameter decreases with increasing temperature until the isotropization temperature is reached and the order parameter falls to zero.

The chains in the nematic domain are free to move both longitudinally and laterally. The low concentration of chain entanglements and the freedom of the chains to move longitudinally give nematic polymers their typical low-melt viscosity (Fig. 6.22).

In a nematic liquid the director field is not uniform unless an external field of electric, magnetic or mechanical nature is applied. The director may vary in a smooth continuous way around certain points called **disclinations**, or more abruptly as in domain walls. Characteristic of the nematic structure are the so-called **schlieren textures** which are readily observed by polarized light microscopy (Fig. 6.23). The dark bands meet at certain points, the disclinations. It should be noted that a dark region indicates that the local molecular director of the optically anisotropic region is parallel to the polarizer or to the analyser, or that the region is optically isotropic.

Disclinations are for liquid crystals what dislocations are for solid crystals. The molecular director 'rotates' about a disclination (point) line. If four dark bands meet, the strength of the disclination is ± 1, and the molecular director is rotating by 360° about the centre of the disclination (Fig. 6.24). Disclinations of strength $\pm 1/2$ (180° rotation of director) are also observed in nematics. They are distinguished by two dark bands meeting in the disclination centre (Fig. 6.23). In fact, disclinations of strength $\pm 1/2$ appear only in nematics. The strength of the disclination is given by:

$$|\text{strength}| = \frac{\text{number of bands}}{4}$$
$$= \frac{\sum \text{angular change of director}}{360°} \quad (6.53)$$

Figure 6.20 Nematic structure with the director indicated by an arrow.

Liquid-crystalline polymers 113

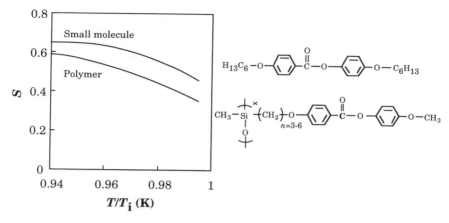

Figure 6.21 Domain order parameter (S) as a function of reduced temperature (T_i being the isotropization temperature, the temperature at which the nematic melt is in equilibrium with the isotropic melt) for a small-molecule- and a side-chain polymer nematic. Drawn after data from Finkelmann and Rehage (1984).

The sign can be experimentally obtained by rotating the crossed polarizer/analyser pair. If the bands radiating out from the singularity are rotating in the same direction as the polarizer/analyser pair, then the sign is positive. Opposite rotations indicate a negative sign.

Figure 6.24 shows the reorientation of the molecular director associated with disclinations of strengths $\pm 1/2$ and ± 1. Disclinations are 'defects' in the sense that each of them can be associated with a certain extra volume, enthalpy and entropy. Two disclinations with the same strength but with different signs may combine and annihilate one another. A uniform director field is then created in this area. This may be a relatively rapid process in low molar mass nematics but very slow in polymers. Thus, heat

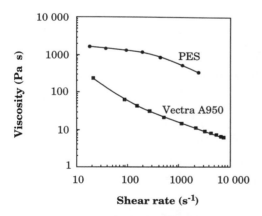

Figure 6.22 Apparent melt viscosity as a function of shear rate at 350°C for polyethersulphone (PES) and liquid crystalline Vectra A950 (copolyester based on p-hydroxybenzoic acid and 2,6-hydroxynaphthoic acid). Drawn after data from Engberg et al. (1994a).

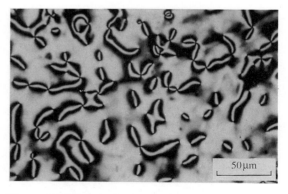

Figure 6.23 Polarized photomicrograph of a nematic liquid crystalline compound, 4-(11-(vinyloxy)undecyloxy)-4-ethoxyphenylbenzoate at 70°C. Note the singularities (disclinations) with two or four radiating dark bands. Photomicrograph by F. Sahlén, Dept of Polymer Technology, Royal Institute of Technology, Stockholm.

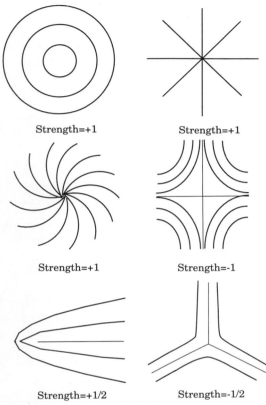

Figure 6.24 Schematic illustration of disclinations of strengths ± 1 and $\pm 1/2$ in liquid crystals with the local directors indicated by the lines. Each type of disclination has a central singularity. Drawn after Donald and Windle (1992, p. 179).

treatment of a 'mobile' nematic polymer leads to a decrease in concentration of disclinations and to a coarsening of the schlieren texture.

The variation of the director field around a disclination is achieved by splaying, twisting or bending (Fig. 6.25). The orientational distortions occur in all types of liquid crystal. The relationship between stress and strain is similar to Hooke's law and each of these distortion cases is associated with an elastic constant: splay (k_{11}), twist (k_{22}) and bend (k_{33}). The elastic constants are also known as the **Frank constants**. In low molar mass nematics, the elastic constants are approximately equal and very small, around 10^{-7} Pa. For polymeric liquid crystals the elastic constants are dependent on molar mass and on the rigidity of the polymer chain. Bend distortion is difficult to accomplish in completely rigid-rod polymers but more easily in polymers containing flexible spacer groups. Splay distortions are, according to de Gennes, unlikely to occur in rigid-rod polymers without the segregation of chain ends to the splay centre.

The cholesteric phase shows, on a local scale, great similarity with the nematic structure. In both cases, order is only orientational. The cholesteric structure is, however, twisted about an axis normal to the molecular director (Fig. 6.26). The twist of the molecular director is spontaneous and originates from the chiral character of the molecules. The cholesteric structure can be obtained by building in chiral elements into a polymer or by the addition of a small-molecule chiral compound to a nematic polymer. The pitch of the helix is temperature-dependent and in lyotropic solutions it also depends on the solvent and on the polymer concentration. For a given combination of solvent and polymer concentration, there is a unique helical pitch for each temperature. Cholesterics are iridescent when the helical pitch is of the same order of length as the wavelength of light. 'Iridescence' means that the wavelength of the reflected light in white light illumination vary with the angle of incidence and observation. The temperature dependence of the helical pitch causes a change in the wavelength of the reflected light with

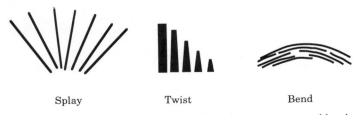

Figure 6.25 Distortion of the director field by splaying, twisting and bending.

Liquid-crystalline polymers 115

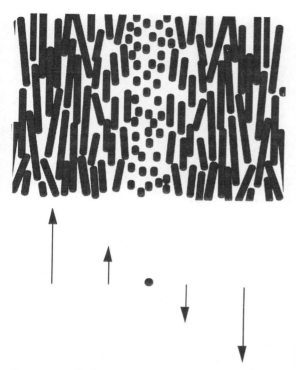

Figure 6.26 Cholesteric structure. Arrows indicate the local mesogenic group director. The central dark circle indicates the perpendicular orientation of the mesogens.

Further cooling of the nematic polymer may result in the formation of a layered mesomorphic structure, the **smectic** (from the Greek word 'smegma', meaning 'soap') structure (Fig. 6.27). A smectic phase possesses both long-range orientational and long-range positional order. The chains are not free to move longitudinally. The viscosity of smectic compounds is consequently very high.

The regular arrangement of the chains (mesogens) in layers is revealed by X-ray diffraction. The X-ray diffraction pattern of a smectic compound always consists of a relatively sharp reflection appearing at a low scattering angle corresponding to the layer thickness, the characteristic repeating distance of the stack (Fig. 6.28).

Smectic A (s_A) and smectic C (s_C) phases display diffuse equatorial wide-angle reflections which indicate that the centres of gravity of the mesogenic groups are irregularly located within the smectic layers. The director is vertical in the diffractograms shown (Fig. 6.28). The typical feature of the scattering pattern originating from the s_A phase is the perpendicular location of the small- and the wide-angle reflections indicating that the mesogenic group director and the smectic layer normal are parallel to each other (Fig. 6.27).

The small-angle reflections originating from the s_C phase are disrupted into reflections symmetrically located around the director at a certain azimuthal angle ϕ, the latter being equal to the angle between the layer normal and the director. There are a great

temperature. Polarized photomicrographs of cholesterics often show parallel lines with a periodicity of half of the helical pitch forming a pattern similar to a fingerprint, the so-called **fingerprint texture**.

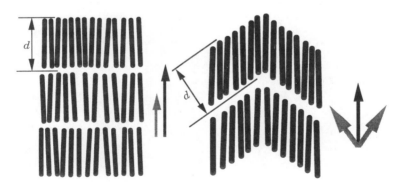

Figure 6.27 Smectic A (left) and one type of smectic C (right) structure. Black arrows indicate director, shaded arrows indicate the directions of the layer normals and the layer thicknesses (*d*) are indicated in the figure. Only the mesogenic groups are shown.

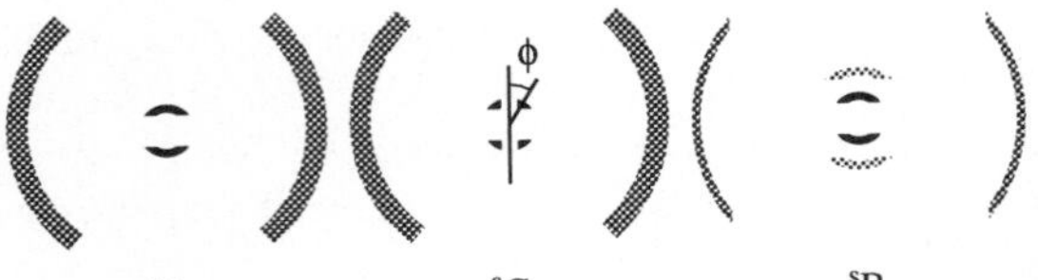

116 The molten state

![X-ray scattering patterns sA sC sB]

Figure 6.28 Typical X-ray scattering patterns for aligned smectic samples. The director is vertical.

number of smectic phases with additional order within the smectic layers, e.g. the smectic B (s_B) phase. The order within the smectic layers is indicated by the sharpness of the wide-angle, equatorial reflection (Fig. 6.28). The particular diffraction pattern shown in Fig. 6.28 is for a so-called hexagonal crystal B (c_B) which is a subgroup of smectic B. It shows a single wide-angle equatorial reflection from the hexagonal packing of the mesogens, and several orders of meridional reflections from the layer spacing. A more complete description of the different smectic structures is given in Demus and Richter (1978), Vertogen and de Jeu (1988) and Donald and Windle (1992).

Polarized photomicrographs of smectic A samples show so-called focal-conic fan textures (Fig. 6.29). Similar but not identical structures are also found in smectic C phases. The origin of these structures is the preference for splay distortion as opposed to the unfavourable twist and bend distortions in these smectics.

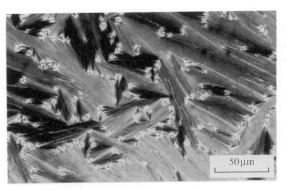

Figure 6.29 Polarized photomicrograph of a polymer possessing smectic A mesomorphism. Photomicrograph by F. Sahlén, Dept of Polymer Technology, Royal Institute of Technology, Stockholm.

It should be noted that the main groups of liquid-crystalline phases presented in Table 6.2 consist of many individual members with only a slight variation in organization. More detailed information is given in Demus and Richter (1978) and Vertogen and de Jeu (1988).

Further cooling of the smectic phase may lead to crystallization or perhaps to the formation of a smectic glass. The semicrystalline state of the rigid-rod polymers is different from that of the flexible-chain polymers described in Chapter 7. The low segmental flexibility of the former prevents chain folding and the crystals should be of the fringed micelle type.

Table 6.2 Main groups of liquid-crystalline structures of main-chain and side-chain polymers

Structure	Orientational order	Positional order	Tilting[a]	Layer order[b] comments
Nematic	yes	no	–	–
Cholesteric	yes	no	–	twisting director
Smectic A	yes	yes	untilted	random
Smectic C	yes	yes	tilted	random
Smectic F	yes	yes	tilted	random
Smectic B	yes	yes	untilted tilted	hexagonal monoclinic
Smectic E	yes	yes	untilted	orthorhombic
Smectic G	yes	yes	tilted	monoclinic

[a] Refers to the angle between molecular director and the layer normal. 'Untilted' means parallel orientation (angle = 0).
[b] Arrangement of the molecular centres of gravity within the layers.

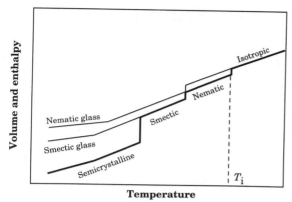

Figure 6.30 Phase transitions in thermotropic liquid-crystalline polymers. The isotropization temperature (T_i) is shown in the diagram.

Figure 6.30 presents a summary of the different possible structures existing in different temperature regions. The phase transitions are dominantly of first-order type. A relatively common thermal behaviour of semiflexible main-chain and side-chain polymers with long spacers is that a smectic structure is formed directly from the isotropic melt without the appearance of any intermediate nematic phase. The development of a smectic phase requires some 'axial' order of the chain. Statistical main-chain copolymers generally do not possess a smectic mesomorphicity. On cooling, they transform into a nematic glass.

The various phase transitions can be revealed by polarized light microscopy which shows changes in the birefringence pattern (texture), by thermal analysis which shows changes in enthalpy or volume, and by X-ray diffraction which shows changes in the local arrangement of atoms. The structural assessment is preferably carried out by X-ray diffraction of oriented (aligned) samples. It is possible to make a preliminary structural assignment by polarized light microscopy from textural observation. A great number of textures of liquid crystals have been collected by Demus and Richter (1978).

Commonly used terms for the mesophases, focusing on their thermodynamical character, are a follows:

- An **enantiotropic** mesophase is observed both on cooling and heating. It is thermodynamically stable within a certain temperature region. Its free energy is at these temperatures lower than those of the isotropic and crystalline (or more ordered liquid-crystalline) phases (Fig. 6.31).
- A **monotropic** mesophase is metastable with respect to the crystalline (or more ordered liquid-crystalline) phase and it appears only under certain conditions on cooling.
- A **virtual** mesophase is potentially possible but thermodynamically less stable than the crystalline (or more ordered liquid-crystalline) phase at the same temperature (Fig. 6.31). The monotropic mesophase is a special case of the virtual mesophase.

6.5.4 CHEMICAL STRUCTURE AND PHASE TRANSITIONS IN LIQUID-CRYSTALLINE POLYMERS

Effect of mesogenic group

The key parameter is the aspect (length-to-width) ratio of the mesogenic group. Tables 6.3 and 6.4 show that

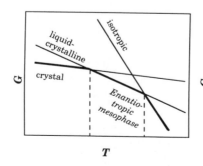
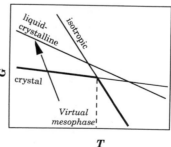

Figure 6.31 Plots of free energy (G) as a function of temperature (T) illustrating enantiotropic and virtual mesophases. The slopes of the lines are equal to the entropies of the different phases.

Table 6.3 Effect of length of mesogenic group on isotropization temperature

$T_i = 160°C$

$T_i = 311°C$

Source: Meurisse et al. (1981).

Table 6.4 Effect of width of mesogenic group on isotropization temperature

$T_i = 354°C$

$T_i = 248°C$

Source: Jo, Lenz and Jin (1982).

relatively small changes in the repeating unit structure may have a profound effect on the isotropization temperature. A polymer with a mesogenic group with a high aspect ratio has a high isotropization temperature. The other main factor is linearity. The introduction of nonlinear cyclic groups, e.g. m-phenylene, causes a pronounced decrease in isotropization and melting temperatures.

Lateral substituents attached to cyclic units of the mesogenic groups increase the width and decrease the aspect ratio for the mesogenic group (Table 6.5 and Fig. 6.32). The shape of the mesogen is the dominant factor controlling the isotropization temperature rather than the polarity and degree of intermolecular interaction. The isotropization temperatures of the polymers with polar halogen substituents are essentially the same as those of the corresponding polymers with the less polar methyl substituent (Table 6.5).

The examples given in this section that demonstrate the dominant effect of the shape anisotropy of the mesogen have dealt only with main-chain polymers. However, the same is also true for side-chain polymers: the isotropization temperature increases with increasing aspect ratio of the mesogenic group.

Effect of flexible group and molar mass

Main-chain and side-chain polymers with longer spacer groups tend to favour smectic mesomorphism, whereas polymers with shorter spacer units more frequently show a nematic order.

The effect of the flexible groups is shown in Fig. 6.33 for a semiflexible main-chain polymer. Notice the pronounced odd–even effect. The polymers with spacer groups having an even number of methylene carbon atoms exhibit greater stability (T_i) and lower entropy (higher entropy change on isotropization) than the corresponding polymers with a spacer group having an odd number of methylene carbon atoms. Notice also that the isotropization temperature (keeping n odd or even) decreases with increasing length of the spacer (Fig. 6.33).

Table 6.5 Effect of lateral substituents

X	Y	T_i (°C)
H	H	294
H	CH$_3$	274
H	Cl	279
H	Br	270
Cl	Cl	255

Source: Lin et al. (1980).

The saw-toothed curves shown in Fig. 6.33 can be given a straightforward explanation. The mesogenic groups have a parallel orientation in a molecule with the spacer groups in an all-trans conformation provided that the number of methylene carbon atoms is even. A molecule with a spacer group with an odd number of methylene carbon atoms in an all-trans state possesses an angular macroconformation, with the mesogens pointing in different directions. The orientational order present in the nematic phase can only be maintained by allowing more conformational disorder in the polymers with an odd-numbered spacer than in those polymers with an even-numbered spacer. This is expressed in the $\Delta S = f(n)$ graphs in Fig. 6.33.

Side-chain polymers constitute a different case. Provided that they have spacer groups longer than six main-chain atoms, they form a smectic phase directly from the melt. Polymers with shorter spacer groups form first a nematic phase and in some cases also a smectic phase at lower temperatures. Polymers with shorter spacer groups, typically with less than six main-chain atoms, show the saw-toothed isotropization–n curve. However, the decrease in isotropization with increasing spacer group length is

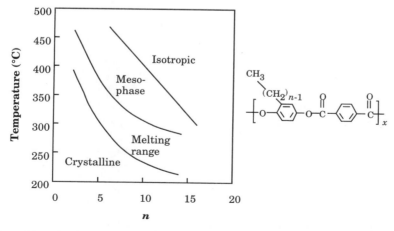

Figure 6.32 Effect of lateral substituent on isotropization temperature and melting temperature range. Drawn after data from Berger and Ballanf (1988).

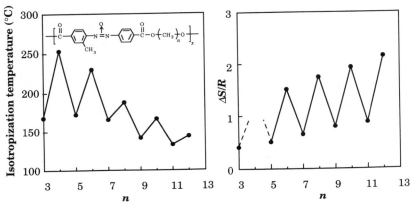

Figure 6.33 Temperature and entropy of isotropization (nematic–isotropic) as a function of spacer length. Drawn after data from Blumstein and Thomas (1982).

not universal and not even typical of the side-chain polymers (Fig. 6.34). The order within the smectic layers, as judged by the enthalpy of the smectic–isotropic transition (Fig. 6.34), increases for these polymers with increasing length of the spacer group. There is a slightly smaller increase in Δs accompanying the increase in Δh which together lead to a small increase in T_{s-i} with increasing n.

Narrow molar mass side-chain polymer samples with a uniformity index of 1.1 can be prepared by living cationic polymerization. The poly(vinyl ether)s show a decrease in Δh_{s-i} (Δs_{s-i}) with increasing degree

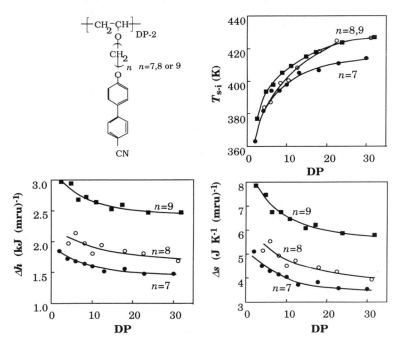

Figure 6.34 Spacer-length and molar-mass dependence of temperature, enthalpy and entropy of isotropization for a series of side-chain poly(vinyl ether)s. The abbreviation 'mru' stands for 'mole repeating units'. Data from Gedde *et al.* (1992). With permission from Butterworth-Heinemann Ltd, UK.

of polymerization (Fig. 6.34), a fact which indicates that the smectic layers are disturbed by the backbone chains in these polymers. Other polymers, e.g. side-chain polysiloxanes, possess essentially constant Δh with increasing degree of polymerization, which may indicate a relatively small degree of interpenetration. This difference in behaviour may be attributed to differences in miscibility of the two 'components' (backbone chain and pendant groups) between the aforementioned classes of polymers.

Effect of copolymerization

Copolymerization is one of the most efficient synthetic techniques to decrease the crystallinity and melting temperature of a given polymer. The melting-point depression occurring in copolyesters (Vectra) based on *p*-hydroxybenzoic acid and 2,6-hydroxynaphthoic acid (HNA) is in fact relatively moderate in comparison with that of other copolymers (Fig. 6.35). The minimum in melting point occurs at about 40 mol% HNA. The decrease in isotropization temperature in the copolyesters is moderate, leading to the desired expansion of the temperature region of a nematic phase.

The X-ray diffraction work of Gutierrez *et al.* (1983) showed the presence of **aperiodic** meridional Bragg reflections in poly(hydroxynaphthoic acid-*co*-

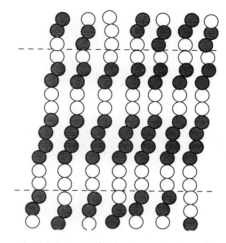

Figure 6.36 Non-periodic layer crystallites of poly(hydroxynaphthoic acid-stat-hydroxybenzoic acid). HBA is indicated by an unfilled circle and HNA by two shaded circles.

hydroxybenzoic acid). They showed that the diffraction pattern could be accounted for by a completely random distribution of comonomers. Windle *et al.* (1985) suggested that the three-dimensional order indicated by the Bragg reflections may be explained by the presence of so-called **non-periodic layer crystallites** (Fig. 6.36). The crystallites consist of a great number of nonperiodic chain segments of essentially the same comonomer sequence distribution. Biswas and Blackwell (1988) assumed, on the other hand, a model that requires only a limited chain register at one point in a sequence and takes no account of differences between monomers or chain sense.

Another example is given in Fig. 6.37, which shows the phase diagram of another copolyester based on *p*-hydroxybenzoic acid (HBA) and ethylene terephthalate. These polymers, first synthesized by Eastman-Kodak, USA, display a nematic mesomorphism at a minimum of 40 mol% of HBA. The melting-point depression is also in this case relatively weak, possibly indicating blockiness in the monomer sequence.

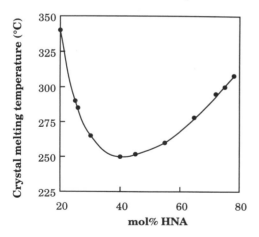

Figure 6.35 Crystal melting temperature as a function of hydroxynaphthoic acid-content of poly(hydroxynaphthoic acid-*co*-hydroxy-benzoic acid) (Vectra). Drawn after data from Calundann and Jaffe (1982).

6.5.5 RHEOLOGY OF LIQUID CRYSTALS

The rheology of liquid crystals is more complex than for ordinary liquids. The anisotropy of the liquid

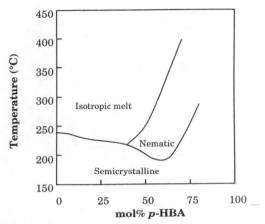

Figure 6.37 Phase diagram of copolymers of ethyleneterephthalate and *p*-hydroxybenzoic acid. Drawn after data collected by Donald and Windle (1992, p. 68).

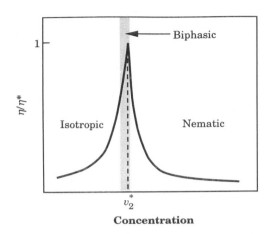

Figure 6.39 Normalized solution viscosity for a solution of a rigid-rod polymer (e.g. poly(γ-benzyl-l-glutamate)). Schematic curve. The critical concentration associated with the formation of the nematic phase is denoted v_2^* and the corresponding viscosity η^*.

crystal makes it necessary to define more than one shear viscosity. Figure 6.38 shows the definition of the three Miesowicz viscosities η_a, η_b and η_c. The measurement of these viscosities requires complete alignment of the domains, which may be possible by application of a magnetic field. It is essential that all disclinations are removed, i.e. that the sample is monodomain-like. It is also possible that the contacting surfaces perturb the alignment.

Most thermotropic and lyotropic nematics, however, are polydomain-like and they contain disclinations, and in these cases only one global shear viscosity is measured.

Let us now go back to the early 1940s, during which Hermans (1946) made important observations about the viscosity of solutions of poly(γ-benzyl-l-glutamate). The viscosity first increased with increasing solute concentration (Fig. 6.39). This is indeed the behaviour of conventional solutions. However, at a certain concentration a dramatic drop in viscosity occurs which coincided with the appearance of the nematic phase. This rheological behaviour was elegantly described by Doi (1982), who used the original Doi–Edwards model (1986) for diluted rigid rods. The latter theory was only valid for dilute and semi-dilute solutions, which prohibited the formation of a nematic phase. However, Doi extended the theory into the concentrated region and

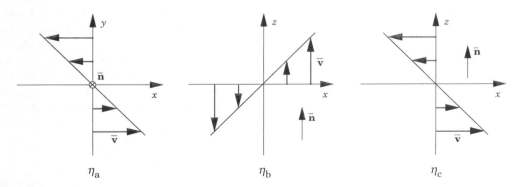

Figure 6.38 Definition of the Miesowisz viscosities. The director is indicated by $\bar{n}$.

the following equation was derived:

$$\frac{\eta}{\eta^*} = \left(\frac{v_2}{v_2^*}\right)\left(\frac{(1-S)^4(1+S)^2(1+2S)(1+3S/2)}{(1+S/2)^2}\right)$$
(6.54)

where S is the order parameter.

Figure 6.39 shows one of the important signatures of nematics, namely their low viscosity. Thermotropic liquid-crystalline polymers such as Vectra are of very low viscosity and complicated, tortuous moulds are readily filled. A generalized shear rate–viscosity curve for liquid-crystalline polymers (nematics) is shown in Fig. 6.40. Shear thinning occurs in both regions I and III. Some nematics only show parts of this curve.

The molar mass dependence of η_0 is significantly stronger for nematics than for entangled polymer melts (η_0 is proportional to $M^{3.4}$). The Doi–Edwards theory for rigid-rod molecules in semi-dilute solutions makes qualitatively correct predictions, although the precise exponent is not known. Figure 6.41 shows the variation in primary and secondary normal stress differences for nematic solutions. The Doi theory can describe the data relatively adequately, with strong negative primary normal stress difference values at intermediate shear rates and an oscillatory behaviour of the second normal stress difference. Marucci (1991) states that the rigid rods tumble at low shear rates and that at higher shear rates a monodomain texture is obtained. It is not clear if thermotropics show the same normal stress difference pattern as lyotropics.

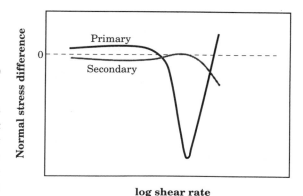

Figure 6.41 Normal stress differences for a solution of poly(γ-benzyl-l-glutamate). Schematic curve. Drawn after data from Magda et al. (1991).

Banded textures are observed in different oriented main-chain liquid crystalline polymers, e.g. in Kevlar fibres and in other lyotropics and thermotropics. The reason for the banding is that molecules after shearing are oriented according to a serpentine. The average direction of the serpentine is parallel to the shear direction. The exact mechanism for the generation of the serpentine structure is not known.

The Leslie–Ericksen theory for flow of nematics is a continuum theory which considers the coupling between velocity field and director field. Details about this important theory are presented in Vertogen and de Jeu (1988).

6.5.6 POLYMER LIQUID-CRYSTALLINITY THEORY

The first steps towards a theory of polymer liquid crystallinity were taken by Ishihara (1951) and Onsager (1949). A few years later, Flory presented a lattice-based theory in which rigid-rod molecules with a certain aspect ratio (x) existed in a certain volume fraction (v_2). Several of the earlier ideas of Onsager were in fact included in the Flory model. It should be noted that the term 'rigid-rod' means completely and permanently rigid. According to the assumptions of the theory, the molecules have no internal flexibility. Flory assumed in his original treatment that the enthalpic contribution to the free energy of mixing was zero. The task was to determine the number of ways (P) in which a population of rigid rods (aspect

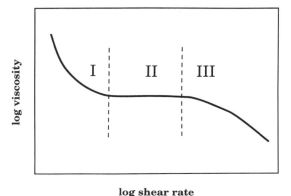

Figure 6.40 Generalized viscosity as a function of shear rate. Drawn after Onagi and Asada (1980).

ratio x) at a particular concentration could be arranged in a given volume. A unique solution is obtained for a given average orientation with respect to the director.

The quantity P can be expressed in the following general terms:

$$P = P_{comb} \cdot P_{orient} \quad (6.55)$$

where P_{comb} is the number of ways of arranging the rigid rods given a certain average orientation ($\bar{y}$, defined in Fig. 6.42) and P_{orient} is the number of additional arrangements which are possible when a range of orientational options are assigned to each rod.

The combinatorial part can be expressed as

$$P_{comb} = \frac{\prod_{i=1}^{n_2} v_i}{n_2!} \quad (6.56)$$

where v_i is the number of ways of arranging polymer molecule i (at a certain orientation; characterized by y) and n_2 is the number of polymer molecules.

The number of ways of placing the first segment of sequence 1 (11) of molecule i is simply $n - x(i - 1)$, where n is the total number of positions in the lattice. This probability is conditional provided that the next segment of the first sequence (12) is vacant, and is given by the ratio

$$\frac{n - x(i - 1)}{n - x(i - 1) + \bar{y}(i - 1)}$$

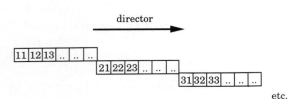

number of steps along director = x
number of lateral steps = y
number of steps in each sequence = x/y

Figure 6.42 Fundamentals of Flory's lattice model showing part of a rigid rod molecule with a total of y lateral steps.

where

$$\bar{y} = \frac{\sum y n_{2y}}{n_2}$$

with n_{2y} the number of rods with misorientation y. The number of ways of arranging sequence 1 is:

$$(n - x(i - 1)) \cdot \left(\frac{n - x(i - 1)}{n - x(i - 1) + \bar{y}(i - 1)} \right)^{(x/y) - 1}$$

The probability that the first segment of sequence 2 is vacant is simply $[n - x(i - 1)]/n$ and the overall conditional probability of the second sequence (and all the rest) is given by:

$$\left[\frac{n - x(i - 1)}{n} \right] \cdot \left[\frac{n - x(i - 1)}{n - x(i - 1) + \bar{y}(i - 1)} \right]^{(x/y) - 1}$$

The total number of different ways of arranging the ith rigid-rod polymer molecule given a certain $\bar{y}$ is simply the product of all the conditional probabilities and the number of ways of placing segment 11:

$$v_i = [n - x(i - 1)] \cdot \left[\frac{n - x(i - 1)}{n} \right]^{y-1}$$

$$\times \left[\frac{n - x(i - 1)}{n - x(i - 1) + \bar{y}(i - 1)} \right]^{(x/y - 1)y}$$

$$= \frac{[n - x(i - 1)]^x}{n^{y-1}[n - x(i - 1) + \bar{y}(i - 1)]^{x-y}}$$

$$\approx \frac{(n - x(i - 1))!(n - (x - \bar{y})i)!}{(n - x)! \, n^{\bar{y}-1}(n - (x - \bar{y})(i - 1))!} \quad (6.57)$$

which, after insertion into eq. (6.56), gives:

$$P_{comb} = \frac{(n_1 + \bar{y}n_2)!}{n_1! n_2! ((n_1 + xn)^{n_2(\bar{y} - 1)}} \quad (6.58)$$

For each average degree of orientation ($\bar{y}$), a number of different arrangements are possible, given in a simplified form according to Flory and Ronca (1979) as:

$$P_{orient} \approx \left(\frac{\bar{y}}{x} \right)^{2n_2} \quad (6.59)$$

Combination of eqs (6.55), (6.58) and (6.59), yields

the following expression:

$$\Delta G_{mix} = -RT \ln(P_{comb} \cdot P_{orient})$$

$$= -RT \ln\left(\frac{(n_1 + \bar{y}n_2)!}{n_1! n_2! (n_1 + xn)^{n_2(\bar{y}-1)}} \cdot \left(\frac{\bar{y}}{x}\right)^{2n_2}\right)$$

$$\approx RT\left(n_1 \ln v_1 + n_2 \ln\left(\frac{v_2}{x}\right) + n_2(\bar{y} - 1)\right.$$

$$- (n_1 + \bar{y}n_2) \ln\left(1 - v_2\left(1 - \frac{\bar{y}}{x}\right)\right)$$

$$\left. - 2n_2 \ln\left(\frac{\bar{y}}{x}\right)\right) \quad (6.60)$$

The free energy of mixing can be expressed as a function of $\bar{y}$ and x by considering that

$$n_1 = \frac{\frac{x}{v_2} - x}{\frac{x}{v_2} + 1 - x} \quad \text{and} \quad n_2 = \frac{1}{\frac{x}{v_2} + 1 - x} \quad (6.61)$$

The contour plots presented in Fig. 6.43 show the free energy as a function of chain orientation expressed in terms of the average side-step number ($\bar{y}$; $\bar{y} = 0$ corresponds to a perfect parallel arrangement of rigid rods) and polymer volume content (v_2).

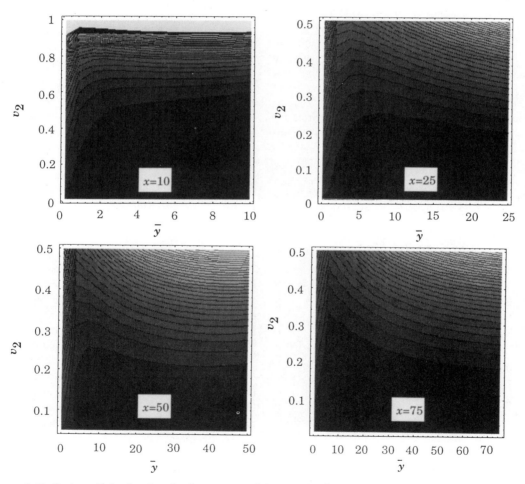

Figure 6.43 Contour plots showing the free energy ($\Delta G/RT$) as a function of volume fraction (v_2) of rigid rod polymer (with different aspect ratios x as shown in the diagrams) and degree of orientation ($\bar{y}$). Higher value of free energy is indicated by lighter colour.

By following the change in free energy at constant v_2 (horizontal shift), a minimum free energy of mixing (corresponding to an anisotropic phase) is obtained at v_2 values greater than a certain critical value v_2^*. This critical value increases strongly with increasing aspect ratio (x) of the rigid-rod molecules (Fig. 6.44).

Flory found the empirical relationship between v_2^* and x given by the following equation, which is plotted in Fig. 6.44:

$$v_2^* = \frac{8}{x}\left(1 - \frac{2}{x}\right) \quad (6.62)$$

It is possible to calculate the free energy of both isotropic (high $\bar{y}$) and anisotropic phases (low $\bar{y}$) and to plot them as a function of v_2. From this plot it is possible to obtain the biphasic region by drawing a common tangent (the two phases have the same chemical potential at equilibrium; see construction of the spinodal in Chapter 4).

Flory added the free energy contribution from the energetic interaction in terms of the Flory–Huggins interaction parameter χ_{12}:

$$\Delta G_{mix} = -RT(\chi_{12}xn_2v_1 + \ln P(\bar{y})) \quad (6.63)$$

where the last term is expressed as eq. (6.60). For good solvents, $\chi_{12} < 0$, the biphasic v_2 region remains narrow, whereas for poor solvents, $\chi_{12} > 0.1$, it broadens considerably (Fig. 6.45).

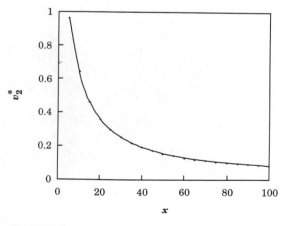

Figure 6.44 The critical (minimum) volume fraction of a rigid-rod polymer at which an anisotropic phase appears as a function of rigid-rod aspect ratio (x).

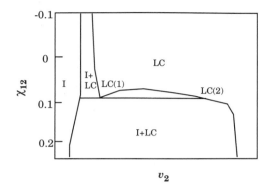

Figure 6.45 Schematic drawing showing phase diagram of rigid-rod polymer mixed with solvents of different solvent power expressed in the Flory–Huggins interaction parameter (χ_{12}). Drawn after Donald and Windle (1992).

The extensive early work of Robinson (1956) on lyotropic solutions of polyglutamates did indeed show phase diagrams very similar to the one presented in Fig. 6.45, hence confirming the predictive power of the Flory theory.

Most of the liquid crystalline polymers have, however, some segmental flexibility, i.e. they are semiflexible and cannot be approximated by a rigid rod. The semiflexible polymers can be represented by a series of rigid-rod moieties connected by flexible (spacer) links. The aspect ratio of the subunits is simply given by the length of the Kuhn segment (section 2.4.4) and the diameter of the molecular cross-section. The aspect ratio referring to the subunits can be inserted into the equations derived by Flory in order to obtain thermodynamic expressions and phase diagrams similar to the one shown in Fig. 6.45. Reasonable agreement is obtained between predictions by the Flory theory and experimental data.

Finally, a few words about other liquid-crystal theories: the mean-field theory of Maier and Saupe (1959, 1960) has been very successful in describing the behaviour of small-molecule liquid crystals, but it has been much less used for polymeric liquid crystals. Other important theories primarily applied to small-molecule liquid crystals are the Landau theory and its extension, the Landau–de Gennes theory. A detailed presentation of these theories, also including the Maier and Saupe theory, is found in Vertogen and de Jeu (1988).

6.6 SUMMARY

Fundamental rheological quantities such as shear viscosity, primary and secondary normal stress coefficients, elongational viscosity and complex viscosity are introduced. Most molten polymers show a reduction in shear viscosity with increasing shear rate (pseudoplasticity). At low shear rates, however, the shear viscosity is practically independent of shear rate. This viscosity value is called the zero-shear-rate viscosity (η_0). At higher shear rates, the viscosity (η) decreases with shear rate ($\dot{\gamma}$) according to a power law expression: $\eta = K\dot{\gamma}^{n-1}$, where K and n are constants, $n < 1$. Polymer melts are visco-elastic non-Newtonian fluids. The response changes from solid-like at short shearing times to liquid-like at long shearing times. These visco-elastic materials may be characterized by their relaxation shear modulus or by their shear creep compliance. From the time dependence of these functions it is possible to calculate useful material parameters, e.g. η_0, the steady-state recoverable shear compliance (J_e^0) and the terminal relaxation time (τ_0).

The molten state of a flexible chain polymer of sufficiently high molar mass is characterized by a chaotic state of interpenetrating, entangled random coil molecules. The time-scales for the rearrangements cover many orders of magnitude, involving rapid local rearrangements and very slow global rearrangements. The slow processes are very important for the rheological properties and they are sensitively dependent on the molecular architecture, i.e. molar mass and degree of chain branching. Quantities like η_0, J_e^0 and τ_0 show an abrupt change in the molar mass dependences at a certain critical molar mass associated with the formation of chain entanglements.

Each chain in the entangled state may be considered as located in a tunnel through the mesh defined by the surrounding chains. The chain is only able to diffuse within its own tunnel. De Gennes (1971), the proposer of this mechanism for self-diffusion, coined the term 'reptation' because of the snake-like character of the motion. He postulated that reptation is the fastest path for rearrangement of the large-scale chain conformation of a linear chain. The theory predicts that relaxation time for the global rearrangements should be proportional to M^3 (M is molar mass),
which is in fair agreement with the experimentally obtained exponent of 3.4. It has been pointed out by Graessley (1982) and de Gennes (1971) that in entangled polymer melts there is a competition between different stress relaxation mechanisms, including reptation and several other processes, e.g. reptation of surrounding chains and tube 'escape'. Better agreement between experimental and calculated viscosity values is achieved when these other relaxation mechanisms are considered.

Polymers, with M smaller than a certain critical value, exhibit negligible chain entanglement and self-diffusion is controlled by the monomeric friction coefficient. Rouse's theory is more applicable to this case; it predicts correctly that the relaxation times associated with global rearrangements should be proportional to M.

Liquid-crystalline polymers consist typically of one-dimensional rod-like groups placed in the main chain or side chain. A great number of liquid-crystalline states have been reported, ranging from those exhibiting only long-range orientational order, nematic and cholesteric phases, to those exhibiting both long-range orientational and positional order, the smectic phases. The stability of the liquid-crystalline phases increases with increasing length-to-width (aspect) ratio of the stiff moieties (mesogens). Flexible and angular groups depress the stability of the mesogens. The phase and thermal transitions appearing in a typical so-called thermotropic liquid-crystalline polymer are:

Low temperature → High temperature
Crystal → Smectic → Nematic → Isotropic liquid

Disclinations are defects existing in liquid crystals involving a localized rearrangement of the molecular director. The strength of the disclination expresses the angular reorientation of the local directors occurring about the centre of the disclination.

The thermodynamics of liquid-crystalline polymers can be relatively adequately described as the Flory lattice theory. This predicts that the stability of the anisotropic phase depends on the aspect ratio of the rigid rods and on the volume fraction of polymer. The theory was originally developed for completely rigid molecules but is also applicable to semiflexible chains. The relevant aspect ratio is here obtained as the ratio of the length of the Kuhn segment and the diameter of the molecular cross-section. The Flory theory is able to predict phase diagrams involving variation of the intermolecular interaction, expressed in terms of a Flory–Huggins interaction parameter.

6.7 EXERCISES

6.1. Elongational flow causes orientation of polymer melts. Discuss the mechanisms for equilibration of the oriented melt for the following polymers: (a) a low molar mass flexible-chain polymer; (b) a high molar mass flexible-chain polymer; (c) a liquid-crystalline (nematic) polymer.

6.2. Present an explanation of the observations relating to the die-swell phenomenon shown in Fig. 6.46.

6.3. Arrange the following nematic liquid-crystalline polymers in order of increasing isotropization temperature.

(a)

$$\left[\!-\!O\!-\!\!\bigcirc\!\!-\!CO\!-\!O\!-\!\bigcirc\!\!-\!\bigcirc\!\!-\!O\!-\!CO\!-\!\bigcirc\!\!-\!O\!-\!R\!-\!\right]_n$$

$R = (-CH_2-)_3$

(b)

$$\left[\!-\!O\!-\!\!\bigcirc\!\!-\!CO\!-\!O\!-\!\bigcirc\!\!-\!\bigcirc\!\!-\!O\!-\!CO\!-\!\bigcirc\!\!-\!O\!-\!R\!-\!\right]_n$$

$R = (-CH_2-)_7$

(c)

$$\left[\!-\!O\!-\!\!\bigcirc\!\!-\!CO\!-\!O\!-\!\bigcirc\!\!-\!O\!-\!CO\!-\!\bigcirc\!\!-\!O\!-\!R\!-\!\right]_n$$

$R = (-CH_2-)_7$

(d)

$$\left[\!-\!O\!-\!\!\bigcirc\!\!-\!CO\!-\!O\!-\!\overset{CH_3}{\bigcirc}\!\!-\!O\!-\!CO\!-\!\bigcirc\!\!-\!O\!-\!R\!-\!\right]_n$$

$R = (-CH_2-)_7$

(e)

$$\left[\!-\!O\!-\!\!\bigcirc\!\!-\!CO\!-\!O\!-\!\bigcirc\!\!-\!O\!-\!CO\!-\!\bigcirc\!\!-\!O\!-\!R\!-\!\right]_n$$

$R = (-CH_2-)_7$

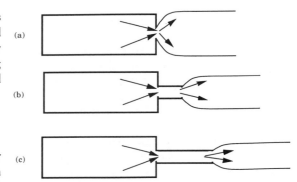

Figure 6.46 Die-swell from three types of extrusion geometry. Drawn after Graessley (1984).

6.4. Draw a diagram of the spatial variation of the director in the structures of the nematic compound shown in Fig. 6.23.

6.5. List the requirements which melt-processable liquid-crystalline polymers need to fulfil. What kind of molecular structure is needed to obtain the desired rheological and thermal properties?

6.6. The X-ray diffraction patterns shown in Fig. 6.47 were obtained for a liquid-crystalline polymer at four different temperatures in order of increasing temperature. Make phase assignments and draw a phase diagram.

6.7. Calculate the order parameter (S) for the sample in Figs 6.47(b) and 6.47(d).

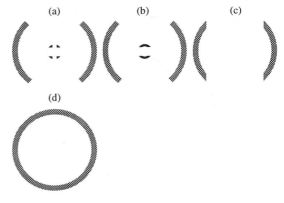

Figure 6.47 Schematic X-ray diffraction patterns obtained at different temperatures.

6.8 REFERENCES

Bawden, F. C. and Pirie, N. W. (1937) *Proc. Roy. Soc. Lond. Ser. B.* **123**, 274.

Berger, K. and Baltanf, M. (1988) *Mol. Cryst. Liq. Cryst.* **157**, 109.

Biswas, A. and Blackwell, J. (1988) *Macromolecules* **21**, 3146.

Blumstein, A. and Thomas, O. (1982) *Macromolecules* **15**, 1264.

Calundann, G. W. and Jaffe, M. (1982) Anisotropic polymers, their synthesis and properties, in *Proceedings of the Robert E. Welsh Conference on Chemical Research XXVI, Houston, Texas, Synthetic Polymers*, p. 247.

Chandrasekhar, S. (1977) *Liquid crystals*, Cambridge University Press, Cambridge.

Cottis, S. G., Economy, J. and Novak, B.E. (1972) US Patent 3 637 595.

de Gennes, P. G. (1971) *J. Chem. Phys.* **55**, 572.

Demus, D. and Richter, L. (1978) *Textures of Liquid Crystals*. VEB Deutscher Verlag für Grundstoffindustrie, Leipzig.

Doi, M. (1982) *J. Polym. Sci., Polym. Phys. Ed.* **20**, 1963.

Doi, M. and Edwards, S. F. (1986) *The Theory of Polymer Dynamics*. Clarendon Press, Oxford.

Donald, A. M. and Windle, A. H. (1992) *Liquid Crystalline Polymers*. Cambridge University Press, Cambridge.

Edwards, S. F. (1977) *Polymer* **9**, 140.

Engberg, K., Strömberg, O., Martinsson, J. and Gedde, U. W. (1994a) *Polym. Eng. Sci.* **34**, 1336.

Engberg, K., Ekblad, M., Werner, P.-E. and Gedde, U. W. (1994b) *Polym. Eng. Sci.* **34**, 1346.

Finkelmann, H., Happ, M., Portugall, M. and Ringsdorf, H. (1978) *Makromolekulare Chemie* **179**, 2541.

Finkelmann, H. and Rehage, G. (1984) *Adv. Polym. Sci.* **60–61**, 99.

Flory, P. J. (1956) *Proc. Roy. Soc.* **234A**, 73.

Flory, P. J. and Ronca, G. (1979) *Mol. Cryst. Liq. Cryst.* **54**, 289.

Gedde, U. W., Jonsson, H., Hult, A. and Percec, V. (1992) *Polymer* **33**, 4352.

Graessley, W. W. (1982) *Adv. Polymer Sci.* **47**, 68.

Graessley, W. W. (1984) Viscoelasticity and flow in polymer melts and concentrated solutions, in *Physical Properties of Polymers*, 2nd edn. (J. E. Mark, ed.). American Chemical Society, Washington, DC.

Gutierrez, G. A., Chivers, R. A., Blackwell, J., Stamatoff, J. B. and Yoon, H. (1983) *Polymer* **24**, 937.

Hermans, P. H. (1946) in *Physics of Cellulose Fibres*, Elsevier, Amsterdam.

Ishihara, A. (1951) *J. Chem. Phys.* **19**, 1142.

Jackson, W. J. and Kuhfuss, H. F. (1976) *J. Polym. Sci., Polym. Chem. Ed.* **14**, 2043.

Jo, B.-W., Lenz, R. W. and Lin, J.-I. (1982) *Macromol. Chemie, Rapid Commun.* **3**, 23.

Jud, K., Kausch, H. H. and Williams, J. G. (1981) *J. Mater. Sci.* **16**, 204.

Kwolek, S. L. (1971) DuPont, US Patent 3 600 350.

Lin, J.-I., Antoun, S., Ober, C. and Lenz, R. W. (1980) *Brit. Polym. J.* **12**, 132.

Magda, J. J., Baek, S.-G., DeVries, K. L. and Larson, R. G. (1991) *Macromolecules* **24**, 4460.

Maier, W. and Saupe, A. (1959) *Z. Naturforschung* **14a**, 882.

Maier, W. and Saupe, A. (1960) *Z. Naturforschung* **15a**, 287.

Marucci, G. (1991) *Macromolecules* **24**, 4176.

Meurisse, P., Noël, C., Monnerie, L. and Fayolle, B. (1981) *Brit. Polym. J.* **13**, 55.

Miller, A. A. (1963) *J. Polym. Sci. Part A* **1**, 1857.

Mitchell, D. R. (1985) *Faraday Discussion Chem. Soc.* **79**, 55.

Onagi, S. and Asada, T. (1980) Rheology and rheo-optics of polymer liquid crystals, in *Rheology*, Vol. 1 (G. Astarita, G. Marucci and L. Nicolais, eds). Plenum, New York.

Onsager, L. (1949) *Ann. N.Y. Acad. Sci.* **51**, 627.

Robinson, C. (1956) *Trans. Faraday Soc.* **52**, 571.

Rouse, P. E. (1953) *J. Chem. Phys.* **21**, 1272.

Vertogen, G. and de Jeu, W. H. (1988) *Thermotropic Liquid Crystals: Fundamentals*, Springer Series in Chemical Physics 45. Springer-Verlag, Berlin.

Windle, A. H., Viney, C., Golombeck, R., Donald, A. M. and Mitchell, D. R. (1985) *Faraday Discussion Chem. Soc.* **79**, 55.

6.9 SUGGESTED FURTHER READING

de Gennes, P. G. (1974) *The Physics of Liquid Crystals*. Clarendon Press, Oxford.

de Gennes, P. G. (1979) *Scaling Concepts in Polymer Physics*. Cornell University Press, Ithaca, NY, and London.

Ferry, J. D. (1980) *Viscoelastic Properties of Polymers*, 3rd edn. Wiley, New York.

Gray, G.W. and Goodby, J. W. (1984) *Smectic Liquid Crystals*. Leonard Hill, Glasgow.

McArdle, C. B. (ed.) (1989) *Side-chain Liquid Crystal Polymers*. Chapman & Hall, New York.

Samulski, E. T. (1993) The mesomorphic state, in *Physical Properties of Polymers*, 2nd edn. (J. E. Mark, ed.) American Chemical Society, Washington, DC.

Woodward, A. E. (1989) *Atlas of Polymer Morphology*. Hanser, Munich, Vienna and New York.

CRYSTALLINE POLYMERS

7.1 BACKGROUND AND A BRIEF SURVEY OF POLYMER CRYSTALLOGRAPHY

Crystalline polymers have a 'reasonably' regular chain structure and a specific preferred chain conformation. The presence of chain defects, e.g. atactic sequences and/or chain branches at high concentrations, makes it impossible for the polymers to crystallize and on cooling they ultimately form a fully amorphous glass. There are exceptional cases of **crystalline** atactic polymers due to side-group crystallization or to the small size of the pendant group. The hydroxyl group in poly(vinyl alcohol) is, for example, sufficiently small and the atactic polymer is crystallizable. Polymers with a small proportion of chain defects crystallize to a lower overall crystallinity than that of the polymer containing no chain defects. The chain defects are normally confined to the amorphous component. Small groups may, however, be housed within the crystals.

It is important to emphasize that the types of polymer dealt with in this chapter have flexible chains. Rigid-rod polymers forming mesomorphic phases are not discussed here. Most of the material presented comes from extensive studies of polyethylene. This polymer should be considered as a 'model' for other flexible-chain polymers and not as a special case.

Let us start by briefly recapitulating the basics of crystallography and pointing out the special features of polymeric crystals. The description of the crystal structure of a polymer is in most respects no different from that of a low molar mass compound. A crystal may be defined as a portion of matter within which the atoms are arranged in a regular, repeated, three-dimensionally periodic pattern. A crystal may be classified into one of seven large subgroups, referred to as **crystal systems**, which are listed in Table 7.1. It should be noted that the unit vectorial distances a, b and c are at the following angles to each other:

$$\alpha = \angle bc$$
$$\beta = \angle ac$$
$$\gamma = \angle ab$$

Periodic arrangements of any motif (e.g. group of atoms) are generated by placing the motif at points located such that each point has identical surroundings. Such infinite arrangements are called **lattices**. Bravais (1850) showed that there are only 14 different ways (so-called Bravais or space lattices) of arranging points in space (Fig. 7.1).

The repeating period of the space lattice that most simply describes the nature of the space lattice when it is repeated by three-dimensional translation is called the **unit cell** of the space lattice. Note the analogy between a, b and c in the unit cell and the corresponding quantities of the crystal system. Cells with only one unique motif are referred to as **primitive**. It is always possible to generate a primitive cell from a given lattice but in many cases end-, face- or body-centred representations are preferred because they may show greater symmetry than the primitive cells.

Crystals exhibit symmetry. A number of different symmetry operations are possible:

- Rotation axis, n-fold, where the motifs are generated using cylindrical coordinates: (r, ϕ), $(r, \phi + 360°/n)$, $(r, \phi + 2 \times 360°/n)$, etc.; n can take values 1, 2, 3, 4 or 6.
- Inversion centre located at $(0, 0, 0)$, where the motifs are located at (x, y, z) and $(-x, -y, -z)$.
- Rotary-inversion axes, which involve a combination of rotation ($\alpha = 360°/n$) and inversion and are indicated by $\bar{n}$, which can take values $\bar{1}, \bar{2}, \bar{3}, \bar{4}$ and $\bar{6}$.
- Mirror planes.
- Screw axes, which involve a combination of translation along the screw axis and a rotation

Crystalline polymers

Table 7.1 Crystal systems

Systems	Axes	Axial angles	Minimum symmetry
Triclinic	$a \neq b \neq c$	$\alpha \neq \beta \neq \gamma \neq 90°$	None
Monoclinic	$a \neq b \neq c$	$\alpha = \gamma = 90°; \beta \neq 90°$	One two-fold rotation axis
Orthorhombic	$a \neq b \neq c$	$\alpha = \beta = \gamma = 90°$	Three perpendicular two-fold rotation axes
Tetragonal	$a = b \neq c$	$\alpha = \beta = \gamma = 90°$	One four-fold rotation axis
Hexagonal	$a = b \neq c$	$\alpha = \gamma = 90°; \beta = 120°$	One six-fold rotation axis
Rhombohedral	$a = b = c$	$\alpha = \beta = \gamma \neq 90°$	One three-fold rotation axis
Cubic	$a = b = c$	$\alpha = \beta = \gamma = 90°$	Three four-fold rotation axes

about the same. It is designated n_δ, where n is the rotation by an angle $\alpha = 360°/n$, and δ is an integer related to the translation component, t: $t = (\delta/n)c$, where c is the length of the unit cell along the screw axis.

- Glide planes, which combine a translation in the plane and a reflection across the plane.

The entire group of symmetry operators that completely describes the symmetry of the atomic arrangements within a crystal is called the **space group**. It has been shown that there are 230 space groups distributed among the 14 Bravais lattice groups.

Another group of symmetry elements is the **point group**, which refers not to a space but to a point. Allowed point-group operators are rotation axes, axes of rotary inversion, inversion centres and mirror planes. Altogether they add up to 32 possible point groups.

A comprehensive list of space and point groups is given in Henry and Lonsdale (1952; 1959; 1962).

In the polymer field, the convention for the choice of the crystallographic directions is that the chain axis

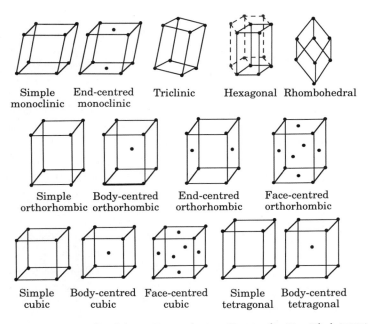

Figure 7.1 Unit cells of the 14 Bravais lattices. Drawn after Van Vlack (1975).

is defined as the c axis, except in the case of the monoclinic cell when the chain axis is the unique axis (b axis).

One central postulate of polymer crystallography is that the conformation of the polymer chains within the crystal is that of the lowest possible energy. Polyethylene has the all-trans conformation as the preferred lowest energy state. Despite this single conformational structure, polyethylene shows two different crystal forms at normal pressure, pointing to the possibility that the all-trans chains may pack differently. The existence of more than one crystal form for a specific compound is referred to as **polymorphism**. For polyethylene, the orthorhombic structure shown in Fig. 7.2 is the most stable. The cross-section of the all-trans stem is not circular, and this leads to the orthorhombic structure. The deviation from hexagonal packing is in fact not very great. This is realized when the central chains in the adjacent (above and below) unit cells are considered. A monoclinic structure has also been reported for polyethylene and is only present in stressed samples. The orthorhombic cell of polyethylene is not primitive. It contains the repeating units of two chains. The chain axis is thus denoted c according to the aforementioned definition.

Isotactic polymers with sufficiently large sidegroups have helical conformations in the crystals. More details of the steric repulsion of the side groups, the cause of the helical conformation, are presented in Chapter 2. Isotactic polypropylene has two helical forms, both of the same conformational energy. The angle of torsion about the CH_2–$CHCH_3$ bond is denoted ϕ_1 and the torsion angle associated with the $CHCH_3$–CH_2 bond is denoted ϕ_2. The two low-energy conformational states are repeats of $\phi_1 = 120°$, $\phi_2 = 0°$ (right-hand helix) and $\phi_1 = 0°$, $\phi_2 = 240°$ (left-hand helix). Note that the trans state is set to $\phi = 0°$. In a crystal of isotactic polypropylene there are four different helices, namely right-hand helices pointing upwards and downwards and left-hand helices also pointing upwards and downwards. It should be noted that an upward, right-hand helix is different from a downward, left-hand helix. It is not surprising that isotactic polypropylene exhibits a number of polymorphs, referred to as α, β and γ. The α cell shown in Fig. 7.3 is one of several possibilities. It has recently become clear that the crystalline structure of α-isotactic polypropylene may show different degrees of disorder in the up and down positioning of the chains. Another important feature of the α structure is that adjacent chains form pairs of left-handed and right-handed helices which pack better than pairs of the same helical type. This is the typical arrangement in crystals of all isotactic polymers with sufficiently large side groups. Isotactic polymers with small pendant groups, e.g. poly(vinyl alcohol) and poly(vinyl fluoride), crystallize in an all-trans conformation.

The γ structure was first discovered in high-pressure crystallized isotactic polypropylene. Later workshowed that the γ phase was also present in baric crystallized samples of low molar mass, typically in the range 1000–2000 g mol^{-1}. It has also been shown that copolymers of propylene with blocks of isotactic

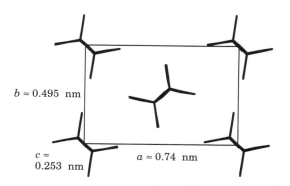

Figure 7.2 View along c (chain) axis of orthorhombic polyethylene crystal.

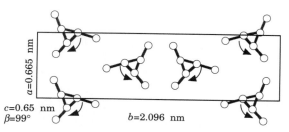

Figure 7.3 Monoclinic cell (α) of isotactic polypropylene from a view along [001]. Drawn after data given in Natta and Corradini (1960).

sequences and minor proportions (4–10 mol%) of ethylene groups develop relatively large fractions of γ phase, also in high molar mass polymers. It was long believed that the γ crystal phase was triclinic with 'ordinary' parallel chains. The X-ray diffraction pattern was, however, never fully explained by the triclinic cell and an orthorhombic cell with the dimensions $a = 0.854$ nm, $b = 0.993$ nm and $c = 4.241$ nm was proposed by Brückner and Meille (1989). The most fascinating aspect of this new finding is that the chains in the proposed cell are not all parallel. The structure consists of alternative layers which are each two chains thick, reminiscent of the α phase, each of the bilayers at an angle of 80° to the adjacent bilayers. This is precisely the angle between the chains in a so-called crosshatch of monoclinic (α) polypropylene (see section 7.3 and Fig. 7.30).

The β (hexagonal) cell was discovered in isotactic polypropylene samples with strongly negative spherulites (see section 7.4 for details of spherulites) crystallized at 128–132°C. The occurrence of the β form was always sporadic. Certain nucleation agents seem to promote the formation of the β structure. In addition to these three crystal structures, isotactic polypropylene also exhibits a so-called smectic or mesomorphic phase, whose presence is favoured by rapid-cooling conditions. It was first reported by Natta and Corradini (1960) and was indicated by an X-ray scattering pattern with two broad peaks centring at 14.8° and 21.2° (2θ, CuKα). Natta and Corradini showed that the density of this phase was 880 kg m^{-3}, compared to a density of 850 kg m^{-3} for fully amorphous polypropylene. Natta suggested that the smectic phase was built up of small bundles of parallel stems of left- and right-handed helices with less order in the direction perpendicular to the chain (helix) axis.

The preferred conformation of poly(oxy methylene) is nearly all-gauche and the crystal structures reported are a trigonal form (I), which is the most stable, and a less stable orthorhombic (II). The chains in the unit cell are of the same handedness, and left- and right-handed molecules evidently appear in different crystal lamellae.

Polymer crystals are always highly **anisotropic**. All tensor-like properties exhibit pronounced direc-tional dependence. The properties along the chain axis are very different from the properties in the transverse directions. The reason for the pronounced anisotropy is that strong covalent bonds connect the atoms along the chain and that the binding forces in the transverse directions originate from the much weaker van der Waals bonds. The modulus for a linear polyethylene single crystal along the chain axis is approximately 300 GPa, which is almost 100 times greater than the moduli in the transverse directions.

The thermal expansion coefficient in the c-axis direction is in fact negative for polyethylene and also for many other polymers (Fig. 7.4). The thermal expansivities along the a and b axes are, however, positive. Polymer crystals are also highly birefringent, with the higher refractive index in the c direction, $\Delta n = \Delta n_c - \Delta n_{a(b)}$, ranging from 0.10 upwards.

In crystals it is necessary to denote planes and direction. This can be done in the conventional way either by Miller's index or by a lattice plane index (Fig. 7.5). Directions are given as the lowest integer vector referring to the coordinate system $x(a)$, $y(b)$

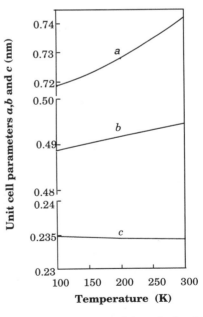

Figure 7.4 Thermal expansion of the orthorhombic cell of polyethylene. Drawn after data from Davis, Eby and Colson (1970).

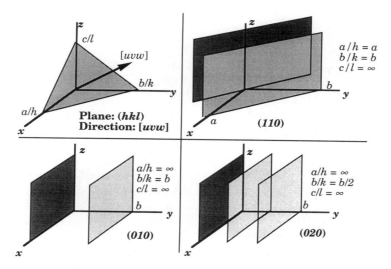

Figure 7.5 Definition of lattice plane index illustrated with three examples.

and $z(c)$. A vector parallel to the chain axis is denoted [001]. Note the square brackets. The method for the notation of planes in the lattice plane index system is illustrated in Fig. 7.5. The first plane intersects the origin of the coordinate system. The next plane intersects the three axes at $x = a/h$, $y = b/k$ and $z = c/l$. The task is to find an integer combination of h, k and l, which is finally presented in parentheses, (hkl). Three examples are given in Fig. 7.5. All planes containing the chain axis, i.e. those parallel to the chain, have the general formula $(hk0)$. The lattice index system indicates not only the orientation of the planes but also the shortest distance between the planes (Fig. 7.5). The set of planes denoted (010) is a subset of (020). The orientation of the two sets of planes is the same but the interplane distances (d_{hkl}) are different: $d_{010} = b$ and $d_{020} = b/2$.

Negative hkl values are indicated by bars, e.g. $(0\bar{1}0)$. Several sets of planes appearing in highly symmetrical crystal structures may be denoted together with brackets of the type $\{hkl\}$, e.g. the following planes in a cubic structure; (100), (010), (001), $(\bar{1}00)$, $(0\bar{1}0)$ and $(00\bar{1})$ are denoted simply $\{100\}$.

Miller's index system is similar to the lattice plane index system but with the difference that the hkl values presented are given as the lowest possible integer values. The Miller's index notation for both the sets of planes shown in Fig. 7.5 with the lattice plane indices (010) and (020) is simply (010). Miller's indices thus provide information only about the orientation of the planes and disregard the interplanar distances involved.

The most densely packed diffraction planes along the chain axis for polyethylene are denoted (002) in lattice plane index notation. The distance between these planes is thus $c/2 \approx 0.127$ nm. In Miller's index notation they are (001).

The **reciprocal space** is a very useful concept that simplifies calculations in crystallography and diffraction. Let $\bar{a}$, $\bar{b}$ and $\bar{c}$ be the translation vectors of the unit cell. A set of vectors $\bar{a}^*$, $\bar{b}^*$ and $\bar{c}^*$, the vectors of the reciprocal cell, exists that fulfils the conditions:

$$\begin{array}{lll} \bar{a} \cdot \bar{a}^* = 1 & \bar{a} \cdot \bar{b}^* = 0 & \bar{a} \cdot \bar{c}^* = 0 \\ \bar{b} \cdot \bar{a}^* = 0 & \bar{b} \cdot \bar{b}^* = 1 & \bar{b} \cdot \bar{c}^* = 0 \\ \bar{c} \cdot \bar{a}^* = 0 & \bar{c} \cdot \bar{b}^* = 0 & \bar{c} \cdot \bar{c}^* = 1 \end{array} \quad (7.1)$$

It can be shown that

$$\bar{a}^* = \frac{\bar{b} \times \bar{c}}{\bar{a} \cdot \bar{b} \times \bar{c}}; \ \bar{b}^* = \frac{\bar{c} \times \bar{a}}{\bar{a} \cdot \bar{b} \times \bar{c}}; \ \bar{a}^* = \frac{\bar{a} \times \bar{b}}{\bar{a} \cdot \bar{b} \times \bar{c}} \quad (7.2)$$

The scalar triple product $\bar{a} \cdot \bar{b} \times \bar{c}$ is equal to the volume of the unit cell, $\bar{a}^*$ is perpendicular to plane

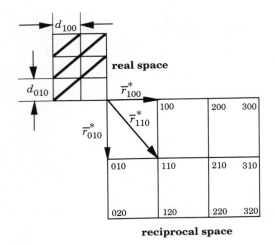

Figure 7.6 Illustration of reciprocal space from a view along [001] of an orthorhombic cell. Drawn after Spruiell and Clark (1980, p. 25).

reciprocal space:

$$\bar{r}^* = h\bar{a}^* + k\bar{b}^* + l\bar{c}^*; \quad \bar{r}^* \perp (hkl) \quad (7.5)$$

and the interplanar spacing (d_{hkl}) can be calculated from:

$$d_{hkl} = \frac{1}{|\bar{r}^*|} \quad (7.6)$$

It is helpful to think of the reciprocal lattice as a representation of the crystal lattice in which the (hkl) planes of the crystal are each represented by a lattice point of the reciprocal lattice. This point in reciprocal space is located in a direction from the origin which is perpendicular to the (hkl) planes in real space (Fig. 7.6).

Crystals are never perfect. They contain various types of defect. Figure 7.7 shows linear (one-dimensional) defects, edge and screw dislocations. The multi-layer growth spiral crystal lamella is due to a screw dislocation with the Burger and dislocation vectors parallel to the c-axis.

Point-like (zero dimensional) defects are also present in polymeric crystals. They arise from the presence of chain ends, kinks (see example in Fig. 7.8) and jogs (molecular defects with collinear stems on each side of the defect).

The presence of molecular (point) defects in polymer crystals may be indicated by an expansion of the unit cell. The unit cell parameters of branched polyethylene have been extensively studied and compared with those of linear polyethylene (Fig. 7.9). The c parameter remains constant and the branched polymer crystals are expanded in the a (mostly) and b directions. Both methyl and ethyl branches cause

bc, $\bar{b}^*$ to plane ac and $\bar{c}^*$ to plane ab. In an orthorhombic cell, the reciprocal cell vectors are parallel to the original cell vectors:

$$|\bar{a}^*| = \frac{1}{|\bar{a}|}; \quad |\bar{b}^*| = \frac{1}{|\bar{b}|}; \quad |\bar{c}^*| = \frac{1}{|\bar{c}|} \quad (7.3)$$

The reciprocal of the reciprocal vectors (cell) is the original cell. Thus,

$$\bar{a} = \frac{\bar{b}^* \times \bar{c}^*}{\bar{a}^* \cdot \bar{b}^* \times \bar{c}^*}; \quad \bar{b} = \frac{\bar{c}^* \times \bar{a}^*}{\bar{a}^* \cdot \bar{b}^* \times \bar{c}^*}; \quad \bar{c}^* = \frac{\bar{a}^* \times \bar{b}^*}{\bar{a}^* \cdot \bar{b}^* \times \bar{c}^*} \quad (7.4)$$

(hkl) in real space is equal to a point $(\bar{r}^*)$ in the

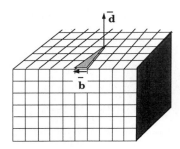 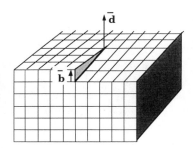

Figure 7.7 Line defects in crystals: edge dislocation (left) and screw dislocation (right). Dislocation vector ($\bar{d}$) and Burger's vector ($\bar{b}$).

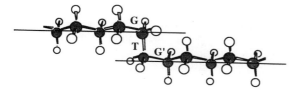

Figure 7.8 Kink (2g1 type); ... GTG' ... in polyethylene; the stems on each side of the defect are parallel but shifted by one lattice unit.

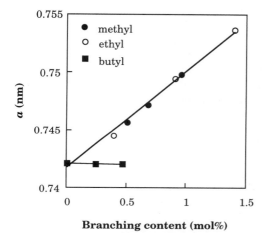

Figure 7.9 The crystallographic *a*-dimension of the orthorhombic cell of branched polyethylene as a function of degree of branching and branch type. Drawn after data from Preedy (1973).

considerable expansion whereas larger pendant groups, propylene or longer homologues, are largely excluded from the crystals. Martinez-Salazar and Baltá-Calleja (1979) have suggested that a fraction of the small branches, methyl and possibly ethyl groups, are included in the crystals in the vicinity of 2g1 kinks. According to this view, the extra 'room' provided by the kinks is not sufficiently large to house the larger pendant groups, and the cell expansion in butyl-branched polyethylene is negligible (Fig. 7.9) and the butyl groups must be confined to the amorphous phase.

7.2 THE CRYSTAL LAMELLA

7.2.1 THE CRYSTAL LAMELLA IN HISTORICAL PERSPECTIVE

It was known from early X-ray diffraction work that polymers never crystallize to 100%. The prevailing view of polymer crystals was that they were fringed micelles (Fig. 7.10). The modern view of **chain folding** was first introduced by Storks (1938). Storks concluded that the chains of semicrystalline *trans*-(polyisoprene) had to fold back and forth. This proposal went by largely unnoticed by the scientific community. Three papers were independently published by Keller (1957), Till (1957) and Fischer (1957) reporting that single crystals were 10 nm thick

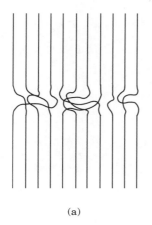
(a)

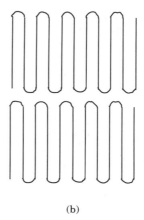
(b)

Figure 7.10 Simple illustrations of two of the central concepts: (a) the historical fringed micelle crystal; (b) the folded chain crystal.

platelets with regular facets and with the chain direction perpendicular to the lamellar surface. Andrew Keller postulated chain folding as the missing link in explaining the fact that the crystal lamellae were very much thinner than the length of the polymer chains (Fig. 7.10).

The story is beautifully told by Keller (1991):

> However, the idea of a morphological hierarchy was alien to the scientific establishment in polymer science at that time. The authorities believed that everything worth knowing can be accounted for by simply considering the statistical behaviour of chain molecules. Crystallisation in particular was seen as a chance coming together of adjacent chain portions forming little micellar bundles but no larger entities.

Andrew Keller entered the H. H. Wills Physics Laboratory in 1955 and was stunned by what he saw:

> The most positive aspect was the extraordinary intellectual ferment coupled with open-mindedness which permeated the whole place... There was no distinction between high and low brow, it was all one intellectual adventure. That is how polymers eventually slotted in between quantum mechanics, dislocations, particle physics, liquid helium, design of new optical instruments and much else... Amongst my 'negative' experiences was first and foremost the nearly total lack of equipment... I cannot deny that the above experimental conditions were frustrating to the extreme, yet they were inducive to make the best use of the little there was and always to concentrate on the essentials, lessons well worth learning... Another negative experience was the total absence of anybody knowledgeable in polymers. As I was still unknown in the field nobody visited me and I had no funds to visit anybody else. Also my access to the polymer literature was highly limited. So I lived and worked for two full years in near complete isolation from the relevant scientific community... Further, when I told him (Sir Charles Frank) that I cannot see how long chains, which I found to lie perpendicular to the basal surface (by combined electron microscopy and diffraction) of layers much thinner than the molecules are long (thickness assessed by electron microscopy shadowing and small-angle X-ray scattering), can do anything else but fold, he said 'of course' and encouraged me to publish immediately... That is how in 1957 in an 'office' filled with fumes, sparks and scattered X-rays, amidst total isolation from, in fact ignorance of the rest of polymer science, single crystals and chain folding were recognised.

7.2.2 FUNDAMENTALS OF CRYSTAL LAMELLA

Most of the examples presented here are taken from the extensive work carried out on polyethylene. They highlight general principles valid for any crystalline polymer. Some of the features are, however, special and unique to the particular polymer, and these cases will be discussed separately.

Single crystals of linear polyethylene prepared from dilute solutions in xylene and similar solvents provided the evidence in favour of chain folding. The large surfaces which contain the chain folds are commonly referred to as the **fold surfaces**. The single crystals shown in Fig. 7.11 exhibit planar, lateral

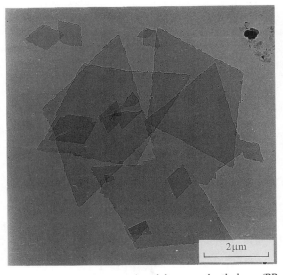

Figure 7.11 Single crystals of linear polyethylene (BP Rigidex 140 60) crystallized from dilute solution in xylene at 74°C. Transmission electron micrograph by A. M. Hodge and D. C. Bassett, University of Reading, Reading, UK.

surfaces, which after careful electron diffraction work, were identified as {110} planes. This led to the recognition of growth **sectors** present in single crystals (Fig. 7.11). The monolayer type of crystal is obtained from very dilute solutions (0.01% of polymer). More concentrated solutions give rise to multi-layer aggregates of crystals. Crystallization from 0.01% linear polyethylene in p-xylene at 70°C gives rise to lozenge-shaped crystals with {110} facets.

Growth at a slightly higher temperature, typically 80°C in p-xylene, resulted in truncated single crystals and two new sectors with {100} facets were observed (Fig. 7.12). The existence of crystallographic facets indicates that the crystallizing molecules are deposited on the crystal surface in a fairly regular manner leading to predominantly regular chain folding. The existence of the faceted single crystals was taken as evidence in favour of particular types of chain folds. In the {110} sectors, folds along the direction [110] were anticipated, whereas for sectors {100}, the chain folding was expected to occur along [010]. In fact, the

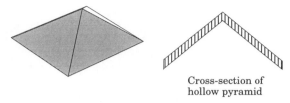

Figure 7.13 Pyramid-shaped single-crystal.

difference in fold types affects the thermal stability of the sectors and it was noted that {100} sectors possess a lower melting point than the {110} sectors.

A frequently occurring feature of the transmission electron micrographs of single crystals of polyethylene is the central corrugations found on the single crystals, which were recognized early and explained as being due to the fact that the crystals are not planar but instead are shaped like hollow pyramids (Fig. 7.13). When the crystals are deposited on the electron microscopy grids, they collapse and this causes the corrugations.

The hollow pyramid shape typical of a single crystal of polyethylene indicates that the chain axis is not parallel with the normal of the lamella. The chain axis is generally at an angle, about 30°, with respect to the lamella normal. The reason for the chain tilt is essentially that a certain type of regular chain fold requires a small vertical displacement of the linear chain in the adjacent position (Fig. 7.14).

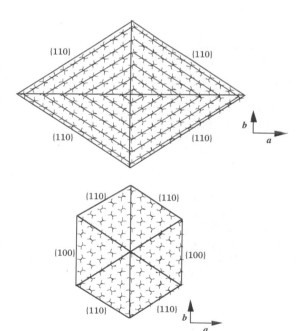

Figure 7.12 Sectorization of polyethylene single crystals. Upper crystal shows only {110} sectors whereas the lower also has {100} sectors.

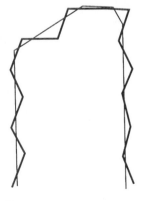

Figure 7.14 Crystallographic (110) fold in linear polyethylene. Note that the lamellar surface and the chain axis are not perpendicular to each other.

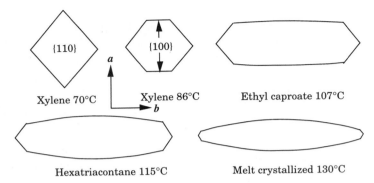

Figure 7.15 Lateral shape of crystals of linear polyethylene crystallized in dilute solutions with different solvents at different temperatures, as indicated in the figure. Drawn after data from Organ and Keller (1985) and Bassett, Olley and Al Reheil (1988).

The crystallographic (110) fold, which was first proposed by Bassett, Frank and Keller (1963), is essentially a path in the diamond lattice. Hence, the hollow pyramid itself is indicative of the dominance of regular chain folding in single crystals. Bassett showed that the fold surfaces in the {110} sectors were parallel to {312} planes and that the fold surfaces in the {100} sectors were parallel to {201} planes. A match between {312} and {201} is obtained only for a certain fixed ratio of {110} and {100} growth. A preference for such a growth ratio was indicated

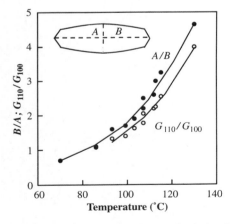

Figure 7.16 Lateral aspect ratio (B/A) and calculated ratio of the linear growth rates perpendicular to {110} and {100}. Drawn after data from Organ and Keller (1985) and Bassett, Olley and Al Reheil (1988).

by Kawai and Keller (1985). Deviation from this ideal shape ratio may occur and is then accompanied by distortions. Crystals grown from the melt exhibit a similar structure, with a roof shape. The latter constitute a part, i.e. a cross-section, of the hollow pyramid. An in-depth discussion of the nature of the fold surface both in single crystals grown from solution and in crystals grown from the melt is given in the following pages.

Organ and Keller (1985) showed that the lateral shape of single crystals of polyethylene grown from solution varied with crystallization temperature (Figs 7.15 and 7.16). The lateral shape of the single crystals showed no systematic correlation with the degree of undercooling. Figure 7.15 also includes more recent findings by Bassett, Olley and Al Reheil (1988) from melt-crystallization at very high temperatures (130°C). Only a minor portion of the sample crystallizes at this temperature and thus the crystallizing species is crystallizing from a dilute solution of crystallizing polymer in a polymer 'solvent'.

Crystals with well-defined {110} and {100} facets are typical of the crystallization of polyethylene at relatively low temperatures, 70 to 90°C from p-xylene (Fig. 7.15). An increase in crystallization temperature causes an increase in relative size and in degree of rounding of the {100} surfaces. At the highest reported temperature (130°C), only a very small part of the periphery of the crystal lamellae is {110} facets. The remaining parts are rounded {100} surfaces. Figure 7.16 shows that the ratio of length along a [100] to length along b [010] increases with increasing crystallization temperature. The curvature of the {100} surface also increases with increasing crystallization temperature. A linear growth rate anisotropy ratio, G_{110}/G_{100}, can be calculated from the crystal shapes (Fig. 7.16) and it is evident that this ratio increases with increasing crystallization temperature.

The temperature dependence of the lateral shape of the crystals has cast important light on the crystallization mechanisms, and Chapter 8 presents a more detailed discussion of this topic.

The rough lateral surfaces shown in Fig. 7.15 constitute a serious problem in the light of some of the current theories of polymer crystallization. Regime I crystallization (a term first introduced by Lauritzen and Hoffman (1960); see Chapter 8) prevailing at 130°C occurs by sequential addition of monolayers. The lateral surfaces are thus expected to be smooth and faceted, in apparent contradiction to the observed, rounded morphology of the lamellar crystals.

Table 7.2 presents the lateral shapes of solution-grown single crystals of a number of other crystalline polymers. The majority of the polymers exhibit faceted crystals and some, e.g. polyoxymethylene, form hollow pyramids. This behaviour is similar to that of polyethylene, and it indicates predominantly regular chain folding in solution-grown crystals of this polymer. The presence of crystallographically smooth side surfaces indicates that folding is more regular but it certainly does not by itself constitute valid proof.

Most of the polymers may under certain conditions show more complex, multi-layer morphologies. The structures presented in Table 7.2 apply to crystals grown exclusively from dilute solutions.

Table 7.2 Lateral shapes of solution-grown single crystals

Polymer	Shape/comments	
Polyoxymethylene	Six-sided (hexagonal hollow pyramids)	⬡
Poly(4-methyl-1-pentene) (isotactic form)	Square-based hollow pyramid	☐
Polytetrafluoroethylene	Irregular hexagonal lamellae	⬡
Poly(1-butene)	Square and hexagonal lamellae	☐ ⬡
Polystyrene (isotactic form)	Hexagonal lamellae	⬡
Poly(ethylene oxide)	Square and hexagonal lamellae	☐ ⬡
Poly(ethylene terephthalate)	Flat ribbons of 30 nm width to spindle-like lamellae	▭
Polyamide 6	Lozenge-shaped lamellae	◇
Polyamide 6,6	Irregular hexagonal lamellae and flat ribbons with H-binding plane parallel to the long direction of the ribbon	⬡ ▭
Polypropylene (isotactic form)	Lath-shaped lamellae	

Sources: Wunderlich (1973); Bassett (1981); Woodward (1989).

A typical feature of the solution-grown single crystals and, in fact, also of crystals grown from the melt, is their low thickness-to-width ratio. Typical values range from 0.01 to 0.001. The equilibrium shape of a crystal can be calculated from the surface energies by searching for the energy minimum at a given crystal volume. High-energy surfaces should be relatively small and the crystal dimensions perpendicular to those surfaces large.

The dimensions of the equilibrium crystal can be obtained by searching for the minimum free energy of the crystal with respect to the melt (ΔG) at a given volume V (Fig. 7.17):

$$\Delta G = V\Delta g^0 + 2L_1L_2\sigma_3 + 2L_1L_3\sigma_2 + 2L_2L_3\sigma_1 \quad (7.7)$$

where Δg^0 is the specific free energy of melting and σ_i are the specific surface free energies. One of the L_i terms can be eliminated from the equation by considering that $V = L_1L_2L_3$ is constant:

$$L_2 = \frac{V}{L_1L_3} \quad (7.8)$$

Insertion of eq. (7.8) into eq. (7.7) yields:

$$\Delta G = V\Delta g^0 + \frac{2V}{L_3}\cdot\sigma_3 + 2L_1L_3\sigma_2 + \frac{2V}{L_1}\cdot\sigma_1 \quad (7.9)$$

By taking the derivatives of ΔG with respect to L_1 and L_3 and setting them equal to zero, the following expression is obtained:

$$\frac{\partial(\Delta G)}{\partial L_1} = L_3\sigma_2 - \frac{V}{L_1^2}\sigma_1 = 0 \Rightarrow \frac{L_1}{\sigma_1} = \frac{L_2}{\sigma_2} \quad (7.10)$$

and

$$\frac{\partial(\Delta G)}{\partial L_3} = L_1\sigma_2 - \frac{V}{L_3^2}\sigma_3 = 0 \Rightarrow \frac{L_2}{\sigma_2} = \frac{L_3}{\sigma_3} \quad (7.11)$$

Combination of eqs (7.10) and (7.11) gives:

$$\frac{L_1}{\sigma_1} = \frac{L_2}{\sigma_2} = \frac{L_3}{\sigma_3} \quad (7.12)$$

Thus, the dimensions of the equilibrium crystal in different directions (i) are proportional to the surface free energies (σ_i) of the perpendicular surfaces (Fig. 7.17).

The specific surface energy of the fold surface (σ_e) of polyethylene is about seven times greater than the specific surface energy of the lateral surfaces (σ). The energy associated with the regular chain folding amounts to 60–70 mJ m^{-2} and the specific surface free energy of the lateral surfaces to 15 mJ m^{-2}. The total specific surface free energy of the fold surface thus amounts to about 90 mJ m^{-2}. The equilibrium ratio of thickness (along the chain axis) to width of a polyethylene crystal is consequently close to 7, which is three to four orders of magnitude greater than experimentally obtained values. It can be concluded that polymer crystals are not in equilibrium. Crystals with this unfavourable thickness-to-width ratio are expected to rearrange when given enough 'thermal stimulation'.

Annealing of monolayer single crystals at sufficiently high temperatures leads to crystal thickening and to the formation of holes, i.e. new lateral surfaces (Fig. 7.18). This 'Swiss cheese' structure typical of annealed monolayer single crystals is thus thermodynamically more stable than the original 'continuous' single crystals. This is because the specific surface energy of the lateral surfaces (approximately 15 mJ m^{-2}) is much less than the specific surface energy of the fold surface (approximately 90 mJ m^{-2}). It should be noted that these holes do not develop on annealing of mats of overlapping single crystals or in melt-crystallized samples.

Polyethylene crystals thicken at temperatures greater than 110°C. Polymer crystals grown under isothermal conditions thicken gradually with time. This has numerous important consequences. For

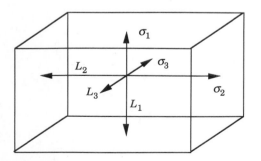

Figure 7.17 Equilibrium shape of crystal with three different surfaces ($i = 1, 2, 3$) with different surface free energies σ_i.

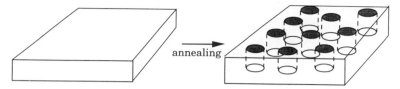

Figure 7.18 Annealing (heat treatment) of a solution-grown monolayer single crystal of polyethylene (schematic drawing).

instance, the recorded melting point of a given polyethylene sample is very dependent on the heating rate in the experiment. Thin crystals have a tendency to thicken during heating prior to melting and for such samples the recorded melting point decreases with increasing heating rate, until finally a constant melting-point value is attained at the highest heating rates. Keller and associates at Bristol showed in studies during the 1980s that the fundamental crystal thickening process is of abrupt and discontinuous character (see Ungar et al. 1985). Crystals doubled or tripled their thicknesses in discrete steps. Spegt and co-workers reported a similar behaviour for narrow fractions of poly(ethylene oxide) (see Spegt 1970). In melt-crystallized samples with a broad crystal thickness distribution this 'integer' crystal thickening is not clearly revealed by experimental data. Experiments on melt-crystallized samples commonly show a linear increase in the average crystal thickness with the logarithm of annealing time.

Polyethylene and a few other polymers form extraordinary thick (μm) crystal lamellae after crystallization at elevated pressure, typically at 4–6 kbar. This was discovered by Wunderlich and co-workers in 1962 (see Geil et al. 1964). Bassett, Block and Piermarini (1974) showed that polyethylene is transformed into a hexagonal phase with appreciable axial disorder at these elevated pressures. The hexagonal phase has been given different names such as 'rotor phase' and 'conformationally disordered crystal'. Most important is that the longitudinal mobility of the chains is extremely high in the hexagonal phase as opposed to that of the conventional orthorhombic phase and hence the crystals grow until or past the length of their molecules. This important phenomenon is further discussed in Chapter 8.

The Bristol group also showed during the 1980s that the thickness (L_c^*) of the first formed crystal depends only on the degree of undercooling ($\Delta T = T_m^0 - T_c$, where T_m^0 is the equilibrium melting point and T_c is the crystallization temperature):

$$L_c^* = \frac{C_1}{\Delta T} + \delta L \qquad (7.13)$$

where C_1 and δL are constants. Experimental data from work on linear polyethylene are shown in Fig. 7.19.

The kinetic crystallization theory of Lauritzen and Hoffman predicts the following expression:

$$L_c^* = \frac{2\sigma T_m^0}{\Delta h^0 \rho_c \Delta T} + \delta L \qquad (7.14)$$

where σ is the fold surface free energy, Δh^0 is the heat of fusion and ρ_c is the crystal phase density. The details of the Lauritzen–Hoffman theory are presented

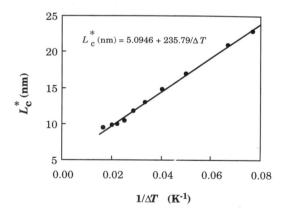

Figure 7.19 Initial crystal thickness plotted as a function of the reciprocal of the degree of undercooling. Drawn after data from Barham et al. (1985).

in Chapter 8. The two expressions, eqs (7.13) and (7.14), have the same dependence on ΔT. The Bristol group showed that the earlier observed significant difference between L_c^* as a function of ΔT for melt- and solution-grown crystals is artificial and due only to differences in rate of crystal thickening at the different absolute temperatures. Thus the initial crystal thickness is controlled solely by ΔT, whereas the crystal thickening rate is controlled by the absolute temperature.

One of the important relationships for crystalline polymers is the Thompson–Gibbs (TG) equation which relates melting point and crystal thickness. The change in free energy on melting (ΔG) is given by:

$$\Delta G = \Delta G^* + \sum_{i=1}^{n} A_i \sigma_i \qquad (7.15)$$

where ΔG^* is the surface-independent change in free energy and σ_i is the specific surface free energy of surface i with area A_i. At equilibrium:

$$\Delta G = 0 \Rightarrow \Delta G^* = \sum_{i=1}^{n} A_i \sigma_i \qquad (7.16)$$

Polymer crystals are lamella-shaped and the two fold surfaces greatly dominate the total surface energy term (Fig. 7.20).

The surface-independent term is equal to:

$$\Delta G^* = \Delta g^* A L_c \rho_c \qquad (7.17)$$

where ρ_c is the density of the crystal phase. Since both Δh and Δs can be regarded as temperature-independent, the specific bulk free energy change (Δg^*) is given by:

$$\Delta g^* = \Delta h^0 - T_m \Delta s^0 = \Delta h^0 \left(1 - \frac{T_m}{T_m^0}\right)$$

$$= \Delta h^0 \left(\frac{T_m^0 - T_m}{T_m^0}\right) \qquad (7.18)$$

If eq. (7.18) is inserted into eq. (7.17), the following equation is obtained:

$$\Delta G^* = \Delta h^0 (T_m^0 - T_m) \frac{A L_c \rho_c}{T_m^0} \qquad (7.19)$$

The total area of the four lateral surfaces is small compared to the area of the fold surfaces and their contribution to the total surface free energy of the crystal can be neglected:

$$\sum_{i=1}^{n} A_i \sigma_i \approx 2\sigma A \qquad (7.20)$$

Combination of eqs (7.16), (7.19) and (7.20) gives:

$$\Delta h^0 (T_m^0 - T_m) \frac{A L_c \rho_c}{T_m^0} = 2\sigma A$$

$$T_m^0 - T_m = \frac{2\sigma A T_m^0}{A L_c \rho_c \Delta h^0} = \frac{2\sigma T_m^0}{L_c \rho_c \Delta h^0}$$

which may be simplified to the Thompson–Gibbs equation:

$$T_m = T_m^0 \left(1 - \frac{2\sigma}{L_c \rho_c \Delta h^0}\right) \qquad (7.21)$$

The Thompson–Gibbs equation predicts a linear relationship between melting point and the reciprocal of crystal thickness (Fig. 7.21). However, in order to obtain the correct crystal thickness from experimental melting-point data, **crystal thickening** has to be inhibited. This can be accomplished by selective crosslinking of the amorphous component by radiation or by controlled rapid heating. Another potential problem, especially relevant to thick crystals, is **superheating**. Melting of a thick crystal is inherently slow and a considerable time must be available to allow equilibrium melting. This non-equilibrium effect is minimized by slow heating.

From density measurements of single crystals, it was recognized early that there is a sizeable fraction

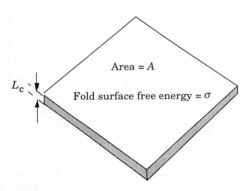

Figure 7.20 Simple model of crystal lamella.

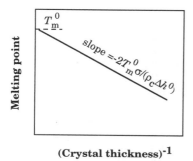

Figure 7.21 Melting point as function of crystal thickness. Schematic curve.

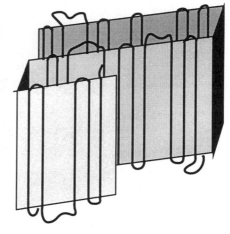

Figure 7.22 Model for the arrangement of a single chain in a solution-grown single crystal. The molecule is located in a plane which is 'super-folded'.

of amorphous material present in any crystalline polymer. This finding indicated that the fold surfaces possessed considerable disorder, i.e. that a significant fraction of the stems do not bend back to enter the crystals at the adjacent positions. Flory (1962) introduced the **random switch-board** model which considered regular, tight folding as rare and statistical re-entry as the typical feature. For many years, the Flory model and regular chain folding model existed without the availability of a critical experiment by which the question could be resolved. More than ten years later, new experimental techniques became available. Small-angle neutron scattering (SANS) of blends of deuterated and protonated polymers, e.g. $(-CD_2-)_n$ and $(-CH_2-)_n$, provided new insight into the fold surface structure. For solution-grown single crystals of linear polyethylene, it was shown that the average radius of gyration $(\langle s \rangle)$ of the molecules was proportional to $M^{0.1}$. This is a considerably weaker molar mass dependence than that of the molecules present in the solution before crystallization ($\langle s \rangle \propto M^{0.5}$). The global dimension of the chains decreases markedly on crystallization. The 'super-folding' model was proposed to take into account data from both SANS and wide-angle neutron scattering (WANS). The plane in which the deuterated molecule is located is super-folded (Fig. 7.22).

Spells, Keller and Sadler (1984) showed by infrared spectroscopy that 75% of the folds in solution-grown single crystals of polyethylene led to adjacent re-entry (tight folds) and that single molecules were diluted by 50% along the {110} fold plane. Both observations are consonant with the super-fold model.

7.2.3 EXPERIMENTAL TECHNIQUES FOR THE STUDY OF LAMELLAR CRYSTALS

The main technique is transmission electron microscopy. Single crystals deposited on a carbon-coated metal grid can be viewed after shadowing with a heavy metal using the bright field technique (Fig. 7.23).

Melt-crystallized samples are semicrystalline with two components; crystalline and amorphous. The natural density difference between the two 'phases' is insufficient to give good contrast in the electron microscope and different techniques are used to enhance the density difference between the

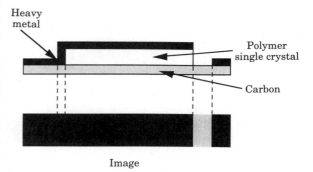

Figure 7.23 Polymer single crystal; contrast obtained by shadowing with a heavy metal (e.g. Au or Pd/Pt).

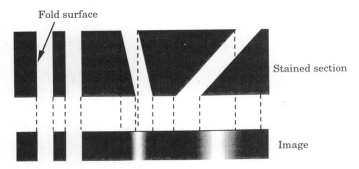

Figure 7.24 Contrast from a stained section with crystals (low density, white areas) and amorphous domains (high density, heavier elements have been added, black areas).

crystalline and amorphous phases. Melt-crystallized polyethylene samples are preferably studied after being etched with chlorosulphonic acid and stained with uranyl acetate. Sulphur, chlorine, oxygen and uranium add selectively to the amorphous component. This method, introduced by the German physicist, Kanig, is only suitable for polyethylene. Unsaturated polymers like polybutadiene or polyisoprene can be stained with OsO_4 which adds selectively to the double bonds located in the amorphous phase. Ruthenium tetroxide (RuO_4) has proven useful for preparing samples with other types of double bond, e.g. polystyrene and polyamides.

Figure 7.24 demonstrates that, in samples prepared according to these techniques, optimum contrast is obtained when the fold surface is parallel to the electron beam. Tilted crystals appear less sharp and areas without lamellar contrast are found over large areas of the electron micrographs. This is one of the disadvantages of these methods. Only a small fraction of the crystals can be viewed at the same time. The use of a tilting stage permits the assessment of more lamellae, i.e. lamellae of other tilt angles are included in the analysis. Another problem is shrinkage of the whole section which may occur in the sample due to dissipation of the electron-beam energy. Samples etched with chlorosulphonic acid for only a short period of time showed significant shrinkage, whereas samples treated with the acid for a long period of time showed only negligible shrinkage.

At the end of the 1970s, Bassett and co-workers introduced a method, using mixtures of concentrated sulphuric acid and potassium permanganate, giving contrast between crystals and amorphous domains, which turned out to be applicable to many different polymers, e.g. polyethylene, polypropylene, poly-(butene-1), polystyrene, poly(aryletherketone)s. Later changes in recipe have involved the introduction of phosphoric acid to avoid the formation of artificial structures. The strong etchant degrades the amorphous phase more quickly than the crystals and the resulting topography is revealed by heavy metal shadowing (Fig. 7.25). Replicates are prepared which are examined in the electron microscope.

The advantage of the permanganic etching method is that all lamellar crystals, including those at a high tilting angle, are revealed. Occasional observation of the lateral shape of the lamellar crystals is possible when the lamellae have a 'flat orientation', as shown in Fig. 7.25 (right-hand crystals).

X-ray diffraction provides information about the crystal thickness in mats of single crystals and in melt-crystallized samples. The periodic variation in (electron) density along a line perpendicular to the fold planes gives rise to constructive interference at very small scattering angles (Fig. 7.26). Thus, small-angle X-ray scattering (SAXS) provides information about the so-called long period, i.e. the sum of the average crystal and amorphous layer thicknesses. The average crystal thickness (L_c) is obtained from the expression:

$$L_c = d \cdot v_c \tag{7.22}$$

where v_c is the volume fraction of crystalline component and d is the SAXS long period.

Wide-angle X-ray diffraction can also be used to determine the crystal size. Small crystals give broad

Figure 7.25 Effect of etching with permanganic acid, creating a surface topography revealing lamellar details of the morphology. Dark areas indicate crystals.

Braggs reflections, and a simple way of assessing the crystal thickness ($\bar{D}_{hkl}$) perpendicular to a given set of (hkl) planes is through the Scherrer equation:

$$\bar{D}_{hkl} = \frac{K\lambda}{\beta \cos\theta} \quad (7.23)$$

where K is the Scherrer shape factor which adopts values close to unity, λ is the wavelength of the X-ray and β is the breadth (in radians, at half the peak value) of the diffraction peak associated with the (hkl) planes. Other causes of the broadening of the diffraction peaks are thermal vibrations and paracrystalline distortions (permanent defects) within the crystals. Hosemann proposed the following equation:

$$\beta_s^2 = \frac{1}{\bar{D}_{hkl}^2} + \frac{\pi^4 g^4}{d_0^2} n^4 \quad (7.24)$$

where β_s is the breadth of the diffraction peak in s units ($s = (2\sin\theta)/\lambda$) after subtracting the instrumental broadening, d_0 is the interplanar distance for the first-order reflection, n is the order of reflection and g is the degree of statistical fluctuation of the paracrystalline distortions relative to the separation distance of the adjacent lattice cell. If β_s^2 is plotted as a function of n^4, the thickness $\bar{D}_{hkl}$ is obtained as the square root of the reciprocal of the intercept.

The crystal thickness can also be estimated by converting melting-point data through application of the Thompson–Gibbs equation (eq. (7.21)). This method is indirect and has to be 'calibrated'. The non-equilibrium nature of polymer crystal melting is another major obstacle. Both crystal thickening and superheating have to be inhibited or at least controlled.

Other methods by which the crystal thickness in polyethylene has been determined are Raman spectroscopy by measurement of the frequency of the longitudinal acoustic mode (which is inversely proportional to the length of the all-trans stems in the crystals), and size exclusion chromatography of samples etched with HNO_3 or O_3. They both degrade the amorphous parts and leaves essentially oligomers with a chain length equal to the all-trans length ($= L_c/\cos\theta$, where θ is the tilt angle).

7.3 CRYSTALS GROWN FROM THE MELT AND THE CRYSTAL LAMELLA STACK

The molecules which add to a growing single crystal from a dilute solution are not in a strong 'competition' with other polymer molecules. Relatively few molecules deposit on the crystal simultaneously. Crystal growth from the concentrated molten state is very different in this respect. The molecules in this case are severely entangled and any polymer molecules add to a specific crystal surface simultaneously. The situation is thus more chaotic and

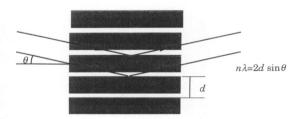

Figure 7.26 Small-angle diffraction from a stack of lamellar crystals (shaded areas) with the repeating distance (long period) d. The condition for constructive interference is expressed in the Bragg equation.

it is not surprising that the single-molecule trajectory of the melt is to some extent preserved in the semicrystalline state.

Melt-grown crystals have the same shape as the solution-grown single crystals in most respects. They are lamella-shaped with a thickness-to-width ratio of 0.01–0.001. A typical feature of melt-crystallized samples, shown in Fig. 7.27, is the **crystal stack**. The lamellae of the stack are almost parallel and the amorphous component is located in the space between the crystal lamellae. The number of lamellae in a stack varies considerably between different samples. Linear polyethylenes of low or intermediate molar mass form stacks with a great many lamellae, whereas branched and/or high molar mass samples exhibit stacks containing only a few crystal lamellae (Fig. 7.27).

The shape of the melt-grown crystals of polyethylene depends on molar mass, degree of chain branching and crystallization temperature. The transmission electron micrographs presented in Fig. 7.27 are from polyethylenes treated with chlorosulphonic acid and uranyl acetate. The crystals appear light and the amorphous regions appear dark. The boundary between a crystal and an amorphous region is sharp only when the fold surface is parallel to the electron beam. The view is dominantly along the radius of the spherulites (see next section) which is parallel to [010]. The growth direction of the crystals is thus predominantly towards the observer. The shape of the crystals of linear polyethylene along this view is straight or roof-shaped (low molar mass samples), C-shaped (intermediate molar mass samples) or S-shaped (high molar mass samples). Branched polyethylenes show mostly C- and S-shaped lamellae.

Careful analysis of the roof-shaped crystal lamellae shows that a roof gable is [010] and that the fold planes on each side of the gable are (h01) planes (Fig. 7.28), the most frequently occurring fold planes are {201}, {301}, {302} combined with {301}. The discovery of these fold planes was made by Keller and Sawada (1964) through electron microscopy of HNO_3-degraded samples. The tilt angle, i.e. the angle between the lamella normal and the chain axis, typically adopts values close to 30°. The analogy between the roof-shaped melt-grown lamellae and the solution-grown hollow pyramids should be noted.

Polymers are always polydisperse with a distribution in molar mass and they may also contain chain branches and different comonomers. Different molecular species crystallize in different stages, i.e. at different temperatures and times. The intermediate and

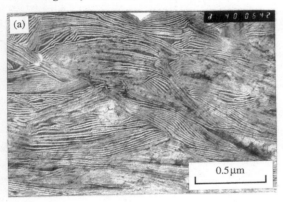

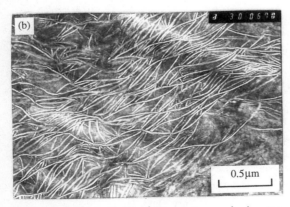

Figure 7.27 Transmission electron micrograph showing lamellar crystals in (a) linear and (b) branched polyethylene. By M. T. Conde Braña, Dept of Polymer Technology, Royal Institute of Technology, Stockholm, Sweden.

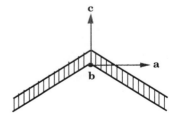

Figure 7.28 Model of roof-shaped lamella with crystallographic directions **a**, **b** and **c**.

the high molar mass component crystallizes early in stacks of thick **dominant** crystals. Small pockets of molten, rejected low molar mass material remain after crystallization of the dominant lamellae. The low molar mass species crystallize in separate crystal lamellae, in stacks of so-called **subsidiary** crystal lamellae. The process leading to separation of high and low molar mass material which accompanies crystallization is referred to as **molar mass segregation**. The electron micrograph presented in Fig. 7.29(a) shows small pockets of segregated, low molar mass material and long and thick dominant lamellae. Some samples show a preference for segregation of the low molar mass species to the spherulite boundaries (Fig. 7.29(b)).

Isotactic polypropylene displays a highly unusual ability to induce epitaxial crystallization of a number of different polymers with their chain axes tilted at large angles, 40–80°, relative to the helix axis direction of the polypropylene substrate. The same polymer shows an epitaxy, i.e. homo-epitaxy, at an angle of 80° (Fig. 7.30). The homoepitaxy is responsible for crosshatching, a structure typical of the monoclinic α structure. This feature causes the lamellar branching and the optical complexity, with both positive and negative spherulites, typical of isotactic polypropylene (section 7.4).

Let us return to the fold surface of melt-grown crystals. The important question is whether regular chain folding is the dominant feature of the fold surface. The problem can be viewed by simple chain-packing considerations. Experiments have shown that the amorphous density is 10–20% lower than the crystal density for most polymers. If all the chains that enter the amorphous phase were to take a random walk before crystal re-entry, the amorphous density would be significantly greater than the crystalline, in disagreement with the experimental data. A significant fraction of the chains must fold back directly in order to account for the low amorphous density. Chain tilting gives fewer amorphous entries per unit area of fold surface.

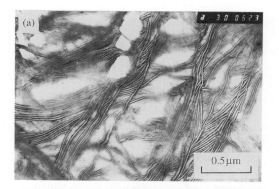

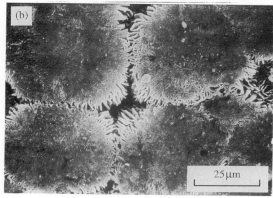

Figure 7.29 (a) Transmission electron micrograph of a binary mixture of low and high molecular mass linear PE. Light areas are due to segregated low molecular mass polymer (M. Conde Braña, Dept. of Polymer Technology, Royal Institute of Technology, Stockholm, Sweden). (b) Scanning electron micrograph of solvent-etched linear PE with removed low molar mass material (U. W. Gedde, Dept. of Polymer Technology, Royal Institute of Technology, Stockholm, Sweden).

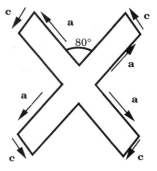

Figure 7.30 Crosshatching of monoclinic isotactic polypropylene; view along [010]. The crystallographic **a** and **c** (helical axis) directions are shown.

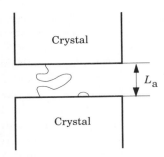

Figure 7.31 Simple model of lamellar stack showing two adjacent crystal lamellae and the two types of chain present: random tie chain and tight, (regular) fold.

It is possible to derive an expression by which the fraction of tight folds can be calculated under the simple assumption that only *two* types of chain leave the crystals, namely perfectly tight folds which occupy no space in the amorphous interlayer and chains performing a random walk in the amorphous interlayer ending in the adjacent crystal (Fig. 7.31). The number of bonds (n) in a typical Gaussian amorphous chain sequence is given by:

$$n = \frac{L_a^2}{Cl^2} \quad (7.25)$$

where L_a is the amorphous layer thickness, which is assumed to be the end-to-end distance, C is a constant for a given polymer–temperature combination and l is the bond length. Entries leading to regular chain folding constitute a fraction (f_{fold}) of the entries that are not contributing to the amorphous weight and the number of chain segments in an 'average type' of amorphous entry is given by:

$$n = \frac{L_a^2}{Cl^2}(1 - f_{fold}) \quad (7.26)$$

The number of chain segments (n^0) in a straight chain is given by:

$$n^0 = \frac{L_a}{l} \quad (7.27)$$

The ratio of the amorphous (ρ_a) to crystalline (ρ_c) density is given by:

$$\frac{\rho_a}{\rho_c} = \frac{n}{n^0} \quad (7.28)$$

The amorphous density is obtained by combining eqs (7.26)–(7.28):

$$\rho_a = \rho_c \frac{L_a}{Cl}(1 - f_{fold}) \quad (7.29)$$

It is known that chain tilting (by an angle θ) is a common feature of polymer crystals. In this case, the number of entries of the fold surface is multiplied by the factor $\cos \theta$, thus:

$$\rho_a = \rho_c \frac{L_a}{Cl}(1 - f_{fold})\cos \theta \quad (7.30)$$

This equation permits the calculation of f_{fold} by insertion of the following values: $\rho_a/\rho_c = 0.85$; $L_a = 5$ nm; $l = 0.127$ nm; $C = 6.85$; $\theta = 30°$. The calculated value of f_{fold} is 0.83; i.e. 83% of all stems are expected to be tightly folded and 17% are expected to be statistical chains in the amorphous interlayer. These figures must, however, be considered in the light of the assumptions made, namely that only tight chain folds and random interlamellar chains are present.

Melt-crystallized binary blends of deuterated (guests) and protonated (host) polymers (linear polyethylene and a few other crystalline polymers) of melt-crystallized samples have also been studied by SANS; samples crystallized at low temperatures to avoid aggregation (segregation) of deuterated molecules, and rapidly cooled samples showed the same value for the average radius of gyration as prior to crystallization. The global dimensions of the chains are essentially the same in the semicrystalline state as in the melt before crystallization. No reliable data are currently available for polymers crystallizing at higher temperatures due to problems associated with the aggregation of deuterated chains. A model which assumes that the local order is high and which invokes a certain fraction of tight chain folding is illustrated in Fig. 7.32. This model describes the state of a rapidly cooled semicrystalline polymer. The single-molecule structure of a sample crystallized more slowly at higher temperatures is less well understood.

The amorphous phase is particularly important for many properties of semicrystalline polymers. The strength and ductility of semicrystalline polymers of high molar mass arise from the presence of

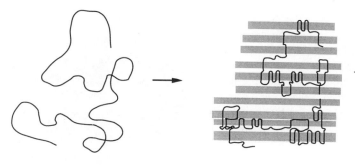

Molecule in melt   Molecule in semicrystalline state

Figure 7.32 Change in conformation of a single chain on crystallization under rapid cooling conditions.

interlamellar tie chains, connecting adjacent crystal lamellae.

It has been suggested, on the basis of density arguments (eq. (7.30)) and other arguments, that a considerable fraction of the chains in the fold surface form relatively tight (regular) folds. Keller and Priest (1968) showed that the chain ends in polyethylene are almost exclusively confined to the amorphous phase. Good estimates suggest that about 90% of the chain ends are located in the amorphous phase. Low molar mass polymers, for linear polyethylene having molar mass values smaller than 3000–4000, consist of extended-chain or once-folded crystals. Higher molar mass polymers form folded-chain crystals with occasional interlamellar tie chains and statistical loops.

The concentration of interlamellar tie chains in a given sample depends on molar mass, which determines the spatial distribution of the chains, and long period, i.e. the sum of crystal and amorphous layer thicknesses. Figure 7.33 shows two very

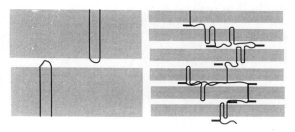

Figure 7.33 Single molecule trajectories showing balance' between molar mass and long period. Note the presence of non-crystallizable chain branches in the right-hand case.

different cases. The low molar mass polymer consists of molecules which are only twice as long as the long period. The chains tend to fold once and there are very few molecular links between adjacent lamellar crystals. The second case shows a high molar mass polymer with occasional non-crystallizable branches. The latter cause a reduction in overall crystallity and long period. The considerable length of the individual molecules allows them to 'travel' over many amorphous interlayers and many tie chains exist in this material.

7.4 SUPERMOLECULAR STRUCTURE

The supermolecular structure is the kind of structure which is typically revealed by optical microscopy. The standard method is in fact polarized light microscopy, but other techniques are also used to reveal these structures, as indicated in the separate presentation of experimental techniques (section 7.5). The size of supermolecular structures ranges from 0.5 μm to several millimetres.

It was perhaps Andrew Keller who introduced the hierarchical morphological scheme starting with the **lamellar (folded-chain) crystal** as the fundamental unit of the **lamellar stacks**, which in turn build up the various **supermolecular structures**, of which the spherulite is the most prominent member. These superstructures are polycrystalline. They consist of a great many lamellar crystals and also many lamellar stacks. This section briefly presents the experimental techniques useful in the assessment of supermolecular structures.

152 *Crystalline polymers*

It may be instructive to start the discussion with spherulites, the best-known supermolecular structure (Fig. 7.32). Spherulites became first known as the circular crystalline objects found in vitreous igneous rocks. The name derives from the Greek word for ball or globe. A spherulite is an 'object' with spherical optical symmetry. Two unique refractive indices may be defined, namely the tangential (n_t) and radial (n_r) refractive indices. For simplicity, let us consider polyethylene. The single crystal or any other oriented structure of this polymer may be considered to be uniaxially birefringent, with the unique direction (largest refractive index) along the chain axis. Negative spherulites with $n_t > n_r$ have a higher proportion of the chains in the circumferential planes than along the radius of the spherulites.

Figure 7.34 shows the two main types of spherulites, non-banded (Fig. 7.34(a)) and banded (Fig. 7.34(b)). The boundaries of the spherulites are non-spherical but smooth after impingement. The boundaries of mature spherulites before impingement appear to be circular (spherical). Examination of the spherulite boundaries by transmission electron microscopy shows more irregular features. Low molar mass species may accumulate in the spherulite boundaries making them very brittle in samples with appreciable fractions of low molar mass species.

The direction of growth of polyethylene spherulites is always close to [010], i.e. the radius of the spherulite is parallel to the crystallographic b axis. Other polymers have other growth directions, e.g. monoclinic isotactic polypropylene grow fastest along the a axis.

Banding often appears in polyethylene samples which have been crystallized at a relatively high degree of supercooling, i.e. at low temperatures. The regular variation in light intensity as a function of radial distance is due to a corresponding variation in orientation of the refractive index ellipsoid. The birefringence is approximately uniaxial. The sample appears isotropic (dark on micrographs) where the long axis of the refractive index ellipsoid is parallel to the light beam, i.e. when the chain-axis director is parallel to the light beam (optical axis) (Fig. 7.35). The transmitted light intensity is at a maximum at locations where the long axis of the refractive index ellipsoid is perpendicular to the light beam (Fig. 7.35). This

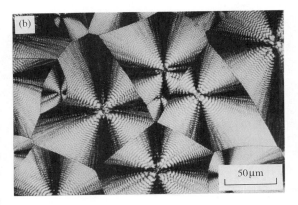

Figure 7.34 Polarized photomicrographs of different polyolefins showing (a) non-banded spherulites (PP); (b) banded spherulites (PE); and (c) axialites (PE). Photomicrographs by S. Laihonen and U. W. Gedde, Dept of Polymer Technology, Royal Institute of Technology, Stockholm.

Figure 7.35 Orientation of refractive index ellipsoids in a banded spherulite.

corresponds to the situation where the normal to the crystal lamellae is perpendicular to the light beam.

It was early suggested that continuously twisting lamellae radiating out from the centre of the spherulites led to the rotation of the refractive index ellipsoid about the radius of the spherulite. Figure 7.36 shows a transmission electron micrograph of a sample with banded spherulites. The same banding is readily revealed in a sample etched with permanganic acid. Circular banded regions with crystal lamellae having a flat orientation are surrounded by circular bands of

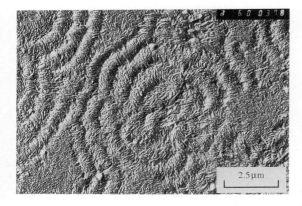

Figure 7.36 Transmission electron micrograph of polyethylene with banded spherulites etched with a mixture of potassium permanganate, sulphuric acid and phosphoric acid (M. Hedenqvist, Dept of Polymer Technology, Royal Institute of Technology, Stockholm, Sweden).

lamellae with the fold surface perpendicular to the sample surface.

It is known that samples displaying banded spherulites also show C- or S-shaped lamellae. Bassett and Hodge (1978) showed that banded spherulites of polyethylene consisted of dominant lamellae whose profiles viewed down [010] were S- or C-shaped. They noticed a uniform chain tilt and c-axis orientation in neighbouring lamellae. These uniform structures were untwisted along the radius of the spherulites a distance of about one-third of the band spacing. Changes in c-axis orientation occurred sharply in screw dislocations of consistent sign and involving only two or three layers of spiral terrace. The single-handedness of these screw dislocations was consistent with observations made on monolayer crystals of polyethylene grown at 413 K which showed giant screw dislocations of similar structure. This was also observed by Stack, Mandelkern and Voigt-Martin (1984) who explained the apparent smooth curving by frequent but abrupt changes in the molecular tilt angle along the width of the lamellae.

An alternative view was presented by Lustiger, Lotz and Duff (1989) and Keith and Padden (1984). Lustiger *et al.* suggested that lamellae in a banded spherulite can be represented by a radially oriented assembly of continuous helicoidally twisted lamellae, i.e. the 'old' view. The lamellar profiles (S or C) can, according to these authors, result from slices that are at oblique angles (not parallel to [010]) through helicoids with straight profiles (from a view along [010]). Keith and Padden proposed, on the other hand, a common origin for S-shaped lamellae and continuously twisting lamellae. They proposed that different degrees of disorder develop at opposite fold surfaces, primarily due to the chain tilting. It was suggested that the resulting differences in surface stress give rise to transient bending moments leading to curved and twisted lamellae.

One of the central questions is how the crystal lamellae starting at a central point can finally form a spherical-shaped spherulite. The spherical shape has to be obtained already by the dominant lamellae. A mechanism for branching and splaying of the dominant lamellae has to be available. Bassett, Olley and Al Reheil (1988) presented transmission electron microscopy data on linear polyethylenes providing

154 Crystalline polymers

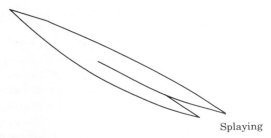

Splaying

Figure 7.37 Spiral development around a screw dislocation producing diverging lamellae in linear polyethylene. The view is essentially along [001]. Drawn after Bassett, Olley and Al Reheil (1988).

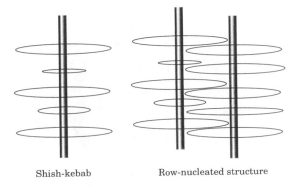

Shish-kebab   Row-nucleated structure

Figure 7.38 Oriented morphologies appearing in polyethylene.

that missing link for polyethylene (Fig. 7.37). The giant screw dislocation generates two lamellar branches that diverge. The branching/splaying mechanisms may well be different for different polymers, e.g. it is likely that the crosshatching phenomenon plays a central role for isotactic polypropylene.

The axialite is a non-spherical and irregular superstructure (Fig. 7.34(c)). Axialites are primarily found in low molar mass polyethylenes, at essentially all crystallization temperatures, and in intermediate molar mass polyethylenes crystallized at higher temperatures, at undercoolings less than 17 K. Very high molar mass samples, typically with molar masses of 10^6 g mol^{-1} or greater, form so-called random lamellar structures. The entanglement effect is so extensive that the mass crystallinity becomes very low and regular lamellar stacking is absent in these samples.

The morphologies discussed up to this point are all formed from isotropic melts. Orientation of the melt causes an increase in the free energy and this itself constitutes an important factor in practical processing. Crystallization may be induced by orientation, the term used for this phenomenon being orientation-induced crystallization. Shish-kebabs are formed from solutions that are subjected to elongational flow, which induces orientation of the solute molecules (Fig. 7.38). A central core of oriented bundles of fibres is formed at first as a direct consequence of the orientation. The shish-kebab consists of a central group of highly oriented fibrils, the shish, from which a great many lamellar crystals have grown, the kebabs. It is known that central fibrils

consist of high molar mass material. A similar structure is formed during extrusion/injection moulding under certain extreme conditions, namely when the melt is subjected to high elongational flow in combination with a high pressure or a high cooling rate. The oriented melt solidifies in a great many fibrous crystals from which lamellar overgrowth occurs. The structure is referred to as row-nucleated (Fig. 7.38). The radiating lamellae nucleating in adjacent fibrils are interlocked, a fact which is considered to be important for the superior stiffness properties of the melt-extruded fibres made by the Bristol group (Keller *et al.*). The fibrillar structure is present in ultra-oriented samples. It consists of highly oriented microfibrils. These microfibrils are sandwiches of alternating sequences of amorphous and crystalline regions. A great many taut interlamellar tie chains are present and the resulting mechanical properties are excellent.

A special kind of morphology found in polymers crystallizing in contact with a nucleating surface is the trans-crystalline structure. The nucleating object may be a flat surface or a fibril. The densely appearing nucleations at the surface of the nucleating object result in a one-dimensional (columnar) growth in a direction parallel to the normal of the surface. The thickness of the trans-crystalline layer depends on the outcome of the competition between surface nucleation and 'bulk' nucleation.

Table 7.3 presents a summary of morphologies found in polyethylene. Many of these structures are also found in a great number of other crystalline polymers. The optical sign of the spherulites is

Table 7.3 Supermolecular structures found in polyethylene

Superstructure	Internal structure	Molecular structure	Crystallization conditions
Axialites (sheaves)	Straight or roof-shaped lamellae	Low molar mass	High crystallization temperature
Banded spherulites	S- or C-shaped lamellae	Intermediate molar mass	Low crystallization temperature
Non-banded spherulites	Straight or weakly 'bent' lamellae	Intermediate molar mass	Intermediate crystallization temperature
Random lamellar structure	Randomly oriented lamellae, short and C-shaped	Very high molar mass	Formed at all temperatures
Shish-kebab structure	Fibrous core with lamellar overgrowth	Intermediate to high molar mass	Oriented crystallization from solution
Row-nucleated structure	Fibrous core with lamellar overgrowth	Intermediate to high molar mass	Oriented crystallization from the melt
Trans-crystalline structure	Columnar growing lamellae	–	Crystallization from a foreign or own nucleating surface

negative for most polymers but occasional examples of positive spherulites are found.

Isotactic polypropylene is a particularly interesting case with regard to supermolecular structure. Both positive and negative spherulites have been reported. The early work of Padden and Keith showed the presence of weakly positive (α-structure) spherulites developing at crystallization temperatures less than 134°C. Crystallization at temperatures above 138°C led to weakly negative (α-structure) spherulites. Crystallization of isotactic polypropylene in the temperature range from 110 to 132°C led to the formation of both α (monoclinic crystal structure) spherulites and highly birefringent β (hexagonal crystal structure) spherulites. The origin of the presence of both negative and positive spherulites is the crosshatching phenomenon. The relative amount of dominant and of daughter lamellae is decisive for which optical sign the spherulites take.

Polyesters, polyamides and polycarbonates occasionally give rise to positive spherulites. All 'even' polyamides (PA 6,6, PA 6,10 and PA 6,12) behave in a similar way. Positive spherulites are found at low temperatures below a certain temperature T_1 (15–20 K below the melting point) and negative spherulites at temperatures between T_1 and T_2 (a temperature only a few kelvin below the melting point). The orientation of the chain axis [001] was preferentially tangential whereas the hydrogen bond planes were radial in both negative and positive spherulites.

7.5 METHODS OF ASSESSING SUPERMOLECULAR STRUCTURE

Polarized light microscopy is the most widely used method to characterize the supermolecular structure (superstructure) of semicrystalline polymers. Small variations in local chain orientation cause variations in the birefringence typically amounting to 0.01 or less. With linear polarized incoming light, a crossed linear polarizer positioned after the objective lens transforms these tiny variations in birefringence to variations in intensity of transmitted light. Figure 7.39 shows the intensity of the transmitted light from

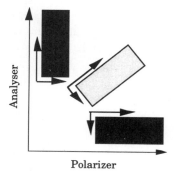

Figure 7.39 Principle behind polarized light microscopy showing the differences in transmitted light intensity for crystals of different orientations with respect to the polarizer/analyser pair. The directions of maximum and minimum refractive indices of the crystals are indicated by the arrows.

156 Crystalline polymers

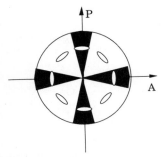

Figure 7.40 Origin of Maltese cross pattern from a negative spherulite. The orientations of the polarizer (P), the analyser (A) and the refractive index ellipses are shown.

three birefringent crystals with different orientations with respect to the crossed polarizer/analyser pair. The intensity of the transmitted light is at a minimum for crystals with their main axes parallel to the polarizer/analyser pair. The maximum transmitted light intensity is obtained for crystals with their main axes at an angle of 45° to the polarizer/analyser pair.

A spherulite is optically a 'spherically' birefringent object with two unique refractive indices: the radial (n_r) and the tangential (n_t). It is convenient to represent the variation in refractive index in the plane by the refractive index ellipse. The length of the major axis of the ellipse is proportional to the maximum refractive index in the plane, whereas the length of the minor axis is proportional to the minimum refractive index. Fig. 7.40 shows a spherulite with the larger refractive index in the tangential direction, i.e. $n_r < n_t$. A spherulite of this type is referred to as **negative**. The direction of the maximum birefringence can be determined by inserting a lambda plate parallel to the main directions of the local birefringence.

Negative spherulites with $n_t > n_r$ can in this manner be distinguished from positive spherulites ($n_t < n_r$). Both types of spherulites show a Maltese cross pattern with a maximum in the intensity of the transmitted light for parts of the spherulites at an angle of 45° to the polarizer/analyser pair.

The principle of small-angle light scattering (SALS) is shown in Fig. 7.41. Typically two types of optical configuration, namely H_v (crossed polarizers) and V_v (parallel polarizers) are used. SALS is a very good technique for the measurement of very small spherulites. The average spherulite radius (R) can be obtained from the scattering angle (θ_{max}) of the maximum scattered intensity (H_v) according to the following expression:

$$R = \frac{4.1\lambda_0}{4\pi n}\left[\sin\left(\frac{\theta_{max}}{2}\right)\right]^{-1} \quad (7.31)$$

where λ_0 is the wavelength of light *in vacuo* and n is the average refractive index of the polymer sample.

Solvent-etching relies on the fact that molecular species with a low melting point are more easily dissolved than the rest of the sample. The solvent removes molecules as whole entities from the sample without breaking any chemical bonds provided that they have segregated during crystallization. Segregated, low molar mass species are concentrated in the spherulite or axialite boundaries. The morphology is revealed by creating a smooth surface by polishing or microtoming, solvent etching, drying and examination in the scanning electron microscope.

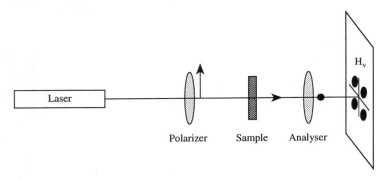

Figure 7.41 Small-angle light scattering with crossed polarizers (H_v).

There are essentially two main etching techniques: (a) vapour etching for only a few seconds; and (b) controlled, isothermal treatment with a liquid solvent for a considerably longer time (several hours). A solvent-etching temperature (T_d) is selected on the basis of the melting point of the segregated species (T_m):

$$T_d = T_m - \Delta T$$

where ΔT is dependent on the solvent power. For p-xylene (as solvent) and polyethylene, ΔT is 31 K. Etching with degrading compounds may also be useful. To this group belong plasma etching and etching with permanganic acid.

Scanning electron microscopy of fracture surfaces may provide useful information about the supermolecular structure, although artificial structures have been reported. These 'pseudospherulites' arise from fractures initiated at spots in front of the main propagating fracture front. These early fractures propagate in a radial manner outwards from the initiation spots and create a spherulite-like topography which can very easily be mistaken for true spherulites.

Low-magnification transmission electron microscopy of chlorosulphonated polyethylene samples also provides information about the supermolecular structure.

7.6 DEGREE OF CRYSTALLINITY

The crystallinity, mass crystallinity or volume crystallinity is the mass or volume fraction of the sample in the crystalline state. It is assumed that only two components exist in the semicrystalline polymer. This postulate may be expressed in the statement that any intensive property (ϕ) is an additative function with contributions from the two components present:

$$\phi = \phi_c w_c + \phi_a (1 - w_c) \quad (7.32)$$

where ϕ_c and ϕ_a are the phase properties and w_c is the mass crystallinity. This equation can be applied to enthalpy of fusion, specific heat, specific volume or any other intensive property.

The most fundamental and direct method of determining the degree of crystallinity is X-ray diffraction. It is based on the principle that the total coherent scattering from N atoms is independent of the state of aggregation. This statement leads to the most fundamental expression of (mass) crystallinity:

$$w_c = \frac{\int_0^\infty s^2 I_c(s) ds}{\int_0^\infty s^2 I(s) ds} \quad (7.33)$$

where I is the total scattered intensity, I_c is the scattered intensity associated with the crystals (Bragg reflections), and $s = (2 \sin \theta)/\lambda$, where λ is the wavelength. There are three complications. First, incoherent scattering has to be subtracted from the total scattering. Second, it is only possible to measure the scattered intensity (I) over a limited range of s. It is then assumed that coherent scattering appearing outside this range is insignificant. Third, the scattering originating from the crystal planes (I_c) includes diffuse components coming from thermal vibrations and paracrystalline defects. It is difficult to separate the crystalline contributions from the amorphous contributions to the total scattering.

There are several practical techniques for the determination of crystallinity. The peak resolution method is based on the approach of Hermans and Weidinger (1950) which was used on cellulose. A 2θ scan including the intensive crystalline peaks and the amorphous halo is made. The diffractogram of the semicrystalline polymer consists of sharp Bragg reflections of the total intensity I_c and an amorphous halo with a total scattered intensity I_a. The mass crystallinity (w_c) is determined from the following equation:

$$w_c = \frac{I_c}{I_c + K I_a} \quad (7.34)$$

where K is a calibration factor. The calibration factor can be set to unity for comparative purposes but, for an absolute determination, the K factor has to be determined by an absolute method, e.g. the Ruland method or from density measurements.

The differential intensity measurement method is suitable for polymers showing a series of strong reflections in the scan. In these cases, it is difficult to make an assessment of the crystalline and amorphous contributions. An amorphous standard and a 'highly crystalline' standard are prepared and scans are taken for the standard samples as well as for the sample of unknown crystallinity. The total scattered intensity is normalized for all three over the 2θ range measured. The differences in the scattered intensities between the unknown sample and the amorphous sample, $I_u - I_a$, and the crystalline and amorphous samples, $I_c - I_a$, are recorded at different increments of 2θ. The mass crystallinity is calculated from the expression:

$$w_c = \frac{\sum_{2\theta}(I_u - I_a)}{\sum_{2\theta}(I_c - I_a)} \quad (7.35)$$

The Ruland (1961) method provides the most fundamental method for crystallinity determination. It recognizes the fact that it is easier to measure the total intensity within the sharp Bragg peaks than the total crystalline intensity also including the 'diffuse' crystalline peaks. The basic equation analogous to eq. (7.33) formulated by Ruland is:

$$w_c = \left[\frac{\int_{s_0}^{s_1} s^2 I_c \, ds}{\int_{s_0}^{s_1} s^2 I \, ds}\right] \cdot K(s_0, s_1, D, \overline{f^2}) \quad (7.36)$$

where K is a 'correction factor' which is a function of the selected s-range $(s_0 - s_1)$, D is the disorder function and $\overline{f^2}$ is the mean-square atomic scattering factor for the polymer. The K factor can be expressed as:

$$K = \frac{\int_{s_0}^{s_1} s^2 \overline{f^2} \, ds}{\int_{s_0}^{s_1} s^2 \overline{f^2} D^2 \, ds} \quad (7.37)$$

where D depends on the type of disorder present in the crystals. For first-order defects (e.g. thermal vibrations), the following equation holds:

$$D^2 = \exp(-ks^2) \quad (7.38)$$

where k is a constant, whereas for disorder of the second kind (paracrystalline defects), the following equation has been used:

$$D^2 = \frac{2 \exp(-as^2)}{1 + \exp(-as^2)} \quad (7.39)$$

where a is a constant. Ruland used eq. (7.38) to calculate K. It is thus possible to calculate K values for different values of k for different s-ranges.

A plot of $s^2 I$ against s^2 is made and corrections are made for incoherent scattering and air scattering. The crystalline contribution is separated from the amorphous by drawing a smooth line between the bases of the peaks. According to Ruland this procedure defines ordered regions larger than 2–3 nm as crystalline. The rest of the procedure is fitting. The results of measurements in different s-ranges should give the same result (the same crystallinity). The adjustable parameter is k, which should be the same for all the measured s-ranges. For more details, including diagrams, see Ruland (1961).

Crystallinity can also be determined from density data and calorimetric measurements. Accurate measurements of the density (with a precision of about 0.2 kg m^{-3}) are carried out in a density gradient column and the mass crystallinity can be obtained from the equation:

$$\frac{1}{\rho} = \frac{w_c}{\rho_c} + \frac{(1 - w_c)}{\rho_a} \quad (7.40)$$

where ρ, ρ_c and ρ_a are the densities of the sample, and the crystalline and amorphous components, respectively. The density of the crystalline component is determined from X-ray unit cell data and the amorphous density is obtained by extrapolation of dilatometric data of molten polymer to lower temperatures. Crystallinity determinations by density measurements are generally in agreement with estimates from X-ray diffraction.

A calorimetric determination of crystallinity can be made by measurement of the heat of fusion using differential scanning calorimetry (DSC). The conversion from heat of fusion to mass crystallinity may be conducted by the total enthalpy method, which is illustrated in Fig. 7.42. It is assumed that only two components exist, and that each of them has a certain

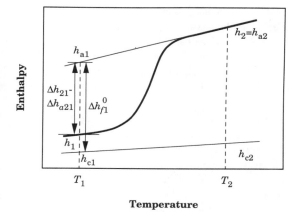

Figure 7.42 The total enthalpy method.

enthalpy, h_a (amorphous) and h_c (crystalline). The enthalpy of the sample at any temperature is given by:

$$h = h_c w_c + h_a (1 - w_c) \quad (7.41)$$

where w_c is the mass crystallinity. Another complication is internal stresses which relax during melting, and this is a process which changes the enthalpy. It is therefore assumed that internal stresses are negligible. The amorphous component is liquid-like, i.e. the amorphous phase enthalpy at $T < T_m$ can be obtained by extrapolation of data from temperatures greater than T_m.

The enthalpy (h_1) at a temperature T_1 well below the melting temperature range is:

$$h_1 = h_{c1} w_{c1} + h_{a1}(1 - w_{c1}) \quad (7.42)$$

where w_{c1} is the crystallinity at T_1, and h_{c1} and h_{a1} are the crystalline and amorphous enthalpies, respectively, at T_1.

At $T_2 (\gg T_m)$, the enthalpy is given by:

$$h_2 = h_{a2} \quad (7.43)$$

The difference in enthalpy of the sample between temperatures T_1 and T_2 is given by:

$$\Delta h_{21} = h_2 - h_1 = [h_{a2} - h_{a1}] + [h_{a1} - h_{c1}]w_{c1}$$
$$= \Delta h a_{21} + \Delta h_{f1}^0 w_{c1} \quad (7.44)$$

i.e.:

$$w_{c1} = \frac{\Delta h_{21} - \Delta h_{a21}}{\Delta h_{f1}^0} \quad (7.45)$$

Degree of crystallinity 159

where $\Delta h_{f1}^0 = h_{a1} - h_{c1}$ is the heat of fusion at T_1. It is important to note that Δh_{f1}^0 is temperature-dependent:

$$\Delta h_{f1}^0 = \Delta h_{T_m^0}^0 - \int_{T_1}^{T_m^0} [c_{pa} - c_{pc}] dT \quad (7.46)$$

where c_{pa} and c_{pc} are respectively the specific heats of the amorphous and crystalline components. Both c_{pa} and c_{pc} are temperature-dependent and can accurately be adapted to third-order polynomial equations.

The energy, which is referred to as $\Delta h_{21} - \Delta h_{a21}$, is obtained by extrapolation of the post-melting scanning base line down to lower temperatures. It is common to find that the extrapolated base line intersects the premelting scanning base line. This temperature is selected as T_1. The area under the curve, starting at T_1 and ending at the well-defined 'end' of the melting peak, then gives the enthalpy $\Delta h_{21} - \Delta h_{a21}$, and Δh_{f1}^0 is calculated according to eq. (7.46).

The total enthalpy method yields crystallinity values in agreement with X-ray and density estimates. Significant deviation was found by Mandelkern (1985) for very high molar mass samples of linear polyethylene. The differences were attributed to the presence of a third, intermediate phase, with mobility in between that of the crystalline and the amorphous components. A problematic case is samples with low thermal stability. Polyvinylchloride is one example. An additional problem with this polymer is also its low crystallinity (5–10%).

The degree of crystallinity depends on the molecular structure (Figs 7.43 and 7.44) and thermal pretreatment (Fig. 7.45). Fully or partially non-crystallizable moieties causes a depression in crystallinity. In both cases, the effect should be proportional to the molar content of the defect (Fig. 7.43). Ethyl branches cause an expansion of the unit cell which is indicative of partial (minor) inclusion of the defects in the crystals. However, at the same time, the crystallinity decreases by 20% per mol% of ethyl groups. It may be concluded that a significant (major) portion of the ethyl groups must be located in the amorphous phase. Larger branches, propyl or longer homologues, are fully non-crystallizable and they

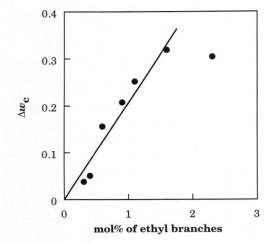

Figure 7.43 Depression in mass crystallinity (Δw_c) as a function of degree of ethyl-group branching. With permission from the Society of Plastics Engineers (Tränkner, Hedenqvist and Gedde (1994)).

cause an even stronger depression in crystallinity. The molar crystallinity depression of smaller groups, carbonyl and methyl groups, is less than that due to ethyl groups.

The crystallinity decreases with increasing molar mass. The very low molar mass polymers crystallize in extended-chain or once- or twice-folded crystals with a very small proportion of amorphous material. Higher molar mass samples crystallize with a larger proportion of statistical chains in the amorphous phase. There are principally two types of statistical chain: statistical loops, i.e. chains returning to the same crystal lamella after a random walk in the amorphous region, and tie chains, i.e. chains which connect adjacent crystal lamellae. Chain entanglements play an important role in the very high molar mass samples. The entanglements effectively depress crystallization of large portions of the molecules, leading to a crystallinity of about 40–50% in linear polyethylene of $\bar{M}_w = 10^6$ g mol^{-1}.

Crystallization at high temperatures generally favours a high final crystallinity (Fig. 7.45). High molar mass polymers are more sensitive to variations in cooling rate/crystallization temperature than low molar mass samples (Fig. 7.45). One of the most dramatic effects is indeed obtained by crystallization of linear polyethylene at very elevated pressures and temperatures. Extended-chain crystals with a thickness of a few micrometres are formed in the pressure region where the hexagonal phase is stable. Such samples approach full crystallinity, i.e. values in the range 95–100% are attained. The chains are very mobile in

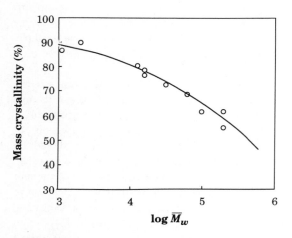

Figure 7.44 Mass crystallinity of linear polyethylene as a function of molar mass. The samples were cooled from the melt at a rate of 10 K min^{-1}. Drawn after data from Tränkner, Hedenqvist and Gedde (1994).

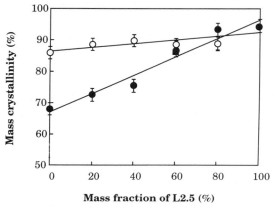

Figure 7.45 Mass crystallinity of binary blends of two linear polyethylenes (L2.5: $\bar{M}_w = 2500$, $\bar{M}_w/\bar{M}_n = 1.15$; L76: $\bar{M}_w = 76\,000$, $\bar{M}_w/\bar{M}_n = 6$) as a function of composition and thermal pretreatment: ○ isothermally crystallized at 396 K; ● cooled at a rate of 400 K min^{-1} from the melt. With permission from Butterworth-Heinemann Ltd, UK (Gustafsson, Conde Braña and Gedde, 1991).

the hexagonal phase and the crystals rapidly approach their equilibrium shape. This is specific for polyethylene and a few other polymers, but high-pressure crystallization in general leads to a high final crystallinity.

The properties of the crystalline and amorphous components in a semicrystalline polymer are different. This statement points to the fact that the degree of crystallinity is a decisive factor for the properties. The stiffness of a semicrystalline polymer increases with increasing crystallinity. More precise data for linear and branched polyethylene are presented in Figs 7.46 and 7.47. At very low temperatures, below the temperature of the onset of the subglass process, the moduli of the two components are very similar and the modulus of the material increases only moderately with increasing crystallinity. At higher temperatures, the phase moduli differ by several orders of magnitude and the modulus of the material is strongly dependent on the crystallinity.

In the case of linear polyethylene, Boyd showed that the data could be adapted to the Tsai–Halpin equation:

$$G = \frac{G_a[G_c + \xi((1 - v_c)G_a + v_cG_c)]}{(1 - v_c)G_c + v_cG_a + \xi G_a} \quad (7.47)$$

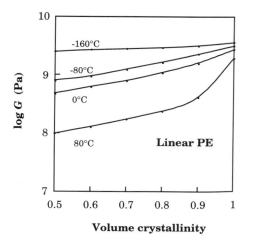

Figure 7.46 Shear modulus (torsion pendulum data at 1 Hz) of linear polyethylenes as a function of crystallinity at different temperatures. From Boyd (1979) who used the data of Illers (1973).

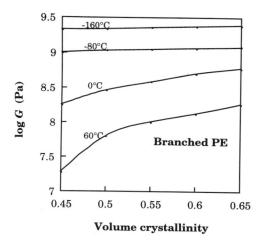

Figure 7.47 Shear modulus (torsion pendulum data at 1 Hz) of branched polyethylenes as a function of crystallinity at different temperatures. From Boyd (1979) who used the data of Illers (1973).

where G is the sample modulus, G_a is the modulus of the amorphous component, G_c is the modulus of the crystalline component, v_c is the volume crystallinity and ξ is the aspect ratio, which for linear polyethylene turned out to take the value 1. The modulus of the amorphous component above its glass transition temperature is several hundred megapascals (found by fitting eq. (7.47) to the data presented in Fig. 7.46), which is significantly higher than the modulus of a typical rubber material. This indicates that the amorphous component is highly constrained by the presence of crystallites.

Branched polyethylene shows a different behaviour (Fig. 7.47). At low temperatures, below the β relaxation (the glass transition of polyethylene), the behaviour is similar to that of linear polyethylene. At higher temperatures, above T_β, the modulus of the amorphous component is crystallinity-dependent.

Transport properties, diffusion coefficient and permeability are also strongly dependent on the degree of crystallinity. The crystalline component is for most polymers impermeable to most small and large molecules. The diffusion coefficient (D) may be described by the following simple equation:

$$D = \frac{D_a(1 - v_c)^n}{B} \quad (7.48)$$

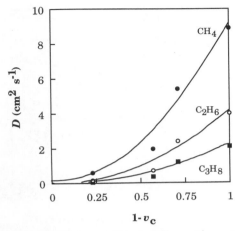

Figure 7.48 Diffusion coefficient of alkanes (shown in the graph) at 25°C as a function of volume fraction of amorphous component in polyethylene. Data from Michaels and Bixler (1961).

Table 7.4 Relaxation processes in semicrystalline polymers

Number of processes	T_{max} (isochronal) $\rightarrow$		
3	γ	β	α
2	β	α_a	

Source: After Boyd (1985a).

where D_a is the diffusion coefficient of the amorphous component, n is an adjustable parameter which takes minimum value 1, and B is the 'blocking factor' that accounts for physical 'crosslinks' between the crystal lamellae. The data presented in Fig. 7.48 are in accordance with eq. (7.48) with n values greater than 1. Reactions such as thermal oxidation, photo-oxidation and radiation-induced crosslinking are confined to the amorphous phase. The rates of these reactions are often found to be proportional to the volume crystallinity.

7.7 RELAXATION PROCESSES IN SEMICRYSTALLINE POLYMERS

Relaxation processes in semicrystalline polymers are complicated. Some of them occur in only one of the components. Others may involve both components. Some of them show a pronounced morphological dependence whereas others show many characteristics with no or only weak dependence on crystallinity and morphology. In isochronal (constant time or constant frequency) experiments between the temperature of liquid nitrogen (−196°C) and the melting point of the polymer, at least two and often three relaxation processes are observed (Table 7.4).

Most of the relaxation processes, the subglass processes (γ or β for polymers with three processes) and the glass transition (β or α_a), can be assigned only to the amorphous component. The molecular interpretation of the subglass processes is reviewed in Chapter 5. The activation energy and the width parameter describing the relaxation time distribution of the subglass process are essentially independent of degree of crystallinity and morphology. This is because the subglass processes are very local, involving only small groups of the main chain or of pendant groups. The swept-out volume associated with the motion is small. The subglass processes are 'short-sighted' and do not sense the structure at a distance of one or two nanometres. The intensity (relaxation strength) of the subglass processes is proportional to the volume fraction of amorphous components. The phase assignment of a relaxation process is preferably achieved by measurement of the relaxation strengths of a series of samples with different crystallinities. Extrapolation of the relaxation strength against crystallinity data to 0% and 100% crystallinity allows the relaxation process to be assigned to the correct phase.

The glass transition, denoted β (or α_a), is almost worth its own chapter. The amorphous component is constrained by the crystals and the glass transition is much broader in a semicrystalline polymer than in a fully amorphous polymer. Polyethylene terephthalate (PETP) can be quenched to a fully amorphous state and it can be crystallized to 60% at elevated pressures. Figure 7.49 shows the dielectric loss process (α_a) and the great difference between two PETP samples of different crystallinity. The amorphous polymer displays a sharper α_a transition than the crystalline. The amorphous polymer shows a clearly asymmetric curve, whereas the semicrystalline polymer shows a symmetric curve.

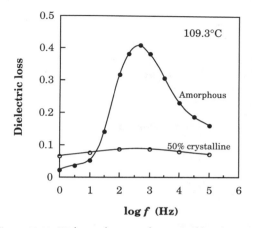

Figure 7.49 Dielectric loss as a function of log frequency at constant temperature, 109.3°C, for two polyethylene terephthalate samples showing the α_a process. Fully amorphous (filled circles) and a sample with 50% crystallinity (open circles). Drawn after Coburn and Boyd (1986).

The glass transition of highly crystalline polymers (linear polyethylene is a prominent example) is very weak and in many cases hardly noticeable. Typical of a highly crystalline polymer is that the small amorphous component consists of relatively short amorphous chain segments which are severely constrained by the crystals. In fact, for linear polyethylene there was a debate as to which of the three relaxation processes, α, β and γ, was the glass transition. The modern view is that the middle process (β), the weakest of the three, is the glass transition. The relaxation strength of the glass transition process is proportional to the volume fraction of amorphous material.

The high-temperature process (α) is present only in polymers showing three relaxation processes. Oxidized polyethylene (one or two C=O groups per thousand CH_2 groups) displays a dielectric α process which appears only in samples with crystallinity. It turns out that the relaxation strength of the dielectric process is proportional to the crystallinity. Another feature of the dielectric α process in polyethylene is that it is extremely sharp. It obeys almost the characteristics of a single time relaxation process. Its temperature dependence obeys the Arrhenius equation. The activation energy increases with increasing lamella thickness up to a certain thickness level above which it remains almost constant.

The mechanical process requires the presence of both a crystalline and an amorphous component. It does not appear in samples above the melting point. Mats of single crystals with essentially no amorphous phase showed no mechanical α process. Figures 7.50 and 7.51 show the mechanisms of the dielectric and mechanical α processes. They are closely related but not identical.

The dielectric process is solely crystalline. It involves a 180° rotation of the dipole and a $c/2$ translation of the chain along its own axis to keep the chain in register with surrounding chains. The 180° twist of the chain is accomplished through a smooth twist which propagates from one side of the crystal to the other. The motion of the twist from one site to the next ($c/2$ translation) involves the passage of an energy barrier of 17 kJ mol^{-1}. The creation energy of the smooth twist with 12 main-chain bonds was calculated at 54 kJ mol^{-1}. There is no strain involved in either of the two states. They are the same in that sense and the dielectric α process lacks mechanical activity. However, the deformation of the amorphous interlayer is accomplished by a mechanically driven twisting/translation of the chain. Thus, this process certainly requires the presence of both the amorphous and the crystalline components.

What are the molecular criteria for the presence of an α process? Polymers with small or no pendant groups, with short repeating units and with weak intermolecular forces are the polymers most likely to have an α process. Examples of polymers belonging to this category are polyethylene, isotactic polypropylene, polyoxymethylene and polyethylene oxide. The pendant phenyl groups of isotactic polystyrene are too large and this polymer does not

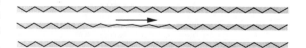

Figure 7.50 Dielectric α relaxation process in linear polyethylene. The smooth twist consisting of 12 carbon atoms out of register propagates through the crystal in the direction of the arrow. Drawn after Boyd (1985b).

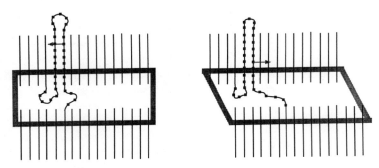

Figure 7.51 Mechanical α relaxation process in linear polyethylene. Drawn after Boyd (1985b).

show an α process. Polyethylene terephthalate has a long repeating unit and cannot form a short twisted structure that shortens the chain by one repeating unit. The energy to form such a defect would be enormous, and it is believed that this is the reason for the absence of an α process in this polymer. Polyamide 6 has, in addition to the long repeating unit, an additional constraint resulting from the strong hydrogen bonds forming layered structures in the crystal. The translation of the chain through the crystal would have to break hydrogen bonds and the energy involved in this motion would be great.

7.8 SUMMARY

The chain structure of semicrystalline polymers is regular. Small proportions of chain defects (local irregularities) lead to a reduced overall degree of crystallinity and a reduction in crystal thickness. The term 'semicrystalline' indicates that no polymer is completely crystalline. A second, disordered, amorphous component is always present. The degree of crystallinity in a given sample can be obtained by X-ray diffraction, density measurements and differential scanning calorimetry. Properties at temperatures between the melting point and the glass transition temperature are particularly sensitive to changes in crystallinity.

A central postulate of polymer crystallography is that the conformation of the polymer chains within the crystals is that of the lowest possible energy. Different crystal structures (so-called polymorphs) of a given compound may arise due to different types of packing of the 'low-energy chains'. The chain axis is selected as the c axis (third coordinate), except in the case of a monoclinic cell in which the chain axis is the unique axis, i.e. the axis which is perpendicular to the other two crystallographic axes. In that case, the chain axis is denoted b. Polymer crystals are always very strongly anisotropic due to the presence of two types of binding forces: strong covalent bonds (along c) and considerably weaker secondary bonds (along a and b).

The morphology of melt-crystallized semicrystalline polymers can be described by the following structural hierarchy: (a) crystal lamellae of folded chains; (b) stacks of almost parallel crystal lamellae, each crystal separated by a thin, amorphous interlayer; (c) supermolecular structures which are polycrystalline structures, e.g. spherulites, axialites, shish-kebabs or trans-crystalline structures. Chain folding was discovered by Keller (1957) and is the dominant feature of solution-grown crystals and also of melt-grown crystals. Single crystals of linear polyethylene grown from dilute solutions at temperatures between 70 and 90°C in xylene or similar solvents are faceted and also tent-like. Distinct growth sectors are present. These observations indicated that chain folding is predominantly regular in solution-grown crystals. Other crystalline polymers show a similar behaviour. Small-angle neutron scattering indicates that crystallization of polymers from dilute solution leads not only to ordering on a very local scale (chain folding) but also to a decrease in the global dimensions of the molecules. Melt-crystallization constitutes a completely different case. The chaotic state of the molecules in the molten state prior to crystallization is to some extent preserved in the semicrystalline state also after crystallization. The mechanism for splaying of the dominant lamellae which must exist in a spherulitic sample may be different in different polymers. Bassett and co-workers found in linear polyethylene splaying of crystal lamellae originating from screw dislocations, and they suggested this as an important step in the formation of spherulites in this polymer.

The thickness of the lamellar crystals (L_c) is small, typically about 10 nm, which leads to a considerable melting-point depression below the equilibrium (infinite crystal thickness) value (T_m^0). The melting point (T_m) is conveniently described by the Thompson–Gibbs equation:

$$T_m = T_m^0 \left[1 - \frac{2\sigma}{\Delta h^0 \rho_c L_c} \right] \quad (7.49)$$

where σ is the specific fold surface free energy, Δh^0 is the heat of fusion and ρ_c is the density of the crystalline component. The thickness of the 'virgin' crystals (L_c^*) is controlled solely by the degree of undercooling ($\Delta T = T_m^0 - T_c$; T_c is the crystallization temperature):

$$L_c^* = \frac{C}{\Delta T} + \delta L_c \quad (7.50)$$

where C and δL_c are constants. The virgin crystals have a thickness-to-width ratio of 0.01 to 0.001, which is many orders of magnitude smaller than the equilibrium value of about 10. Polymer crystals become gradually thicker when given enough 'thermal stimulation'. The rate of this process is controlled by the absolute temperature and is independent of ΔT.

Two or three relaxation processes occur in semicrystalline polymers. The low-temperature (γ or β) process is a subglass process occurring in the amorphous phase. The medium or high temperature process (β or α_a) is associated with the glass–rubber transition of the amorphous component. The glass transition is very weak, and in many cases difficult to find, in highly crystalline polymers like linear polyethylene. A certain class of polymers shows a high-temperature relaxation process denoted α, which is a combined crystalline and amorphous process. Reorientation of the chain by a 180° twist of the molecule in the crystals and a certain axial translation of the chain are the basis of the dielectric α process, whereas the mechanical process involves shearing of the amorphous interlayer which is facilitated by chain slippage (the 'dielectric process'). The α process is absent in polymers with bulky side groups and/or strong intermolecular interaction.

7.9 EXERCISES

7.1. The mass crystallinity of a single crystal is 85%. Carry out a simplified calculation of the degree of tight, regular folding, assuming that the amorphous chains are performing a random walk in the amorphous, fold-surface component.

7.2. Calculate the melting points of linear polyethylene crystals having thickness 100 nm, 50 nm, 10 nm and 5 nm. Use the following data: $\Delta h^0 = 293$ J g^{-1}, $\rho_c = 1000$ kg m^{-3}, $T_m^0 = 415$ K, and $\sigma = 90$ mJ m^{-2}, and assume that the Thompson–Gibbs equation is valid.

7.3. Derive an expression for f_{fold}, considering also the finite length of a polymer molecule. Assume also that the chain ends are all located in the amorphous phase. Then draw a plot of $f_{fold} = f(M)$ using the following values: $\rho_a = 0.85\rho_c$; $L_a = 5$ nm; $l_{cc} = 0.127$ nm; $C = 6.85$; $\theta = 30°$. Plot $f_{fold} = f(\theta)$ using the following values: $\rho_a = 0.85\rho_c$; $L_a = 5$ nm; $l_{cc} = 0.127$ nm; $C = 6.85$; $M = \infty$. Plot $f_{fold} = f(L_a)$ using the following values: $\rho_a = 0.85\rho_c$; $\theta = 30°$; $l_{cc} = 0.127$ nm; $C = 6.85$; $M = \infty$.

7.4. Segregation of low molar mass species leads to a finely dispersed system of phase-separated domains of high and low molar mass species. Explain why.

7.5. Negative and positive spherulites are readily distinguished by polarized light microscopy. How? Also explain the differences in morphology between the two.

7.6. Explain the optical origin of banded spherulites. Write a short summary of the current ideas describing the underlying morphological structure.

7.7. Large defects cannot be accommodated in the crystalline phase. Calculate (a) the statistical distribution of chain defects; (b) the melting-point distribution in polyethylene with 1 mol% of chain defects. Use the following data: $\Delta h^0 = 293$ J g^{-1}, $\rho_c = 1000$ kg m^{-3}, $T_m^0 = 415$ K, and $\sigma = 90$ mJ m^{-2}.

7.8. What is the chain orientation around a nucleating fibre?

7.9. Figure 7.45 shows that the diffusion coefficient is **not** a linear function of crystallinity. Suggest a possible explanation.

7.10. Isotactic polypropylene shows polymorphism. X-ray diffraction shows the presence of α and γ crystals in 50/50 proportions in poly(i-propylene-stat-ethylene) (6.6 mol%) crystallized at 393 K. The measured enthalpy change associated with melting was $\Delta h_{21} - \Delta h_{a21} = 44 \text{ J g}^{-1}$ and the onset of occurred at 395 K. The heat of fusion at the equilibrium melting point (460.7 K) is 206 J g^{-1} for the α phase and 165 J g^{-1} for the γ phase. The specific heats of the crystalline and amorphous components were given by Gaur and Wunderlich (1981) as follows:

Temperature (K)	c_{pc} (J mol^{-1} K^{-1})	c_{pa} (J mol^{-1} K^{-1})
150	37.68	39.53
200	47.47	50.53
250	56.53	61.95
300	70.39	84.11
350	82.12	109.2

Calculate the fractions of α and γ crystalline material in the sample.

7.11. Calculate the SAXS long period and the average crystal thickness from the SAXS data shown in Fig. 7.52. Use the following data: mass crystallinity $(w_c) = 0.60$; $\rho_c = 1000 \text{ kg m}^{-3}$; $\rho_a = 855 \text{ kg m}^{-3}$, wavelength $(\lambda) = 0.15$ nm.

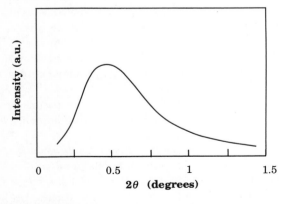

Figure 7.52 SAXS data for a medium-density polyethylene sample. Intensity is plotted on an arbitrary units scale.

7.10 REFERENCES

Barham, P. J., Chivers, R. A., Keller, A., Martinez-Salazar, J. and Organ, S. J. (1985) *J. Mater. Sci.* **20**, 1625.

Bassett, D. C. (1981) *Principles of Polymer Morphology*. Cambridge University Press, Cambridge.

Bassett, D. C. and Hodge, A. M. (1978) *Proc. Roy. Soc. A* **359**, 121.

Bassett, D. C., Frank, F. C. and Keller, A. (1963) *Phil. Mag.* **8**, 1739.

Bassett, D. C., Block, S. and Piermarini, G. J. (1974) *J. Appl. Phys.* **45**, 4146.

Bassett, D. C., Olley, R. H. and Al Reheil, I. A. M. (1988) *Polymer* **29**, 1539.

Boyd, R. H. (1979) *Polym. Eng. Sci.* **19**, 1010.

Boyd, R. H. (1985a) *Polymer* **26**, 323.

Boyd, R. H. (1985b) *Polymer* **26**, 1123.

Bravais, A. (1850) *J. Ecole Polytech.* **19**, 1–128.

Brückner, S. and Meille, S. V. (1989) *Nature* **340**, 455.

Coburn, J. and Boyd, R. H. (1986) *Macromolecules* **19**, 2238.

Davis, G. T., Eby, R. K. and Colson, J. P. (1970) *J. Appl. Phys.* **41**, 4316.

Fischer, E. W. (1957) *Z. Naturf.* **12a**, 753.

Flory, P. J. (1962) *J. Am. Chem. Sci.* **84**, 2857.

Gaur, U. and Wunderlich, B. (1981) *J. Phys. Chem., Ref. Data* **10**, 1051.

Geil, P. H., Anderson, F. R., Wunderlich, B. and Arakawa, T. (1964) *J. Polym. Sci. A* **2**, 3707.

Gustafsson, A., Conde Braña, M. T. and Gedde, U. W. (1991) *Polymer* **32**, 426.

Henry, N. F. M. and Lonsdale, K. (1959) *International Tables for X-ray Crystallography*. Vol. I. Kynoch Press, Birmingham.

Henry, N. F. M. and Lonsdale, K. (1959) *International Tables for X-ray Crystallography*. Vol. II. Kynoch Press, Birmingham.

Henry, N. F. M. and Lonsdale, K. (1962) *International Tables for X-ray Crystallography*. Vol. III. Kynoch Press, Birmingham.

Hermans, P. H. and Weidinger, A. (1950) *J. Polym. Sci.* **5**, 565.

Illers, H. H. (1973) *Kolloid Z. Z. Polym.* **251**, 394.

Kawai, T. and Keller, A. (1985) *Phil. Mag.* **11**, 1165.

Keith, H. D. and Padden, F. J. Jr (1984) *Polymer* **25**, 28.

Keller, A. (1957) *Phil. Mag.* **2**, 1171.

Keller, A. (1991) 'Chain-folded crystallisation of polymers from discovery to present day: a personalised journey' in *Sir Charles Frank FRS, OBE, An Eightieth Birthday Tribute*, Chambers, R. G. (ed.). Adam Hilger, Bristol.

Keller, A. and Priest, D. J. (1968) *J. Macromol. Sci.* **B2**, 479.

Keller, A. and Sawada, F. (1964) *Macromol. Chem.* **74**, 190.

Lauritzen, J. I. Jr and Hoffman, J. D. (1960) *J. Res. Nat. Bur. Std.* **64a**, 73.

Lustiger, A., Lotz, B. and Duff, T. S. (1989) *J. Polym. Sci., Polym. Phys. Ed.* **27**, 561.

Mandelkern, L. (1985) *Polym. J.* **17**, 337.

Martinez-Salazar, F. J. and Baltá-Calleja, F. J. (1979) *J. Crystal Growth* **48**, 282.

Michaels, A. S. and Bixler, H. J. (1961) *J. Polym. Sci.* **50**, 413.

Natta, G. and Corradini, P. (1960) *Nuovo Cimento Suppl.* **15**, 40.

Organ, S. J. and Keller, A. (1985) *J. Mater. Sci.* **20**, 1571.

Preedy, J. E. (1973) *Br. Polym. J.* **5**, 13.

Ruland, W. (1961) *Acta Crystallogr.* **14**, 1180.

Spegt, P. (1970) *Makromol. Chem.* **139**, 139.

Spells, S. J., Keller, A. and Sadler, D. M. (1984) *Polymer* **25**, 749.

Spruiell, J. E. and Clark, E. S. (1980) Unit cell and crystallinity, in *Methods of Experimental Physics*, Vol. 16, Part B (R. A. Fava, ed.). Academic Press, New York.

Stack, G. M., Mandelkern, L. and Voigt-Martin, I. G. (1984) *Macromolecules* **17**, 321.

Storks, K. H. (1938) *J. Amer. Chem. Soc.* **60**, 1753.

Till, P. H. (1957) *J. Polym. Sci.* **24**, 30.

Tränkner, T., Hendenqvist, M. and Gedde, U. W. (1994) *Polym. Eng. Sci.* **34**, 1581.

Ungar, G., Stejny, J., Keller, A., Bidd, I. and Whiting, M. C. (1985) *Science* **229**, 386.

Van Vlack, L. H. (1975) *Elements of Materials Science and Engineering* (3rd edn). Addison-Wesley, Reading, MA.

Woodward, A. E. (1989) *Atlas of Polymer Morphology*. Hanser, Munich, Vienna and New York.

Wunderlich, B. (1973) *Macromolecular Physics*, Vol 1: *Crystal Structure, Morphology, Defects*. Academic Press, New York and London.

7.11 SUGGESTED FURTHER READING

Baltá-Calleja, F. J. and Vonk, C. G. (1988) *X-ray Scattering of Synthetic Polymers*. Polymer Science Library 8. Elsevier, Amsterdam.

Fava, R. A. (ed.) (1980) *Methods of Experimental Physics*, Vol. 16, Part B: *Polymers: Crystal Structure and Morphology*. Academic Press, New York.

CRYSTALLIZATION KINETICS 8

8.1 BACKGROUND

The crystallization of flexible-chain polymers is a fascinating branch of polymer physics. The central paradigm that is a part of all descriptions and theories of polymer crystallization is chain folding, discovered by Keller (1957) (see Chapter 7). Polymer crystallization theories have arisen largely from theories developed earlier for small-molecule crystallization. The principle of chain folding has been implemented in these theories. Some of the growth theories for polymer crystallization, e.g. that proposed by Lauritzen and Hoffman (1960), have been in use for over 30 years. Important phenomena such as crystal thickening and molecular fractionation are essentially untreated by currently existing growth theories.

This chapter presents first some fundamental aspects of nucleation, and second the general Avrami equation, which is frequently used to describe overall crystallization. The growth theories of Lauritzen and Hoffman and Sadler and Gilmer are discussed in sections 8.4.2 and 8.4.3. Molecular fractionation and orientation-induced crystallization are dealt with in sections 8.5 and 8.6.

Crystallization involves both diffusion of the crystallizable units to the crystal front and nucleation. Short-range diffusion occurs more rapidly at higher temperatures. All diffusive motions are completely 'frozen in' at temperatures below the glass transition temperature. When the diffusing molecule reaches the crystal boundary, it has to form a stable nucleus. The conditions for stability are described by nucleation theory. The increase in free energy is due to the positive contribution from the surface energies ($\sigma_i A_i$, where σ_i is the specific surface energy of surface i) which, at temperatures below the equilibrium melting point, opposes the negative contribution from the crystallization free energy ($\Delta g V_{\text{crystal}}$, where Δg is the specific change in free energy and V_{crystal} is the volume of the nucleus):

$$\Delta G = \Delta g V_{\text{crystal}} + \sum_i A_i \sigma_i \qquad (8.1)$$

Let us, for the purpose of demonstration, select a particularly simple case: a spherical crystal. The change in free energy on crystallization (ΔG) is then given by:

$$\Delta G = \frac{4\pi r^3}{3} \cdot \Delta g + 4\pi r^2 \sigma \qquad (8.2)$$

where r is the radius of the spherical crystal and σ is specific free energy of the surface. The radius of the sphere (r^*) associated with the free energy barrier is obtained by setting the derivative of ΔG with respect to r equal to zero:

$$\frac{\partial \Delta G}{\partial r} = 4\pi r^{*2} \Delta g + 8\pi r^* \sigma = 0$$

$$r^* = -\frac{2\sigma}{\Delta g} \qquad (8.3)$$

The temperature dependence of this equation lies in Δg:

$$\Delta g = \frac{\Delta h^0 \Delta T}{T_m^0} \qquad (8.4)$$

where Δh^0 is the heat of fusion (per unit volume), T_m^0 is the equilibrium melting point, $\Delta T = T_m^0 - T_c$ is the degree of supercooling and T_c is the crystallization temperature. Equation (8.4) is valid provided that Δh^0 and the entropy of fusion (Δs^0) are temperature-independent, which is a good approximation in a limited temperature range near the equilibrium melting temperature. Insertion of eq. (8.4) into eq. (8.3) yields:

$$r^* = -\frac{2\sigma T_m^0}{\Delta h^0 \Delta T} \qquad (8.5)$$

170 Crystallization kinetics

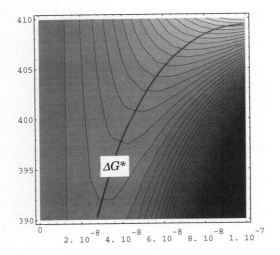

Figure 8.1 Free energy as a function of size of nucleus and crystallization temperature; $\Delta h^0 = 300 \times 10^6$ J m^{-3}; $\sigma = 0.3$ J m^{-2}; $T_m^0 = 418$ K. Higher value of free energy is indicated by the lighter colour.

It should be noted that Δh^0 is negative and that the radius of the critical nucleus increases with decreasing degree of supercooling. By inserting eq. (8.5) into eq. (8.2), the free energy barrier (ΔG^*) can be derived:

$$\Delta G^* = \frac{4\pi r^{*3}}{3} \cdot \Delta g + 4\pi r^{*2}\sigma$$

$$= \frac{4\pi(-2\sigma T_m^0)^3}{3(\Delta h^0 \Delta T)^3} \cdot \frac{\Delta h^0 \Delta T}{T_m^0} + \frac{4\pi(-2\sigma T_m^0)^2 \sigma}{(\Delta h^0 \Delta T)^2}$$

(8.6)

which can be simplified to:

$$\Delta G^* = \frac{4\pi\sigma^3(T_m^0)^2}{(\Delta h^0)^2 \Delta T^2}\left[-\frac{8}{3} + 4\right] = \frac{16\pi\sigma^3(T_m^0)^2}{3(\Delta h^0)^2 \Delta T^2} \quad (8.7)$$

Figure 8.1 shows ΔG schematically as a function of the size of the nucleus at different temperatures.

Nucleation occurs more readily at lower crystallization temperatures because of the lower critical nucleus size and the lower free energy barrier associated with the process (Fig. 8.1). Different types of nucleation are possible (Fig. 8.2). Primary nucleation involves the formation of six new surfaces, whereas the secondary and tertiary nucleation involve fewer, four and two, respectively. The free energy barrier is highest for primary nucleation and lower for secondary and tertiary nucleation. Homogeneous, primary nucleation occurs very seldom. Both calculations and experimental data show that 50–100 K of supercooling is needed to achieve true homogeneous nucleation. Instead, crystallization is in all 'practical' cases initiated at foreign particles, i.e. nucleation is heterogeneous.

Crystal growth occurs by secondary and tertiary nucleation (Fig. 8.2). The initial step is the formation of a secondary nucleus, which is followed by a series of tertiary nucleation events.

From a consideration of both diffusive transport and nucleation, the following general temperature (T_c) dependence is obtained for the overall crystallization rate ($\dot{w}_c$):

$$\dot{w}_c = C \exp\left(-\frac{U^*}{R(T_c - T_\infty)}\right) \exp\left(-\frac{K_g}{T_c(T_m^0 - T_c)}\right) \quad (8.8)$$

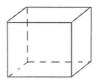

Primary nucleation **Secondary nucleation** **Tertiary nucleation**
    (*n*=6)   (*n*=4)   (*n*=2)

Crystal growth (view along the chains)

Figure 8.2 Main types of nucleation; *n* is the net number of newly formed surfaces.

where C is a rate constant, U^* is an (energy) constant, R is the gas constant, T_∞ is a temperature at which all segmental mobility is frozen in, and K_g is a kinetic constant for the secondary nucleation. The first factor (C) depends on the segmental flexibility and the regularity of the polymer: $C = 0$ for an atactic polymer, and C is low for very inflexible polymers such as isotactic polystyrene. The second factor expresses the temperature dependence of the rate of the short-range transport of the crystallizing segments according to the William–Landel–Ferry (WLF) equation (see Chapter 5). It is zero at $T_c = T_\infty$. The second factor may be expressed differently using an exponential expression for self-diffusion by reptation (see Chapter 6). The third factor expresses the temperature dependence of the nucleation rate. It is zero at $\Delta T = 0$, i.e. at $T_c = T_m^0$. Figure 8.3 shows the overall crystallization rate as a function of temperature according to eq. (8.8). The bell-shaped crystallization rate curve is general for all types of materials, including metals, inorganic compounds and polymers.

8.2 THE EQUILIBRIUM MELTING TEMPERATURE

8.2.1 UNDILUTED SYSTEMS

The equilibrium melting temperature (T_m^0) is central to most crystallization theories. The degree of supercooling (ΔT) is defined with reference to T_m^0 according to the following equation:

$$\Delta T = T_m^0 - T_c \qquad (8.9)$$

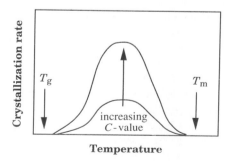

Figure 8.3 General curve of crystallization rate as a function of crystallization temperature.

where T_c is the crystallization temperature. The equilibrium melting temperature refers normally to a crystal of infinite thickness. Homopolymers of intermediate or high molar mass can grow crystals of practically infinite thickness. Let us take an illustrative example: the extended-chain length of a polyethylene of $M = 100\,000$ g mol^{-1} is $100\,000/14 \times 0.127$ nm ≈ 900 nm. The melting-point depression arising from the finite crystal thickness (900 nm) is in this case negligible, as is realized after insertion of the appropriate data into the Thompson–Gibbs equation (see Chapter 7). Low molar mass homopolymers can obviously grow crystals of only limited thickness. The lowest free energy state is the crystal of maximum thickness, i.e. the extended-chain crystal. The equilibrium melting point of the low molar mass polymers is thus the melting point of the extended-chain crystals. Copolymers constitute an even more complicated case. The 'foreign' moieties with a 'statistical' placement in the polymer chains may be fully non-crystallizable or they may be included to some extent in the crystals. The equilibrium crystal thickness is also in this case limited. The equilibrium melting temperature can be determined by a number of different methods. The methods presented are, strictly speaking, applicable only to polymers of intermediate to high molar mass:

1. The melting point (T_m) of samples with a well-defined crystal thickness (L_c) can be measured and the data extrapolated to $L_c^{-1} = 0$ using the Thompson–Gibbs equation:

$$T_m = T_m^0 \left[1 - \frac{2\sigma}{\Delta h^0 \rho_c L_c} \right] \qquad (8.10)$$

where ρ_c is the crystal density, Δh^0 is the heat of fusion (per unit mass) and σ is the specific fold surface free energy. The Thompson–Gibbs equation is derived in Chapter 7. A few polymers, polyethylene among them, can be crystallized at elevated pressures to form extended-chain crystals. These micrometre-thick crystals melt at temperatures very close indeed to T_m^0. Superheating effects are dominant in extended-chain crystals, whereas crystal thickening is the main feature of crystals of normal thickness (see Chapters 7 and 10).

2. Data for the enthalpy (Δh) and entropy (Δs) of fusion can be obtained for relatively small molecules (oligomers) and the equilibrium melting point is obtained by extrapolation of Δh and Δs data as a function of degree of polymerization (x) to infinite x:

$$T_m^0 = \lim_{x \to \infty} \frac{\Delta h(x)}{\Delta s(x)} \quad (8.11)$$

The Broadhurst equation (applicable to linear polyethylene) is probably the best-known equation of this kind:

$$T_m = 414.3 \left[\frac{x - 1.5}{x + 5.0} \right] \quad (8.12)$$

where x is the number of repeating units of the polymer.

3. The diagram shown in Fig. 8.4 is often referred to as a Hoffman–Weeks plot. It demonstrates that the curve of T_m versus T_c for virgin crystals of a certain δL_c intersects the graph of $T_m = T_c$ at the equilibrium melting temperature. Curve b (Fig. 8.4) shows the melting point of the initially formed crystals with $\delta L_c = 3$ nm ($L_c^* = 2\sigma/\Delta g + \delta L_c$, where L_c^* is the initial crystal thickness, and $2\sigma/\Delta g$ is the minimum crystal thickness possible at the particular crystallization temperature). More mature crystals with thickening ratios $\beta = L_c/L_c^* > 1$ show rectilinear plots with slopes equal to $1/\beta$; L_c is the thickness of the mature crystal.

The melting point T_m of a crystal which has thickened by a factor β is given by the Thompson–Gibbs equation:

$$T_m = T_m^0 \left[1 - \frac{2\sigma}{\Delta h^0 \rho_c L_c^* \beta} \right] \quad (8.13)$$

The melting point of the virgin crystal is equal to the crystallization temperature under the simplistic assumption that $\delta L_c = 0$:

$$T_c = T_m^0 \left[1 - \frac{2\sigma}{\Delta h^0 \rho_c L_c^*} \right] \quad (8.14)$$

Equation (8.14) can be rearranged to give:

$$L_c^* = \frac{2\sigma T_m^0}{\Delta h^0 \rho_c (T_m^0 - T_c)} \quad (8.15)$$

and eq. (8.13), after insertion of eq. (8.15), gives:

$$T_m = T_m^0 \left[1 - \frac{2\sigma}{\Delta h^0 \rho_c \beta \frac{2\sigma T_m^0}{\Delta h^0 \rho_c (T_m^0 - T_c)}} \right]$$

$$= \frac{T_c}{\beta} + T_m^0 \left[1 - \frac{1}{\beta} \right] \quad (8.16)$$

The extrapolation of T_m–T_c data to $T_m = T_c = T_m^0$ is, strictly speaking, only valid in the case of data from samples with crystals of a constant β value.

Table 8.1 presents data on the equilibrium melting point of a few selected polymers. The high melting point of PA 6,6 is due to its high enthalpy of fusion, which is due to strong secondary bonds (hydrogen bonds between amide groups), whereas the high melting point of PTFE is due to its low entropy of fusion. At a high temperature, PTFE crystals show considerable segmental mobility which leads to a relatively small increase in entropy on melting. The high melting point of POM is due to the high enthalpy of fusion of this polymer. Part of this originates from intermolecular interactions involving the oxide groups and part from intramolecular sources. PEO has a much

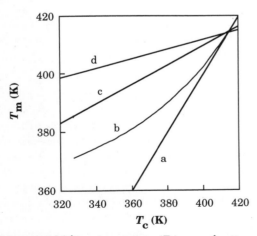

Figure 8.4 Melting temperature (T_m) as a function of crystallization temperature (T_c) for polyethylene: (a) $T_m = T_c$; (b) crystals with $\delta L_c = 3$ nm; (c) crystals with a thickening ratio of 3 ($\delta L_c = 0$); (d) crystals with a thickening ratio of 6 ($\delta L_c = 0$).

Table 8.1 Equilibrium melting-point data for a few selected polymers[a]

Polymer	T_m^0 (K)	Δs^0 (J Kmol)$^{-1}$ [a]	Δh^0 (kJ mol^{-1})[b]	CED (kJ mol^{-1})[c]
PE	414.6	9.91 (1)	4.11 (1)	4.18
PTFE	600	5.69 (1)	3.42 (1)	3.35
iPP	460.7	7.55 (2)	2.31 (3)	4.74
POM	457	10.70 (2)	4.98 (2)	5.23
PEO	342	8.43 (3)	2.89 (3)	4.88
PA 6,6	553	10.2 (12)	4.85 (14)	11.7

Source: Wunderlich (1980, pp. 72–73).
[a] Per mole of flexible chain units; the number of flexible units per repeating unit is given in parentheses.
[b] Per mole of interacting groups: the number of interacting groups per repeating unit is given in parentheses.
[c] Cohesive energy density per mole of interacting groups.

lower melting point than POM primarily due to the low enthalpy of fusion. The fundamental thermodynamical relationship between the quantities presented in Table 8.1 is as follows:

$$T_m^0 = \frac{\Delta h^0 \cdot x}{\Delta s^0 \cdot y} \quad (8.17)$$

where x is the number of interacting groups per repeating unit, and y is the number of flexible units per repeating unit.

The equilibrium melting-point data refer to samples of infinite molar mass. Samples of intermediate or high molar mass show only a negligible melting-point depression with respect to that of a polymer of infinite molar mass. The equilibrium melting point of a low molar mass polymer is, however, significantly depressed.

The most stable crystal of a polymer with x repeating units is the one which has no chain folds. All the crystalline chains of the equilibrium crystal should be fully extended. The melting point of this crystal is denoted $T_m(0, x)$. The zero indicates that there are no folds at the crystal surface. The change in free energy at the equilibrium melting temperature is zero:

$$\Delta h[T_m(0, x)] - T_m(0, x) \cdot \Delta s[T_m(0, x)] = 0 \quad (8.18)$$

It is, according to Hoffman et al. (1975), reasonable to assume that Δh and Δs are constant near the melting point, and this reasoning leads to the following expression for the crystal thickness (L_c):

$$L_c > \frac{2\sigma' T_m(0, x)}{\Delta h(T_m(0, x)) \cdot [T_m(0, x) - T_c]} \quad (8.19)$$

It is argued by Armitstead and Goldbeck-Wood (1992) that Δh and Δs are not true bulk properties, but that they instead both include effects of the chain ends. They therefore vary with crystal thickness. The surface free energy factor (σ') can be regarded as the contribution to the surface tension originating from only the folds. The other part of the fold surface free energy is included in $\Delta h(T_m(0, x))$ and $\Delta s(T_m(0, x))$.

A different approach, proposed by Buckley and Kovacs (1976), that expresses the chain-end effects separately leads to the expression:

$$\Delta g = \Delta h^0 - T\Delta s^0 - kT \cdot \frac{\ln(Cx)}{xv} \quad (8.20)$$

where Δh^0 and Δs^0 are the quantities referring to infinite molar mass (i.e. true bulk properties). C is a constant related to the segmental flexibility and v is the volume of the repeating unit. The third term in eq. (8.20) comes from the entropy of localization due to pairing of chain ends. The following expression, analogous to the Thompson–Gibbs equation, but including the chain-end effect, is derived from eq. (8.20):

$$T_m = T_m^0 \left[1 - \frac{2\sigma}{\Delta h^0 L_c} - \frac{RT_m^0 \ln x}{x \Delta h^0} \right] \quad (8.21)$$

8.2.2 DILUTE SYSTEMS

The equilibrium melting point, i.e. the melting point of an infinitely thick crystal, is lowered in the presence of a low molar mass diluent. The equilibrium melting point of a crystalline polymer of uniform chain length (degree of polymerization = x) can be derived from the Flory–Huggins equation (see Chapter 4) yielding the following expression:

$$\frac{1}{T_m} - \frac{1}{T_m^0} = \left(\frac{R}{\Delta h_u}\right)\left(\left(\frac{V_u}{V_1}\right)v_1 \\ + \frac{1}{x}\left(v_2 + \frac{1}{x - \zeta_0 + 1}\right) \\ - \left(\frac{V_u}{V_1}\right)\chi_1 v_1^2\right) \quad (8.22)$$

where T_m^0 is the equilibrium melting point of the pure polymer, T_m is the equilibrium melting point of the polymer in the presence of the diluent (component 1), Δh_u is the heat of fusion per mole of repeating units, V_u is the molar volume of the repeating unit of the polymer, V_1 is the molar volume of the diluent, v_1 and v_2 are the volume fractions of diluent and polymer, ζ_0 is the number of repeating units along the equilibrium crystal thickness and χ_{12} is the interaction parameter. Figure 8.5 shows schematically the depression of the melting point caused by the diluent.

If we consider only polymers with very large x, the second term in eq. (8.22) can be neglected and the following equation holds:

$$\frac{1}{T_m} - \frac{1}{T_m^0} = \left(\frac{R}{\Delta h_u}\right)\left(\frac{V_u}{V_1}\right)(v_1 - \chi_1 v_1^2) \quad (8.23)$$

The equilibrium melting point is indeed very difficult to determine. When crystallized at elevated pressures (500 K and 435 MPa), polyethylene forms so-called extended-chain crystals. These extraordinarily thick crystals melt at temperatures very near the equilibrium melting point. Cormier and Wunderlich (1966) determined the dissolution temperatures for such samples in a variety of solvents and, by fitting eq. (8.23) to the experimental data, the interaction parameter values listed in Table 8.2 were obtained.

It was shown in the same report that the effect of the finite size of 'ordinary' crystals on their melting points ($T_m(L)$; dilute solutions) could be described by the Thompson–Gibbs equation:

$$T_m(L) = T_m^0 \text{(solvent)} \left(1 - \frac{2\sigma}{\Delta h^0 L \rho_c}\right) \quad (8.24)$$

where T_m^0 (solvent) is the equilibrium dissolution temperature in the solvent concerned and the other quantities are as explained previously. Cormier and Wunderlich showed that σ remained constant at 83 ± 10 mJ m^{-2} for different diluents. Figure 8.6 shows schematically the effect of diluent and crystal thickness on the crystal melting point.

The dissolution temperature (T_d) of a given crystal in a given solvent (dilute solution) is, according to eq. (8.25), proportional to the melting point (T_m):

$$T_d = \left(\frac{T_d^0}{T_m^0}\right) T_m \quad (8.25)$$

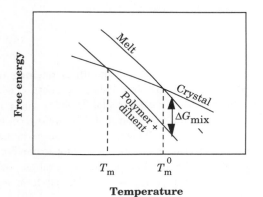

Figure 8.5 The effect of free energy lowering by dilution (ΔG_{mix}) on the melting point.

Table 8.2 Dissolution temperatures of extended-chain crystals of polyethylene and interaction parameter values according to eq. (8.23)

Solvent	Dissolution temp.[a] (K)	χ_{12}
None	414.6	
Decane, $C_{10}H_{22}$	394.6	0.43
Octadecane, $C_{18}H_{38}$	401.2	0.28
Hexatriacontane, $C_{36}H_{74}$	406.3	0.11

Source: Cormier and Wunderlich (1966).
[a] Final polymer concentration in solution was 0.0001 (w/w), i.e. $v_1 = 0.9999$.

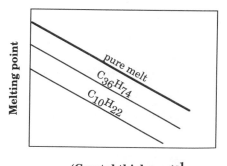

Figure 8.6 Melting point (dissolution point) as a function of reciprocal crystal thickness and solvent.

where T_d^0 is the equilibrium dissolution temperature in the dilute solution, and T_m^0 is the equilibrium melting point in the pure polymer.

8.3 THE GENERAL AVRAMI EQUATION

8.3.1 THEORY

The general Avrami equation is applicable to any type of crystallization. It is not restricted to polymers. It describes the time evolution of the overall crystallinity. The pioneer work was conducted during the 1930s and 1940s by Evans, Kolmogoroff, Johnson and Mehl, and Avrami. Wunderlich (1978) concludes that 'without the parallel knowledge of the microscopic, independently proven mechanism, the macroscopic, experimentally derived Avrami equation and the Avrami parameters are only a convenient means to represent empirical data of crystallization'. However, interest in the Avrami equation has been renewed in the development of simulation programmes for injection moulding which also includes solidification simulation. The simple form of the Avrami equation makes it a suitable model for the solidification of polymer melts in these packages.

Expressions are derived in this section for a few selected cases and it is shown that the derived equations have a certain common mathematical form. This is expressed in the general Avrami equation. Figure 8.7 illustrates the fundamentals of the model. It is assumed that crystallization starts randomly at different locations and propagates outwards from the nucleation sites. The problem which is dealt with can be stated as follows. If raindrops fall randomly on a surface of water and each creates one leading expanding circular wave, what is the probability that the number of waves which pass a representative point P up to time t is exactly c? The problem was first solved by Poisson in 1837 and the resulting equation is referred to as the **Poisson distribution**:

$$p(c) = \frac{\exp(-E)E^c}{c!} \quad (8.26)$$

where E is the average value of the number of passing waves. The probability that no fronts pass P is given by:

$$p(0) = \exp(-E) \quad (8.27)$$

To illustrate the general Avrami equation, a particularly simple case is selected: **athermal** nucleation followed by a spherical free growth in three dimensions. All nuclei are formed and start to grow at time $t = 0$. The spherical crystals grow at a constant rate $\dot{r}$. It is an established fact that crystallization from a relatively pure melt occurs at a constant linear growth rate. All nuclei within the radius $\dot{r}t$ from

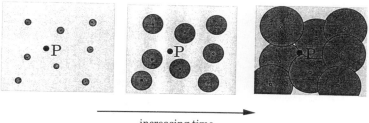

increasing time

Figure 8.7 Snapshots illustrating the growth of circular crystals and the fundamentals of the Avrami model.

point P have formed spherical waves which pass the arbitrary point P during time t. The average number of crystal fronts (E) passing under these conditions is given by:

$$E(t) = \tfrac{4}{3}\pi(\dot{r}t)^3 g \qquad (8.28)$$

where g is the volume concentration of nuclei.

The probability $p(0)$ is equivalent to the volume fraction $(1 - v_c)$ of the polymer which is still in the molten state:

$$p(0) = 1 - v_c \qquad (8.29)$$

where v_c is the volume fraction of crystalline material. Combination of eqs (8.27)–(8.29) yields:

$$1 - v_c = \exp(-\tfrac{4}{3}\pi \dot{r}^3 g t^3) \qquad (8.30)$$

A slightly more complex case involves **thermal nucleation**. The nuclei are here formed at a constant rate both in space and time, similar to 'normal' rain. Let us select the case of three-dimensional growth at a constant linear rate. The number of waves (dE) which pass the arbitrary point (P) for nuclei within the spherical shell confined between the radii r and $r + dr$ is given by:

$$dE = 4\pi r^2 \left(t - \frac{r}{\dot{r}}\right) I^* \, dr \qquad (8.31)$$

where I^* is the nucleation density (number of nuclei per cubic metre per second). The total number of passing waves (E) is obtained by integration of dE between 0 and $\dot{r}t$:

$$E = \int_0^{\dot{r}t} 4\pi r^2 I^* \left(t - \frac{r}{\dot{r}}\right) dr = \frac{\pi I^* \dot{r}^3}{3} t^4 \qquad (8.32)$$

which, after insertion into eqs (8.27) and (8.29), gives:

$$1 - v_c = \exp\left(-\frac{\pi I^* \dot{r}^3}{3} t^4\right) \qquad (8.33)$$

Crystallization based on different nucleation and growth mechanisms can be described by the same general formula, the general Avrami equation:

$$1 - v_c = \exp(-Kt^n) \qquad (8.34)$$

where K and n are constants typical of the nucleation and growth mechanisms (Table 8.3). Equation (8.34) can be expanded according to $\exp(-Kt^n) \approx 1 - Kt^n + \ldots$ and, for the early stages of crystallization, where there is little restriction of crystallization due to impingement, the so-called **Göler equation** is applicable:

$$v_c = Kt^n \qquad (8.35)$$

The Avrami exponent (n) increases with increasing 'dimensionality' of the growth (Table 8.3). Diffusion-controlled growth reduces the value of the exponent by a factor of 1/2 compared with the

Table 8.3 Derived Avrami exponents (n: eq. (8.34)) for different nucleation and growth mechanisms

Growth geom.	Athermal[a]	Thermal[a]	Thermal[b]
Linear problem			
Line	1	2	1
Two-dimensional			
Circular	2	3	2
Three-dimensional			
Spherical	3	4	5/2
Fibrillar	≤1	≤2	
Circular lamellar	≤2	≤3	
Solid sheaf	≥5	≥6	

[a] Free growth; $\dot{r}$ = constant.
[b] Diffusion control; $\dot{r} \propto 1/\sqrt{t}$.

corresponding 'free' growth case. There are certain limitations and special considerations for polymers with regard to the Avrami analysis:

1. The solidified polymer is always only semicrystalline. The crystallinity behind the crystal front is never 100%. This is taken into consideration by modifying eq. (8.30) to:

$$1 - \frac{v_c}{v_{c\infty}} = \exp(-Kt^n) \quad (8.36)$$

where $v_{c\infty}$ is the finally reached volume crystallinity, i.e. the volume crystallinity attained behind the crystal front (Fig. 8.8).

2. The volume of the system studied changes during crystallization. This can be taken into consideration by the following modification of eq. (8.34):

$$1 - v_c = \exp\left(-K\left[1 - v_c\left(\frac{\rho_c - \rho_l}{\rho_l}\right)\right]t^n\right) \quad (8.37)$$

where ρ_c is the density of the crystal phase and ρ_l is the density of the melt.

3. The nucleation is seldom either simple athermal or simple thermal. A mixture of the two is common.
4. Crystallization always follows two stages: (a) **primary** crystallization, characterized by radial growth of spherulites or axialites; (b) **secondary** crystallization, i.e. the slow crystallization behind the crystal front caused by crystal thickening, the formation of subsidiary crystal lamellae and crystal perfection. Secondary crystallization is slow and the initial rapid crystallization is normally dominated by primary crystallization.

Crystallization kinetic data, obtained by differential scanning calorimetry (DSC) and dilatometry, can be analysed if the primary data are transformed to volume crystallinity as a function of crystallization time. The constants in the Avrami equation are obtained by taking the double logarithm of eq. (8.36):

$$\ln\left[-\ln\left(1 - \frac{v_c}{v_{c\infty}}\right)\right] = \ln K + n \ln t \quad (8.38)$$

More information about the DSC and dilatometric methods can be found in Chapter 10. The crystallinity obtained by DSC is often given as mass crystallinity (w_c) which can be converted to a volume fraction (v_c) as follows:

$$v_c = \frac{\dfrac{w_c}{\rho_c}}{\dfrac{w_c}{\rho_c} + \dfrac{(1-w_c)}{\rho_a}} = \frac{w_c}{w_c + \dfrac{\rho_c}{\rho_a}(1-w_c)} \quad (8.39)$$

Figure 8.9 shows a typical Avrami plot. The Avrami equation represents data at low degrees of conversion. The fit at higher degrees of conversion is poor. The Avrami exponent is obtained from the slope of the initial linear part of the curve.

8.3.2 EXPERIMENTAL OBSERVATIONS

Fractions of linear polyethylene have been studied by several researchers (Kovacs 1955; Ergoz, Fatou and Mandelkern 1972). Low molar mass fractions

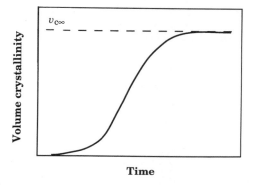

Figure 8.8 Development of crystallinity under isothermal conditions.

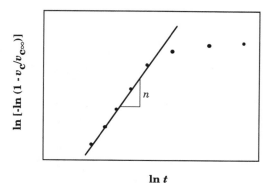

Figure 8.9 Avrami plot showing that the crystallization data at low degrees of conversion follow the Avrami equation (eq. 8.38)).

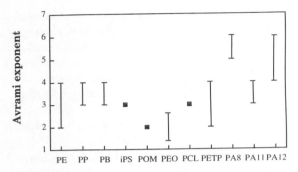

Figure 8.10 Data collected by Wunderlich (1978). (PCL = polycaprolactone.)

$\bar{M}_c \leq 10\,000$, show data which follow the Avrami equation with an exponent equal to 4. Fractions of intermediate molar mass, $10\,000 < \bar{M}_w < 1\,200\,000$, have an Avrami exponent near 3. High molar mass samples, $\bar{M}_w \geq 3\,000\,000$, exhibit kinetics with $n = 2$. The tendency for the Avrami exponent to decrease with increasing molar mass is presumably due to differences in morphology and crystal growth mechanisms between samples of different molar mass.

The low molar mass samples display axialitic (sheaf-like) morphology. A relatively high value of n is expected on the basis of the data presented in Table 8.3. The reduction in n for the fractions of intermediate molar mass is consonant with the spherulitic morphology observed in these samples. The crystallization of high molar mass polymers is strongly influenced by chain entanglements and the slow and incomplete crystallization leads to small, uncorrelated crystals, i.e. to a so-called random lamellar structure. A low value of n is expected for such a 'low-dimensional' growth.

Figure 8.10 presents a summary of Avrami exponent data obtained for different polymers.

8.4 GROWTH THEORIES

8.4.1 INTRODUCTION

Growth theories can be divided into **equilibrium** theories and **kinetic** theories. Equilibrium theories explain some features of the crystal thickness. They predict the existence of two minima in free energy, one at a finite crystal thickness and the other at infinite thickness. The crystal thickness associated with the minimum in free energy increases with increasing temperature up to a certain critical temperature. The experimentally determined relationship between crystal thickness and degree of supercooling is not predicted. Equilibrium theories are today usually ignored by the polymer scientific community.

There is currently a consensus among scientists in the field that kinetic factors control growth rate and morphology. Kinetic theories acknowledge that the end state is not that with the lowest possible free energy. The growth rate depends on the crystal thickness. Crystals with a range of crystal thicknesses, all greater than the minimum value $2\sigma/\Delta g$, are formed. The resulting crystal thickness distribution is determined by the relationship between crystallization rate and crystal thickness at the particular temperature. A maximum in growth rate is obtained at a crystal thickness which is greater than $2\sigma/\Delta g$ by a 'small' term δL_c. Kinetic theories are principally of two types: **enthalpic nucleation** theories and **entropic** theories. Nucleation theories, e.g. the Lauritzen–Hoffman theory, assume that the free energy barrier associated with nucleation has an energetic origin. The Sadler–Gilmer theory regards the free energy barrier as predominantly entropic. Kinetic theories predict the temperature dependence of growth rate, initial crystal thickness (L_c^*) and other morphological parameters.

8.4.2 THE LAURITZEN–HOFFMAN THEORY

Theory

The Lauritzen–Hoffman (LH) theory has been the dominant growth theory for polymer crystallization for the last 20 years. New experimental data have led to the revision and development of the theory during recent years. The original theory of Lauritzen and Hoffman (1960) is presented here in detail. Later modifications of the theory are outlined in the subsequent text. Finally, criticisms of the LH theory including the later developed versions are mentioned and discussed.

The LH theory provides expressions for the linear growth rate (G), i.e. the rate at which spherulites or axialites grow radially, as a function of degree of supercooling ($\Delta T = T_m^0 - T_c$, where T_m^0 is the equilibrium melting point and T_c is the crystallization

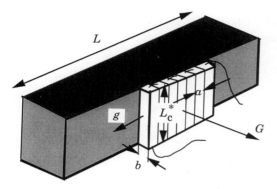

Figure 8.11 A growing crystal lamella according to the LH theory.

Figure 8.11 illustrates the model. It is assumed that the crystal lamellae at the growth front grow at the same rate as the macroscopically observed linear growth rate. The nucleation, being either secondary or tertiary, controls the crystal growth together with the short-range diffusion of the crystallizing units.

This presentation of the LH theory considers a very simple case, namely a homopolymer of intermediate or high molar mass. Fold length fluctuation is not considered in this simple treatment. A secondary nucleus is first formed and it spreads out laterally at the rate g. The thickness of the stem along the growth (G) direction is b. It is not necessary for the steady-state structure of the growing crystals to be smooth. In fact two of the three regimes of crystallization that are defined by the LH theory are characterized by a surface which contains several patches on to which stems are deposited.

Figure 8.12 shows the energy map for this process in more detail. The first step, the deposit of the first stem, occurs at rate A_0 and involves the lateral surface free energy ($2bL_c\sigma_L$) reduced by a fraction of the free energy of crystallization ($\psi abL_c\Delta g$). Note that Δg in this case is defined as being positive for $T_c < T_m^0$. The remainder of the free energy of crystallization (($1 - \psi)abL_c\Delta g$) for the first stem is released on the

temperature). The linear growth rate is experimentally determined by polarized light microscopy. A typical experiment involves first heating a 2–20 μm thick polymer film to a temperature well above the melting point and then rapidly cooling it to a constant temperature after which a number of photomicrographs are taken at suitable time intervals. The growth of the spherulites (axialites) is measured directly from the photomicrographs. A linear increase with time of the spherulite (axialite) radii is almost always observed.

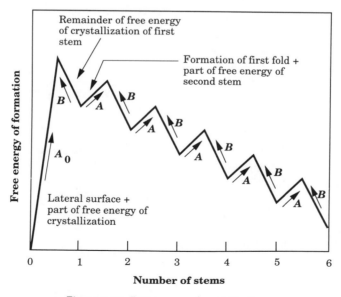

Figure 8.12 Energy map of crystallization.

180 Crystallization kinetics

other side of the first maximum. The next process involves the formation of the first fold ($2ab\sigma$) which is accompanied by the release of a fraction of the free energy of crystallization ($\psi abL_c\Delta g$). On the rear side of the second maximum, the rest of the free energy of crystallization for the second stem is released. Later crystallizing stems exhibit the same energy barriers as the second stem (Fig. 8.12).

The rate of deposit of the first stem is:

$$A_0 = \beta \exp\left[-\frac{2bL_c\sigma_L - \psi abL_c\Delta g}{kT_c}\right] \quad (8.40)$$

where

$$\beta = \frac{kT_c}{h} J_1 \exp\left[-\frac{U^*}{R(T_c - T_\infty)}\right] \quad (8.41)$$

This expression (for β) is derived from the WLF equation and h is Planck's constant, J_1 is a constant, U^* is a constant (dimension J mol^{-1}), and T_∞ is the temperature at which diffusion is stopped. Later versions of the LH theory use another expression derived from de Gennes's theory for self-diffusion (see Chapter 6) for the β parameter:

$$\beta = \left(\frac{kT_c}{h}\right)\left(\frac{1}{\bar{M}_z}\right)\exp\left[-\frac{\Delta E_r}{RT_c}\right] \quad (8.42)$$

where ΔE_r is the activation energy for transport of a molecule across the phase boundary by reptation.

The rate equations of the subsequent steps are:

$$A = \beta \exp\left[-\frac{(2ab\sigma - \psi abL_c\Delta g)}{kT_c}\right] \quad (8.43)$$

$$B = \beta \exp\left[-\frac{((1-\psi)abL_c\Delta g)}{kT_c}\right] \quad (8.44)$$

The ratios A_0/B and A/B are independent of ψ. The absolute rate constants are, however, dependent on ψ.

The crystallizing polymer will add 1, 2, ..., n stems to the substrate surface. The steady-state constant flux of segments (S) will find its solution in the following balance equations:

$$S = N_0A_0 - N_1B = N_1A - N_2B = \cdots$$
$$= N_vA - N_{v+1}B \quad (8.45)$$

The following equations are obtained by rearranging eq. (8.45):

$$N_{v+1} = \left(\frac{A}{B}\right)N_v - \frac{S}{B} \quad (8.46)$$

$$N_{v+1} = \left(\frac{A}{B}\right)^2 N_{v-1} - \frac{S}{B}\left(1 + \frac{A}{B}\right) \quad (8.47)$$

Further substitution of N_{v-1} finally leads to:

$$S = N_0A_0\left(1 - \frac{B}{A}\right) \quad (8.48)$$

This equation can be intuitively understood: if N_0A_0 surface nuclei start to grow, the fraction of these which will melt is related to the ratio of the backward to the forward process, i.e. B/A.

If eqs (8.40)–(8.44) are inserted into eq. (8.48), the following expression is obtained:

$$S(L_c) = N_0\beta\left[\exp\left(-\frac{(2bL_c\sigma_L - \psi abL_c\Delta g)}{kT_c}\right)\right.$$
$$\left.\times\left[1 - \exp\left(-\frac{(2ab\sigma - abL_c\Delta g)}{kT_c}\right)\right]\right] \quad (8.49)$$

This quantity is thus the rate of crystallization and its dependence on crystal thickness (L_c) and temperature (T_c). The average crystal thickness L_c^* can be derived from the expression:

$$L_c^* = \frac{\int_{L_c=2\sigma/\Delta g}^{\infty} L_c S(L_c)\, dL_c}{\int_{L_c=2\sigma/\Delta g}^{\infty} S(L_c)\, dL_c} \quad (8.50)$$

The lower limit is obtained by considering that the lowest possible crystal thickness should correspond to a melting point which is the same as the crystallization temperature:

$$L_{c\min} = \frac{2\sigma}{\Delta g} = \frac{2\sigma T_m^0}{\Delta h^0 \rho_c(T_m^0 - T_c)} \quad (8.51)$$

which after rearrangement gives:

$$T_c = T_m^0\left[1 - \frac{2\sigma}{\Delta h^0 \rho_c L_{c\min}}\right] \quad (8.52)$$

Equation (8.52) is identical to the Thompson–Gibbs equation. Note that Δh^0 is mass related heat of fusion (in J kg^{-1}) and that $T_c = T_m$. Integration of eq. (8.50) gives the expression:

$$L_c^* = \frac{2\sigma}{\Delta g} + \frac{kT}{2b\sigma_L} \cdot \frac{\left(2 + (1 - 2\psi)a \cdot \frac{\Delta g}{2\sigma}\right)}{\left(1 - \frac{a\Delta g \psi}{2\sigma_L}\right) \cdot \left(1 + \frac{a\Delta g(1 - \psi)}{2\sigma_L}\right)} \quad (8.53)$$

For $\psi = 1$ this reduces to:

$$L_c^* = \frac{2\sigma}{\Delta g} + \frac{kT}{2b\sigma_L} \cdot \frac{\left(\frac{4\sigma_L}{a} - \Delta g\right)}{\left(\frac{2\sigma_L}{a} - \Delta g\right)} = \frac{2\sigma}{\Delta g} + \delta L_c \quad (8.54)$$

δL_c becomes infinite when

$$\Delta g = \frac{2\sigma_L}{a}$$

The degree of supercooling (ΔT_s) corresponding to the singularity can be derived as follows:

$$\frac{\Delta h^0 \rho_c \Delta T_s}{T_m^0} = \frac{2\sigma_L}{a} \Rightarrow \Delta T_s = \frac{2\sigma_L T_m^0}{a \Delta h^0 \rho_c} \quad (8.55)$$

A calculation of ΔT_s using data for linear polyethylene (and $\psi = 1$) yields the value 55 K.

The singularity (or δL-catastrophe as it is sometimes called) may be avoided by setting $\psi = 0$:

$$L_c^* = \frac{2\sigma}{\Delta g} + \frac{kT}{2b\sigma_L} \cdot \frac{\left(\frac{4\sigma_L}{a} + \Delta g\right)}{\left(\frac{2\sigma_L}{a} + \Delta g\right)} = \frac{2\sigma}{\Delta g} + \delta L_c \quad (8.56)$$

The LH theory does not include any elements of crystal thickening. All crystals are 'virgin', i.e. unthickened. The LH theory introduced three growth regimes: I, II and III. The corresponding growth-rate equations are now derived.

Regime I growth

Regime I growth is characterized by the fact that the secondary nucleation step controls the linear growth rate (G). The lateral growth rate (g) is significantly greater than the rate of formation of secondary nuclei (i):

$$g \gg i \quad (8.57)$$

The whole substrate is completed and covered by a new monolayer (Fig. 8.13). Monolayers are added one by one and the linear growth rate (G_I) is given by:

$$G_I = biL \quad (8.58)$$

where b is the monolayer thickness, i is the surface nucleation rate (nuclei per length of substrate per second), L is the substrate length $= n_s a$, n_s is the number of stems on the substrate and a is the stem width. Equation (8.58) is a particular solution to the more general Frank equation of nucleation. It should be noted that iL is the rate of the formation of secondary nuclei on a given substrate. Each secondary nucleation leads to the formation of a monolayer of thickness b. The linear growth rate must thus be the product of iL and b.

The total rate of crystallization, i.e. the total flux (S_T) in the system of N totally vacant stems (event surface nuclei/time) is

$$S_T = iNa \quad (8.59)$$

$$S_T = \left(\frac{1}{L_u}\right) \int_{2\sigma/\Delta g}^{\infty} S(L_c) \, dL_c \quad (8.60)$$

where L_u is the length of the repeating unit. Combination of eqs (8.58) and (8.59) gives:

$$G = bS_T \frac{n_s}{N} \quad (8.61)$$

Figure 8.13 Regime I growth. Each square corresponds to the cross-section of a stem.

Integration of eq. (8.60) yields:

$$S_T = \left(\frac{N_0 \beta}{L_u}\right) P \exp\left(\frac{2ab\sigma\psi}{kT_c}\right) \exp\left(-\frac{4b\sigma\sigma_L}{\Delta g kT_c}\right) \quad (8.62)$$

where

$$P = RT_c \left[\frac{1}{2b\sigma_L - \psi ab\Delta g} - \frac{1}{2b\sigma_L - (1-\psi)ab\Delta g}\right]$$

Lauritzen and Hoffman (1960) used in their original work the WLF equation (details given in Chapter 5) for expressing the short-range diffusion factor (β):

$$\beta = \left(\frac{kT_c}{h}\right) J_1 \exp\left(-\frac{U^*}{R(T_c - T_\infty)}\right) \quad (8.63)$$

Finally, combination of eqs (8.61)–(8.63) yields:

$$G_I = G_{0I} \exp\left(-\frac{U^*}{R(T_c - T_\infty)}\right) \exp\left(-\frac{4b\sigma\sigma_L}{\Delta g kT_c}\right) \quad (8.64)$$

where

$$G_{0I} = b\left(\frac{kT_c}{h}\right) J_1 \exp\left(\frac{2ab\sigma\psi}{kT}\right)$$

It should be noted that $P \approx L_u$. However, as pointed out earlier, in the later versions of the LH theory the transport factor is derived from de Gennes's theory (the reptation model, see Chapter 6) and β is given by:

$$\beta = \left(\frac{kT_c}{h}\right) \cdot \left(\frac{1}{\bar{M}_z}\right) \exp\left(-\frac{\Delta E_r}{RT_c}\right) \quad (8.65)$$

where ΔE_r is the activation energy for reptation.

Regime II growth

Regime II growth occurs by multiple nucleation (Fig. 8.14). The secondary nucleation rate is in this case more rapid than in regime I, i.e. $g < i$.

Sanchez and DiMarzio (1971) and Frank (1974) showed independently that the linear growth rate (G_{II}) can be expressed as follows:

$$G_{II} = b\sqrt{ig} \quad (8.66)$$

where g is the lateral growth rate $= a(A - B)$, given

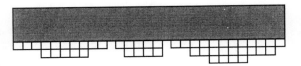

Figure 8.14 Regime II growth. Each square corresponds to the cross-section of a stem.

by:

$$g = a\beta\left[\exp\left(-\frac{2ab\sigma}{kT_c} + \frac{\psi abL_c\Delta g}{kT_c}\right) - \exp\left(-\frac{(1-\psi)abL_c\Delta g}{kT_c}\right)\right]$$

$$\approx a\beta \exp\left(-\frac{2ab\sigma(1-\psi)}{kT_c}\right) \quad (8.67)$$

The reasoning of Sanchez and DiMarzio was briefly as follows: an isolated stem undergoes nucleation at time $t = 0$ and it grows laterally in the two opposite directions. At a given time t, the length of the substrate covered by an additional monolayer is proportional to $2gt$. The rate at which new nuclei form on this patch is simply $2gti$, where i is the nucleation rate. The number of nuclei formed during the incremental time dt is given by:

$$N(t)\,dt = 2gti\,dt \quad (8.68)$$

The total number of nuclei formed during a time period t is:

$$\int_0^t N(t)\,dt \propto \int_0^t 2git\,dt = git^2 \quad (8.69)$$

The average time $\langle t \rangle$ to form a new nucleus on a growing patch is:

$$gi\langle t \rangle^2 \propto 1 \Rightarrow \langle t \rangle \propto \frac{1}{\sqrt{ig}} \quad (8.70)$$

The rate at which new layers are formed is given

by:

$$G_{II} = \frac{b}{\langle t \rangle} = b\sqrt{ig} \qquad (8.71)$$

which is identical to eq. (8.66). The continuum theory of Frank leads to the same result.

If eqs (8.59), (8.62) and (8.67) are inserted into eq. (8.66), an expression for the growth rate of regime II is obtained:

$$G_{II} = G_{0II} \cdot \exp\left(-\frac{U^*}{R(T_c - T_\infty)}\right)\exp\left(-\frac{2b\sigma\sigma_L}{\Delta g k T_c}\right) \qquad (8.72)$$

where

$$G_{0II} = b\left(\frac{kT_c}{h}\right)\exp\left(\frac{(2\psi - 1)ab\sigma}{kT_c}\right) \qquad (8.73)$$

Regime III growth

Regime III growth occurs by prolific multiple nucleation (Fig. 8.15). The niche separation in this case is the same size as the stem width. It can be shown that the growth rate (G_{III}) is given by:

$$G_{III} = biL \qquad (8.74)$$

The following expression is then derived for regime III growth. It is noted that the nucleation factor, the last exponent in eq. (8.75), is the same as that for regime I (cf. eq. (8.64)):

$$G_{III} = G_{0III} \cdot \exp\left(-\frac{U^*}{R(T_c - T_\infty)}\right)\exp\left(-\frac{4b\sigma\sigma_L}{\Delta g k T_c}\right) \qquad (8.75)$$

where $G_{0III} = C(kT_c/h)$.

Figure 8.15 Schematic description of regime III growth. Each square corresponds to the cross-section of a stem.

Experimental data

The LH theory compares favourably with experimental data in two important aspects: the temperature dependence of both the initial crystal thickness (L_c^*) and the linear growth rate (G), given by eqs (8.53), (8.64), (8.72) and (8.75). The following equations can be fitted to the experimental data:

$$L_c^* = \frac{C_1}{\Delta T} + C_2 \qquad (8.76)$$

and

$$G = \beta \exp\left(-\frac{K_g}{T_c \Delta T f}\right) \qquad (8.77)$$

where C_1, C_2, K_g and f are constants. Plots of (log G)/β as a function $1/T_c\Delta T$ consist of lines with relatively abrupt changes in the slope coefficients (Fig. 8.16). The regions of linear behaviour are denoted regimes.

Table 8.4 presents data for linear polyethylene with some characteristics of the growth regimes.

It should be noted that:

$$\Delta g = \Delta h^0 - T_c \Delta s^0 = \Delta h^0 \left[1 - \frac{T_c}{T_m^0}\right]$$

$$= \frac{\Delta h^0}{T_m^0}[T_m^0 - T_c] = \frac{\Delta h^0 \Delta T}{T_m^0} \qquad (8.78)$$

and insertion of eq. (8.78) into eqs (8.64), (8.72) and (8.75) results in the expressions for K_g shown in Table 8.4. Note that the dimension of Δh^0 is here J m^{-3}.

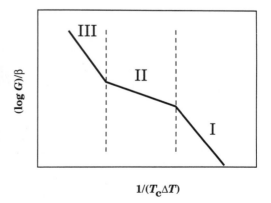

Figure 8.16 Growth rate regimes (schematic curve).

Table 8.4 Growth-rate equations and growth-rate data of linear polyethylene

	Regime I	Regime II	Regime III
$K_g{}^a$	$\dfrac{4b\sigma\sigma_L T_m^0}{\Delta h^0 k}$	$\dfrac{2b\sigma\sigma_L T_m^0}{\Delta h^0 k}$	$\dfrac{4b\sigma\sigma_L T_m^0}{\Delta h^0 k}$
ΔT (K)	< 17	17–23	> 23
$\dfrac{g}{i}$	≫ 1	< 1	≪ 1
Supermolecular structure	axialitic	spherulitic	spherulitic

$^a\ G = \beta\ \exp(-K_g/T_c\Delta Tf)$.

Figure 8.17 presents data for linear polyethylene. The transition from regime I to regime II crystallization was reported by Hoffman et al. (1975) as being sharp and accompanied by a distinct change in supermolecular structure from axialitic to spherulitic. The LH theory predicts that the ratio of the slopes (K_g) in the regimes I and II growth regions is 2:1. A list of references providing data on regime transitions was compiled by Hoffman and Miller (1989).

The transition from regime I to regime II can be predicted by the so-called Z test originally proposed by Lauritzen (1973). The quantity Z is defined by:

$$Z = \frac{iL^2}{4g} \qquad (8.79)$$

where L is the length of the substrate. Regime I is predicted to occur when $Z < 0.1$ and regime II when $Z > 1$. The substrate length at the temperature for the transition from regime I to regime II crystallization has been estimated on the basis of eq. (8.79). In the early work, L values of the order of 1 μm were suggested. However, refined analysis later led Hoffman and Miller (1989) to suggest values of the order of 0.1 μm. This is a very important question. As pointed out in Chapter 7, recent data obtained by electron microscopy showed that linear polyethylene crystals growing at 130°C (regime I) exhibit a rough growth surface. This is clearly inconsistent with the original suggested length of the substrate (1 μm). Hoffman overcomes the problem by suggesting that the surface is microfaceted, with facets as small as 0.02 μm.

Figure 8.18 shows early data from Hoffman et al. (1975) for fractions of linear polyethylene. The transition between regimes I and II growth occurs at a supercooling of 17.5 ± 1 K in the molecular mass range displayed in the figure. The K_g values for the same samples are shown in Fig. 8.19. The ratio $K_g(I)/K_g(II)$ is indeed very close to 2, as predicted by the LH theory.

From the K_g values, it is possible, using eqs (8.64) and (8.72), to calculate the product of the surface energies, $\sigma\sigma_L$ (Fig. 8.20). The surface energy product increases strongly with increasing molar mass up to

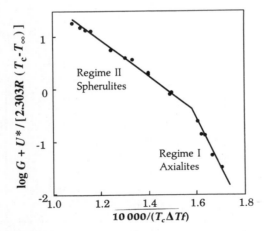

Figure 8.17 Linear growth rate data for a linear polyethylene sharp fraction: $M_w = 66\,000$. Drawn after data of Rego Lopez et al. (1988).

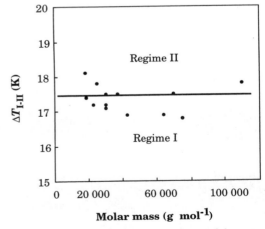

Figure 8.18 Degree of supercooling (ΔT_{I-II}) of the regime I–II transition for linear PE fractions plotted as a function of molecular mass. Drawn after data of Hoffman et al. (1975).

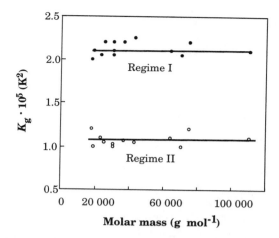

Figure 8.19 K_g values recorded for fractions of linear PE. Drawn after data from Hoffman et al. (1975).

a molar mass value of about 20 000 g mol^{-1}, above which it is approximately constant. It may be assumed that σ_L is independent of molar mass. The increase in the surface energy product with increasing molar mass can then be assigned to the fold surface. It is known from melting point–crystal thickness data for high molar mass samples using the Thompson–Gibbs equation that σ is about 95 mJ m^{-2}. This allows the determination of σ_L, which takes the value

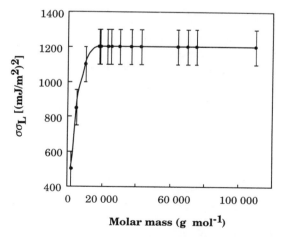

Figure 8.20 Surface free energy product from linear growth rate data for fractions of linear PE. Drawn after data from Hoffman et al. (1975).

13 $\pm$ 2 mJ m^{-2}. It should, however, be pointed out that this determination is self-consistent in that the value obtained is calculated using an equation with parameters (constants) which are defined by the theory. An independent determination of σ_L has not yet been carried out.

The LH theory has been under constant modification during its 30-year life. Crystallization of multicomponent chains and fold-length fluctuations are phenomena which were addressed in different revisions during the 1960s. Diffusion plays an important role in crystallization, and the modern view of reptation was implemented by Hoffman, Guttman and DiMarzio (1979). This modification was driven by the suggestion made by Yoon and Flory (1977) that regular, tight chain folding was prohibited by the long times needed to rearrange the entangled molecules. Hoffman and co-workers showed, however, that the self-diffusion rate was sufficiently high to ensure a sizeable proportion of tight chain folding. This development also led to a modification of the growth-rate equation. The linear growth rate in all three regimes turned out to be proportional to $1/\bar{M}_z$. This expression was thus introduced into the pre-exponential factor of the growth-rate equations. The last modification addressed the problem of the curved edges of crystals melt grown at high temperatures, clearly within the regime I temperature region. Figure 8.21 shows the model of Mansfield, Miller and Hoffman (see Mansfield 1988; Hoffman and Miller 1989). The elliptical shape of the crystal is in fact microfaceted with steps of about 100 nm. The linear growth rate becomes approximately equal to the lateral growth rate (g).

Criticism of the LH theory has grown steadily during the last ten years. It is difficult to present a comprehensive and short review of the critique. Dosière, Colet and Point (1986) measured the persistence length (the average distance between adjacent secondary nucleation sites) by a decoration technique and obtained a maximum value of 200 nm. This led to much lower g values than those predicted by the LH theory. Point, Colet and Dosière (1986) calculated the persistence length from growth-rate data of solution-crystallization of linear polyethylene within the framework of the LH theory. The L values obtained were clearly unrealistic.

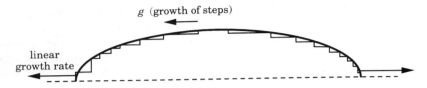

Figure 8.21 Elliptical profile of melt-grown crystal (130°C) of linear polyethylene with steps along the two long axes.

The persistence length is determined by the ratio of the lateral growth rate (g) to the secondary nucleation rate (i). The LH theory predicts a growth-rate transition, essentially because g changes only moderately with temperature whereas i changes strongly. This means that the persistence length is strongly dependent on the temperature. It has been suggested by other authors that both i and g have the same temperature dependence and that the kinetic (persistence) length is essentially temperature-independent. There are alternative explanations of the change in slope accompanying the regimes I–II transition. Point and Dosière (1989), after criticism of the LH theory (and of developments of it), suggest other causes for the change in the slopes: molar mass segregation, a pronounced temperature dependence of the interfacial energies, viscosity effects, and temperature dependence of the nucleation processes.

8.4.3 THE SADLER–GILMER THEORY

The Sadler–Gilmer theory, i.e. entropic theory, has its roots in the crystallization behaviour of small molecules. At temperatures greater than a certain so-called roughening temperature (T_r), macroscopically rounded and usually microscopically rough crystal surfaces are generated, and this is taken as evidence that free energy barriers to large fluctuations are very small under these conditions. At temperatures below T_r, growth occurs through invariantly faceted surfaces. Kinetic roughening occurs, however, at very low temperatures. The same temperature trend is in fact found for polyethylene. Solution-grown single crystals grown from good solvents at low temperatures are always faceted (Chapter 7). Crystallization at higher temperatures in poorer solvents yield single crystals of slightly rounded {100} sectors. Crystallization at very high temperatures leads to leaf-shaped crystals. Bassett, Olley and Al Reheil (1988) were able to view isolated crystals of polyethylene as being melt-crystallized at 130°C, i.e. within the regime I crystallization temperature region. Figure 8.22 shows the rounded features with small tips of {110} facets of these crystals compared with a crystal formed at a lower crystallization temperature showing only {110} growth facets.

The growth rate for small molecules at temperatures above T_r is proportional to ΔT, whereas experiments clearly show that the logarithm of linear growth rate is proportional to $1/(T_c\Delta T)$. Sadler (1983) took that as an indication that some free energy barrier must also be present above T_r. The barrier cannot be energetic because the attachment of new stems occurs without any change in the surface area. He then proposed that the free energy barrier has an entropic origin. The multitude of possible chain conformations lead to an **entropic** growth barrier. This is the essence of the entropic barrier theory, hereinafter called the Sadler–Gilmer theory.

The Sadler–Gilmer theory not only is consistent with the observations made by electron microscopy (Fig. 8.22), but also predicts a temperature dependence of the linear growth rate and of the crystal thickness

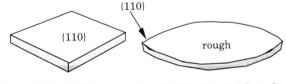

$\Delta T=45°C$; $T_c=70°C$ (p-xylene) $\Delta T=15°C$; $T_c=130°C$ (melt)

Figure 8.22 Shape of polyethylene crystals crystallized at conditions specified in the figure.

in accordance with basic phenomenological laws:

$$G \propto \exp\left(-\frac{K_g}{T_c \Delta T}\right)$$

$$L_c^* = \frac{C}{\Delta T} + \delta L_c$$

(8.80)

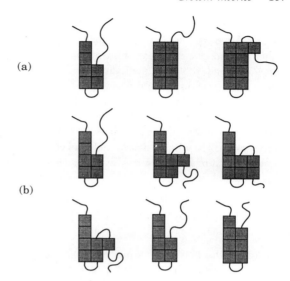

Sadler's reasoning was that the interaction energy between a whole stem and the underlying substrate is large compared with kT, and roughening will not occur. If, however, a small part of a stem is considered, the interaction energy is small and possibly comparable with kT, and roughening is a possibility. Sadler found that six CH_2 units formed a group with an interaction energy which was equal to $0.6kT$. It is known that a simple cubic face is rough under such conditions. The main concepts in the Sadler–Gilmer model are 'blind attachment', 'pinning' and 'detachment'. As a chain molecule attaches to a rough crystal surface, it does not necessarily choose a conformation which is suitable for later stages of growth. The stem length may lie below the limit of thermodynamic stability. The chain may form chain folds and loops which prohibit further growth on to these sites. These sites are said to be 'pinned'. The growth of the crystal becomes frustrated, blocked by these non-viable chain conformations. The advance of the growth may occur only by ongoing on- and off-flux of the chain segments which gradually sort out chain conformations suitable for incorporation into the body of the crystal. These sorting-out processes are illustrated in Fig. 8.23.

The Sadler–Gilmer model considers both forward and backward reactions and the ratio of the two (k_-/k_+) can be expressed as:

$$\frac{k_-}{k_+} = \exp\left(\frac{2\varepsilon}{kT_m^0} - \frac{m\varepsilon}{kT_c}\right)$$

(8.81)

where ε is the interaction energy, m is the number of neighbouring sites, T_m^0 is the equilibrium melting point and T_c is the crystallization temperature. The chain connectivity, a property of all polymers, leads to pinning according to a set of rules which limits the number of sites available for attachment and detachment. A very simple two-dimensional approach to the phenomenon is shown in Fig. 8.24. Attachment

Figure 8.23 Different types of growth: (a) chain folds are only formed when the strip reaches a certain length (LH theory); (b) chain folds are formed also at intermediate positions leading to unfavourable conformation (pinning). The 'sorting' process leads to an accumulation of 'good' conformations.

and detachment are only allowed at the outermost position (surface layer).

The kinetics expressions can be generated by two different methods. One is Monte Carlo simulation, by which a growing crystal is generated and the exact

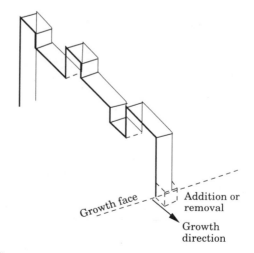

Figure 8.24 Two-dimensional representation of growth of crystal.

path is followed in real time. Data for G and L_c^* can be generated by running many simulations. In the other method, rate equations of the different steps are generated and solved numerically. The results are given as average values and no information is given about the growth 'mechanism'. Monte Carlo simulations performed by Sadler and Gilmer (1984) and Goldbeck-Wood and Sadler (1989) yielded thickness and growth-rate data for different crystallization temperatures consonant with experimental data (eq. (8.80)).

8.4.4 CRYSTALLIZATION VIA METASTABLE PHASES?

It is known that polyethylene may exist in a hexagonal phase at elevated pressure. The hexagonal phase is stable within a certain temperature and pressure region (Fig. 8.25). A lowering of the pressure at constant temperature leads to a phase transformation from hexagonal to orthorhombic structure. The hexagonal phase of polyethylene allows rapid crystal thickening which, if sufficient time is given to the process, leads to the formation of extended-chain crystals without any chain folds. Crystal thickening of orthorhombic crystals is a considerably slower process.

Trans-poly(1,4-butadiene) has two crystal forms, a monoclinic phase which is stable at low temperatures and a hexagonal phase which is stable at high temperatures. The hexagonal phase of this polymer is similar to the hexagonal phase of polyethylene: it

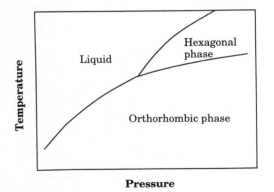

Figure 8.25 Temperature–pressure phase diagram of polyethylene. Schematic drawing.

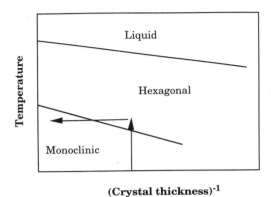

Figure 8.26 Temperature–crystal thickness phase diagram for trans-poly(1,4-butadiene). Schematically drawn after data from Rastogi and Ungar (1992). Arrows show the route of Rastogi and Ungar's experiment.

favours also rapid crystal thickening. Rastogi and Ungar (1992) reported the following important findings. By first heating the sample, trans-poly(1,4-butadiene), at normal pressure, they noticed a transition from monoclinic to hexagonal structure at a certain temperature. The sample was then held at constant temperature, still at normal pressure, which caused rapid crystal thickening and, interestingly, at a certain crystal thickness, a transition from hexagonal to monoclinic structure. Figure 8.26 shows schematically the phase diagram that they were able to construct based on their observations. The hexagonal phase is only stable for thin crystals and because of the rapid crystal thickening, the monoclinic phase is formed at a certain crystal thickness.

Keller et al. (1994) now suggest that this scheme may well apply to other polymers, e.g. polyethylene. Crystallization may under certain conditions start and proceed through the mobile hexagonal phase, which permits rapid crystal thickening, with the orthorhombic phase being formed when the crystals reach a certain critical thickness, the latter being controlled by the crystallization temperature. Computations by Keller et al. (1994) showed that there is a strong likelihood that this scheme applies to melt-crystallization of polyethylene also at normal pressure. However, solution-crystallization of the same polymer would, according to these calculations, not involve the hexagonal phase.

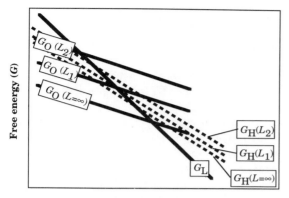

Figure 8.27 Schematic representation of the free energy as a function of temperature at constant pressure; M = melt; O = orthorhombic phase; H = hexagonal phase; note that $L_1 > L_2$. Drawn after Keller et al. (1994).

The thermodynamic arguments in favour of crystal-thickness-dependent phase stability are illustrated in Fig. 8.27. The hexagonal (H) phase is clearly only unstable at infinite crystal thickness. For the crystal thickness denoted L_1, there is a triple point, i.e. the three phases have the same free energy at the temperature at the three-line intersection. Thinner crystals ($L_c < L_1$) are more stable in the hexagonal phase than in the orthorhombic (O) phase within a temperature region from the triple point to the intersection between G_O and G_H (at that particular crystal thickness). The chain mobility of the hexagonal phase causes rapid crystal thickening, i.e. a vertical shift downwards in the diagram (Fig. 8.27), and at a certain stage a transition from hexagonal to orthorhombic structure is expected.

The thermodynamic criterion for the existence of a triple point may be expressed as follows:

$$\left(\frac{\sigma_H}{\Delta h_H}\right) < \left(\frac{\sigma_O}{\Delta h_O}\right) \quad (8.82)$$

where σ_H and σ_O are the fold surface free energies of the hexagonal and orthorhombic phases and Δh_H and Δh_O are the heat of fusion of the hexagonal and orthorhombic phases.

The most general statement that can be deduced from the scheme is that phase transformation starts with the phase which is stable down to the smallest size, irrespective of whether this is stable or metastable when fully grown. Further experimental data on polyethylene are evidently soon to be presented and their implications for theory are awaited with interest.

8.4.5 CRYSTAL GROWTH THEORIES: STATUS AS AT 1994

The presentation of growth theories in this chapter has followed the historical development in an almost chronological order. New experiments, particularly those revealing the lateral shape of crystals, have cast new light on the crystallization process and forced theory to change accordingly. Theoretical activities are currently intensive and it is likely that new modifications of 'old' theories and even completely new ideas will arise. For more information on the most recent developments, see Armitstead and Goldbeck-Wood (1992), Point and Villars (1992), and Mansfield (1988; 1990). Another current development of great significance, due to Keller et al. (1994), is that relating to 'metastable' phases (e.g. the hexagonal phase in polyethylene) possibly existing during the initial stages of crystallization also at normal pressure (section 8.4.4). Perhaps a new paradigm for polymer crystallization will take shape in the coming years.

8.5 MOLECULAR FRACTIONATION

8.5.1 INTRODUCTION

Crystallization of most polymers is accompanied by the separation of different molecular species, a process referred to as **molecular fractionation**. In linear polyethylene, fractionation occurs due to differences in molar mass. The low molar mass material crystallizes at low temperatures in subsidiary lamellae located between the dominant lamellae and in the spherulite boundaries (see Chapter 7).

The first direct evidence of molecular fractionation in polyethylene was provided by Bank and Krimm (1970). The first extensive studies were reported by Mehta and Wunderlich (1975). The data obtained by Mehta and Wunderlich indicated that, at each crystallization temperature, there exists a critical molar mass (M_{crit}) such that the molecules of molar mass greater than M_{crit} are able to crystallize at this temperature, whereas molecules of molar mass less than M_{crit} are unable to crystallize. Fractionation was found to be relatively sharp in terms of molar mass. Figure 8.28 shows that M_{crit} increases with increasing crystallization temperature.

The lower limit of segregation is set by the hypothetical equilibrium of crystallization. It is assumed that dynamic equilibrium is achieved between fully extended-chain crystals and the surrounding melt. At equilibrium, the molecular length of the crystallizable species corresponds closely to the lamellar thickness, and molecules too short or too long introduce defects and increase the free energy and are thus rejected from the crystal. The equilibrium melting point of a given molecular species is dependent not only on its molar mass but also on the molar masses of the other different species present in the blended melt:

$$\frac{1}{T_m} - \frac{1}{T_m^0(M)} = \frac{R}{\Delta H}$$
$$\times [-\ln v_p + (\bar{x} - 1)(1 - v_p) - \bar{x}\chi(1 - v_p)^2] \qquad (8.83)$$

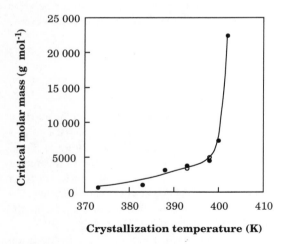

Figure 8.28 Critical molar mass of melt-crystallized linear polyethylene as a function of crystallization temperature. Filled circles: data for a broad molar mass sample: $M_n = 8500$ g mol^{-1}; $M_w = 153\,000$ g mol^{-1} of Mehta and Wunderlich (1975); open circles: data for a sample with $M_n = 12\,900$ g mol^{-1}; $M_w = 108\,000$ g mol^{-1} of Gedde, Eklund and Jansson (1983).

where T_m is the temperature of the melting–crystallization considered, $T_m^0(M)$ is the equilibrium melting–crystallization temperature of the pure species of the molar mass considered, R is the gas constant, v_p is the volume fraction in the melt of the crystallizing species, ΔH is the molar heat of fusion, χ is the interaction parameter and $\bar{x} = \sum v_i x_i / \sum v_i$ is the volume fraction of crystallizing species with respect to all other species in the blend. It is possible to calculate an equilibrium critical molar mass for each temperature of crystallization from eq. (8.83) and from the molar mass distribution data of the polymer considered. Wunderlich and Mehta (1974) showed that the experimental and theoretical data were in agreement at high ΔT. The equilibrium critical molar mass calculated was, however, significantly lower than the experimental data for low ΔT. Wunderlich and Mehta therefore suggested that each molecule must undergo a **molecular nucleation** before crystallization. Wunderlich and Mehta claimed that fractionation is governed not by equilibrium considerations but rather by the size of the molecular nucleus under the given conditions. The free energy change on folded-chain crystallization of a molecule on a substrate is given by:

$$\Delta G = vabL_c\Delta g + 2bL_c\sigma_L + 2vab\sigma + 2ab\sigma_{ce} \qquad (8.84)$$

where v is the number of crystallizing stems of a given molecule, σ_{ce} is the extra free energy associated with each chain end and σ and σ_L are the specific surface free energies of the fold surface and the lateral surface, respectively. The quantities a and b are defined in Fig. 8.11. It was suggested by Zachmann (1967) that the major part of σ_{ce} is due to the entropy reduction of the non-crystallized cilia. The size of the critical nucleus (L_{crit}) can be calculated on the basis of eq. (8.84) to be:

$$L_{crit} = \frac{4\sigma\sigma_L b(T_m^0)^2}{(\Delta h^0)^2\Delta T^2} + \frac{2\sigma_{ce}T_m^0}{\Delta h^0\Delta T} + \frac{2kT_cT_m^0}{ab\Delta h^0\Delta T} \qquad (8.85)$$

The first term of eq. (8.85) dominates at small ΔT, whereas the second and third terms dominate at higher ΔT. Wunderlich and Mehta (1974) showed that eq. (8.85) was in agreement with experimental data without the introduction of any adjustable parameters. One of the weak points concerns σ_{ce}, because no independent measurement of σ_{ce} has been made. The

σ_{ce} value used by Wunderlich and Mehta (1974) was 100 mJ m^{-2}, which is only an estimate.

It is instructive to consider a binary mixture of polymers of different molar mass. Let us assume that they are linear polyethylene sharp fractions with significantly different chain lengths. Figure 8.29 illustrates the behaviour. The temperature T_1 is the upper crystallization temperature for the high molar mass (H) polymer. Between temperatures T_1 and T_2 only H crystallizes. Temperature T_2 is the upper crystallization temperature for the low molar mass (L) polymer. Temperatures T_1 and T_2 can be obtained from the graph of critical molar mass against upper crystallization temperature presented in Fig. 8.28. At temperatures below T_2, both polymers crystallize but in separate crystal lamellae. Some limited cocrystallization (crystallization of both polymers in the **same** crystal lamella) occurs at very low crystallization temperatures.

Polymers with branching and tacticity exhibit not only molar mass segregation but also segregation phenomena relating to the structural irregularities. The crystallization temperature range is shifted towards lower temperatures with increasing degree of chain branching. The multicomponent nature of branched polyethylene arises from the fact that the chain branches are randomly positioned on the polymer backbone chain. Segregation is thus never sharp.

Blends of linear and branched polyethylene normally crystallize in two stages. The components crystallize separately provided that they are of similar molar mass. A diagram similar to that presented in Fig. 8.29 may also be constructed for mixtures of linear and branched polyethylene. The linear polyethylene crystallizes at the highest temperatures, forming regular-shaped (dominant) crystal lamellae. The branched polymer crystallizes at lower temperatures in finer, S-shaped lamellae located between the stacks of the dominant lamellae. Crystallization at temperatures below T_2 involves mostly separate crystallization although some cocrystallization may occur, particularly at very low crystallization temperatures. Although linear and branched polyethylene are chemically very similar, the Bristol group (Barham et al. 1988) reported that they may phase-separate in the molten state.

A special case was recently reported by Gedde (1992) and Conde Braña and Gedde (1992): binary mixtures of a low molar mass linear polyethylene ($\bar{M}_w = 2500$ g mol^{-1}; $\bar{M}_w/\bar{M}_n = 1.15$) and higher molar mass branched polyethylenes ($\bar{M}_w \approx 100\,000$ g mol^{-1}; degree of branching $= 0.5$–1.2 mol%) cocrystallized dominantly. The crystallization temperature region of the two components is almost identical. The depression in crystallization temperature with respect to that of high (or intermediate) molar mass polymer is due to the finite molar mass of the linear polymer and to the presence of chain defects (branches) along the chain in the branched polymer. Figure 8.30 presents the strongest evidence for cocrystallization. The transmission electron micrographs of the chlorosulphonated sections (see Chapter 7) show a gradual change from a shorter S or C shape of the crystal lamellae in the pure branched polymer to straight and longer, occasionally roof-shaped crystal lamellae in blends with a high concentration of linear polymer. A monotonous decrease in thickness of the amorphous interlayers with increasing content of linear polyethylene is strikingly apparent in the micrographs shown in Fig. 8.30. This is due to the

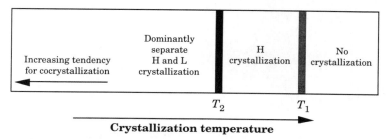

Figure 8.29 Schematic diagram showing the crystallization behaviour of binary mixtures of two linear polyethylene fractions of mass-average molar masses 2500 g mol^{-1} (L) and 66 000 g mol^{-1} (H), respectively. Drawn after data from Gedde (1992).

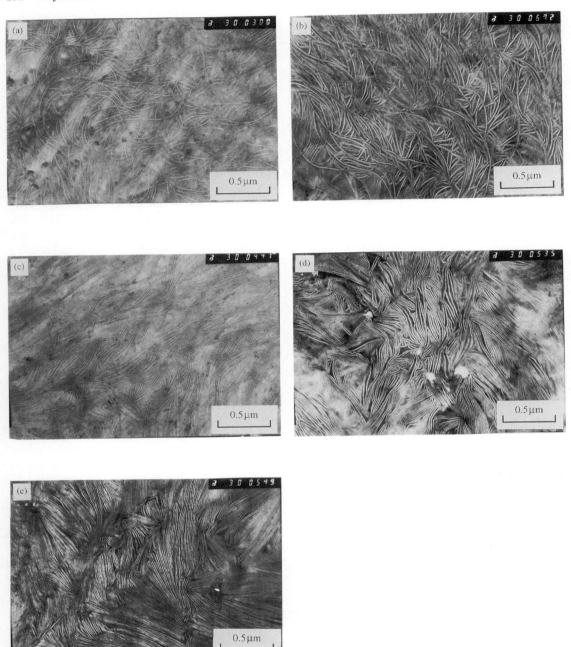

Figure 8.30 Transmission electron micrographs of chlorosulphonated sections of L2.5/BE1.5 crystallized at 387.2 K for 0.7 h and then cooled at a rate of 80 K min^{-1} to room temperature: (a) 0% L2.5; (b) 20% L2.5; (c) 40% L2.5; (d) 60% L2.5; (e) 80% L2.5. From Conde Braña and Gedde (1992), with permission from Butterworth-Heinemann Ltd, UK.

fact that the linear polymer crystallizes in either an extended or a once-folded conformation. The average length of the amorphous chain segments is very small compared to that of the branched polymer. The gradual decrease in average amorphous layer thickness and the narrowing of the amorphous thickness distribution with increasing concentration of linear polymer are strong evidence in favour of cocrystallization of the linear and branched components.

8.5.2 METHODOLOGY FOR REVEALING SEGREGATION AND COCRYSTALLIZATION IN PE BLENDS

The question addressed here is whether the components crystallize separately or together. DSC is probably the most popular method. In this case the melting-point distribution $w(T_m)$ is measured and is transformed into a crystal thickness distribution $w(L_c)$ using the Thompson–Gibbs equation. Let us select a sample with a broad, continuous molar mass distribution. The sample is crystallized under isothermal conditions for different periods of time. Each isothermal treatment is followed by a rapid cooling of the sample to a temperature at which crystallization is no longer active. The melting endotherm is recorded after each cycle of isothermal and rapid cooling treatment, and it consists of two melting peaks. The high temperature peak (HTP) is from the melting of the species crystallizing under the isothermal conditions, whereas the low temperature peak (LTP) is from crystals formed during the cooling stage (Fig. 8.31). The relative size of the LTP decreases monotonously with increasing isothermal crystallization time until a constant value is reached (Fig. 8.31). This crystal fraction constitutes the fraction of the sample which is unable to crystallize at the particular crystallization temperature.

The low-temperature melting material can in 'favourable' cases be isolated by solvent extraction and analysed by methods such as size exclusion chromatography.

Another interesting case concerns the 'solid-state' miscibility of binary mixtures. Crystallization at 392 K followed by rapid cooling to room temperature of a 50/50 blend of two linear polyethylene narrow fractions of $\bar{M}_w = 2500$ g mol^{-1} and 66 000 g mol^{-1}, respectively, led to the melting thermogram shown in Fig. 8.32. Peaks I and II originate from the isothermally crystallized material; peak I from the 66 000 g mol^{-1} material and peak II from the 2500 g mol^{-1} material. Peak III comes from the material crystallizing during the cooling phase, which consists almost exclusively of low molar mass material.

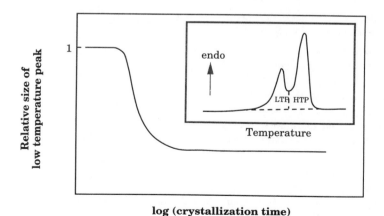

Figure 8.31 Fraction of low temperature peak as a function of isothermal crystallization temperature for a sample with a broad molar mass distribution. Insert: melting trace of sample crystallized under isothermal conditions followed by rapid cooling.

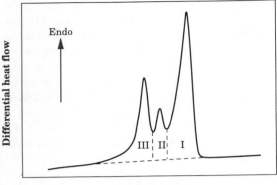

Figure 8.32 Schematic melting thermogram of a binary linear polyethylene blend. Drawn after data presented by Rego-Lopez et al. (1988).

Small-angle X-ray scattering (SAXS) yields information about the long period (crystal plus amorphous thickness) distribution. The presence of two coexisting periodicities indicates segregation. Similar information can be obtained by Raman spectroscopy (see Chapter 7).

Polarized light microscopy can in some cases be useful. Norton and Keller (1984) observed coarse, large spherulites consisting of linear polyethylene and a fine-grained material consisting essentially of segregated branched material that was formed upon subsequent cooling. Segregation occurs, however, in most samples on such a 'local' scale that polarized light microscopy is not a useful technique for making an assessment. Polarized light microscopy can, however, reveal brittleness in the spherulite boundaries and can thus provide indirect evidence of segregation of low molar mass species to the spherulite boundaries.

Transmission electron microscopy provides the most direct information about segregation phenomena. A good example is shown in Fig. 8.30. The lamellar structures of the pure components are compared with the lamellar structure of the blended sample. Morphological features such as lamella thickness, width and form, and amorphous layer thickness provide valuable information for the assessment of segregation. It is useful to combine DSC and transmission electron microscopy. The former technique provides information about the overall segregation, which facilitates the interpretation of the transmission electron micrographs. Scanning electron microscopy of samples treated with a solvent so that only the segregated species are removed from the sample (assessed by DSC confirming the complete removal of the low-temperature melting peak) is another useful technique for morphological characterization of segregation.

8.6 ORIENTATION-INDUCED CRYSTALLIZATION

This phenomenon can, from a thermodynamical point of view, be described according to Fig. 8.33. The molecules align (orientate) under the influence of a mechanical field (shear or elongation). The degree of orientation depends both on the deformation rate and on the relaxation times involved. The relaxation time is a system-response parameter which depends strongly on molar mass. During shear or elongational flow, there is a competition between the applied field and the Brownian motion of the molecules which essentially aims at disorder and isotropicity. The relaxation time for a high molar mass sample is considerably higher than that for a low molar mass sample. That means that orientation more readily occurs in a high molar mass sample than in a low molar mass sample. It is known that orientation causes an increase in the free energy. This is an essential feature of the theory of rubber elasticity (Chapter 3).

The increase in the equilibrium melting point is thus due to the increase in free energy of the melt, the latter being ideally entropy-driven:

$$T_m^0 = \frac{h_m - h_c}{s_m - s_c} = \frac{\Delta h^0}{s_m^0 - \frac{R}{2}\left(\lambda^2 + \frac{2}{\lambda} - 3\right) - s_c^0}$$

$$= \frac{\Delta h^0}{\Delta s^0 - \frac{R}{2}\left(\lambda^2 + \frac{2}{\lambda} - 3\right)} \quad (8.86)$$

where λ is the molecular 'draw ratio', R is the gas constant and Δh^0 and Δs^0 are the quantities referring to the isotropic melt. The second term in the denominator is from the statistical mechanical theory of rubber elasticity (for further details, see

Chapter 3). The effective degree of supercooling at a given crystallization temperature (T_c) increases due to orientation:

$$\Delta T = T_m^0 - T_c = \cfrac{\Delta h^0}{\Delta s^0 - \cfrac{R}{2}\left(\lambda^2 + \cfrac{2}{\lambda} - 3\right)} - T_c$$

$$= \frac{\Delta h^0}{\Delta s_0}\left[\cfrac{\Delta s^0}{\Delta s^0 - \cfrac{R}{2}\left(\lambda^2 + \cfrac{2}{\lambda} - 3\right)}\right] - T_c$$

$$T_m^0(\lambda = 1)\left[\cfrac{1}{1 - \cfrac{\cfrac{R}{2}\left(\lambda^2 + \cfrac{2}{\lambda} - 3\right)}{\Delta s^0}}\right] - T_c \quad (8.87)$$

The kinetics of orientation-induced crystallization has been treated by Janeschitz-Kriegl (1992), Eder, Janeschitz-Kriegl and Liedauer (1992) and Haudin and Billon (1992), essentially using the Avrami equation.

8.7 SUMMARY

Crystallization involves the short-distance diffusion of crystallizable units and the passage of a free energy barrier of enthalpic or entropic origin. The overall crystallization rate ($\dot{w}_c$) can be described according to the equation:

$$\dot{w}_c = C\exp\left(-\frac{U^*}{R(T_c - T_\infty)}\right)\exp\left(-\frac{K_g}{T_c\Delta T}\right) \quad (8.88)$$

where C is a rate constant, U^* is an (energy) constant, R is the gas constant, T_c is the crystallization temperature, T_∞ is a temperature at which all segmental mobility is frozen in, K_g is a kinetic constant for the secondary nucleation, $\Delta T = T_m^0 - T_c$ is the degree of supercooling and T_m^0 is the equilibrium melting point. The first factor (C) depends on the segmental flexibility and on the regularity of the polymer: $C = 0$ for an atactic polymer and C is low for very inflexible polymers such as isotactic polystyrene. The second factor expresses the temperature dependence of the rate of the short-range transport of the crystallizing segments. The third factor expresses the temperature dependence of the nucleation rate.

The overall crystallization can also be compactly described by the Avrami equation, which for polymers takes the form:

$$1 - \frac{v_c}{v_{c\infty}} = \exp(-Kt^n) \quad (8.89)$$

where v_c is the volume crystallinity at time t, $v_{c\infty}$ is the long-time (final) volume crystallinity value, and K and n are constants. The Avrami exponent (n) is dependent on the nucleation mechanism and crystal growth geometry/mechanism. Wunderlich (1978) concludes that 'without the parallel knowledge of the microscopic, independently proven mechanism, the macroscopic, experimentally derived Avrami equation and the Avrami parameters are only a convenient means to represent empirical data of crystallization'. However, a renewed interest in the Avrami equation comes from the development of simulation programmes for injection moulding which also includes solidification simulation. The simple form the Avrami equation makes it suitable for the representation of the solidification of polymer melts in these packages.

The phenomenological laws that any crystal growth theory must address are:

$$G \propto \exp\left(-\frac{K_g}{T_c\Delta T}\right) \quad (8.90)$$

$$L_c^* = \frac{C}{\Delta T} + \delta L_c \quad (8.91)$$

where G is the linear growth rate (e.g. radial growth rate of spherulites). K_g is a kinetic constant, T_c is the crystallization temperature, $\Delta T = T_m^0 - T_c$ is the degree of supercooling, T_m^0 is the equilibrium melting point, L_c^* is the initial crystal thickness, and C and δL_c are constants. Typical for most polymers is that they show different growth-rate regimes (with different constants in this equation). These regimes are referred to as I, II and III, respectively. Chain folding is the third component of the paradigm. Later important observations that have been generally accepted concern the lateral shape of the crystals. Crystallization at lower temperatures leads to faceted crystals, whereas crystallization at high temperatures leads to elliptic or leaf-shaped crystals.

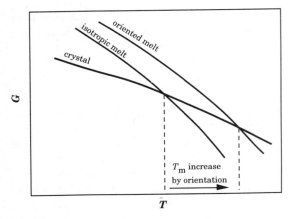

Figure 8.33 Schematic curves of free energy as a function of temperature for crystal, isotropic and oriented melts.

Kinetic crystal growth theories can meet most of these requirements. These theories state that the crystallization rate is strongly dependent on crystal thickness. Crystals below a certain minimum thickness ($2\sigma/\Delta g$, where σ is the fold surface free energy and Δg is the 'volume-related' change in crystallization free energy) are unstable (growth rate is zero) and the free energy barrier for the formation of very thick crystals is very high (growth rate is small). The Lauritzen–Hoffman (LH) theory assumes an enthalpic (nucleation type of) barrier. This theory has undergone a number of modifications during its 30-year life, and its most recent revision was forced by the data concerning the lateral habit. The Sadler–Gilmer theory, which is more recent than the LH theory, is also a kinetic theory but suggests that the free energy barrier to growth is dominantly entropic. Crystal thickening, crystal perfection and molecular fractionation are not explicitly treated by the kinetic theories.

Keller et al. (1994) proposed a new scheme for polymer crystallization, also most probably applicable to melt-crystallization of polyethylene at normal pressure. They proposed that crystallization starts with the formation of thin hexagonal phase crystals, which in this crystal size range are more stable than the orthorhombic crystals. The high mobility of the chains in the hexagonal phase leads to rapid crystal thickening which proceeds until the orthorhombic phase becomes more stable than the hexagonal phase,

i.e. a crystal-size-induced phase transformation occurs. Only limited experimental evidence has so far been presented, but more is said to be coming.

Rejection of low molar mass or 'defect' molecular species during crystallization is referred to as molecular fractionation. Wunderlich and Mehta (1974) showed that at each crystallization temperature there exists a critical molar mass (M_{crit}) such that the molecules of a molar mass greater than M_{crit} are able to crystallize at this temperature, whereas molecules of a molar mass less than M_{crit} are unable to crystallize. The lower limit of segregation is set by the hypothetical equilibrium of crystallization. It is assumed that dynamic equilibrium is achieved between extended-chain crystals and the surrounding melt. At equilibrium, the molecular length of the crystallizable species corresponds sharply to the lamella thickness and molecules which are too short or too long introduce defects and increase the free energy and are thus rejected from the crystal. It is possible to calculate an equilibrium critical molar mass for each temperature of crystallization. Wunderlich and Mehta showed that the experimental and theoretical data were in agreement at high ΔT. The equilibrium critical molar mass calculated was, however, significantly lower than the experimental value at the lower ΔT. It was therefore suggested by Wunderlich that every molecule must undergo molecular nucleation before crystallization. The size of the critical nucleus (L_{crit}) is given by:

$$L_{crit} = \frac{4\sigma\sigma_L b(T_m^0)^2}{(\Delta h^0)^2 \Delta T^2} + \frac{2\sigma_{ce} T_m^0}{\Delta h^0 \Delta T} + \frac{2kT_c T_m^0}{ab\Delta h^0 \Delta T} \quad (8.92)$$

where σ_L is the surface free energy of the lateral surfaces, σ_{ce} is the surface free energy associated with the chain ends, b is the thickness of the stem along the growth direction, and a is the corresponding thickness in the direction perpendicular to the growth direction and the chain axis. The first term of eq. (8.86) dominates at small ΔT, whereas the second and third terms dominate at higher ΔT. Wunderlich and Mehta showed that this equation was in agreement with experimental data without the introduction of any adjustable parameters.

Shearing or elongational flow of a crystallizable melt causes orientation (alignment) of the molecules and an increase in the free energy of the melt. Crystallization of an oriented melt occurs more readily than of the corresponding isotropic melt. This process is referred to as orientation-induced crystallization.

8.8 EXERCISES

8.1. Figure 8.34 shows the three possible temperature dependencies of the overall crystallization rate. Explain.

8.2. The time for solidification of the polymer melt constitutes a significant fraction of the total cycle time for injection moulding. Suggest possible actions which can be taken to shorten the cycle time for a component made of isotactic polypropylene.

8.3. Explain in simple thermodynamic terms why a stretched band of natural rubber crystallizes.

8.4. Fit the general Avrami equation to the following data obtained by DSC at two different temperatures for a highly branched polyethylene sample:

110°C		115°C	
Time (s)	Δh^a (J g^{-1})	Time (s)	Δh^a (J g^{-1})
30	1.9	180	1.6
60	7.5	300	5.1
90	14	450	8.4
120	17	600	12.7
150	18.5	780	14.2
180	20.4	1200	16.5
300	21.5	1500	17.1
600	22.7	1800	17.3
1020	24.2	3600	17.7
1800	23.3	7200	18.6

^{a} Exothermal (crystallization) heat evolved at time t.

8.5. Explain the break in the plot of log G against $1/T_c \Delta T$.

8.6. How can homogeneous nucleation in a polymer be studied?

8.7. How can molecular fractionation be studied? Discuss possible methods of studying phase separation in the melt prior to crystallization.

8.8. Is it possible to cocrystallize high and low molar mass fractions of 'essentially' the same polymer (i.e. with the same repeating unit)?

8.9. Describe different methods for the determination of the surface free energies of the fold surface (σ) and of the lateral surface (σ_L).

8.10. Fit the LH growth-rate equation with $T_m^0 = 410$ K to the following linear growth-rate data obtained by polarized light microscopy of a branched polyethylene:

Temperature (K)	Linear growth rate (μm s^{-1})
379.2	0.51
380.2	0.56
381.2	0.5
382.2	0.404
383.1	0.234
383.1	0.244
384.1	0.206
384.2	0.172
385.6	0.091 9
386.2	0.07
387.2	0.029 8
388.3	0.027 8
389.3	0.015 4
390.2	0.009 26
391.1	0.006 2
392.3	0.002 9
393.2	0.000 777

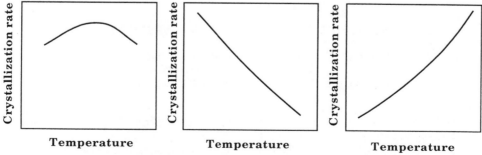

Figure 8.34 Schematic curves showing the overall crystallization rate as a function of temperature.

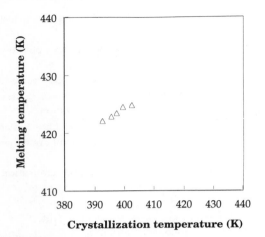

Figure 8.35 Melting peak temperature as a function of crystallization temperature of isotactic polypropylene crystallized at different temperatures to only 5% crystallinity.

8.11. Determine the equilibrium melting point from the data presented in Fig. 8.35.

8.9 REFERENCES

Armitstead, K. and Goldbeck-Wood, G. (1992) *Adv. Polym. Sci.* **100**, 220.
Bank, M. I. and Krimm, S. (1970) *J. Polym. Sci., Polym. Lett.* **8**, 143.
Barham, P. J., Hill, M. J., Keller, A. and Rosney, C. C. A. (1988) *J. Mater. Sci. Lett.* **7**, 1271.
Bassett, D. C., Olley, R. H. and Al Reheil, I. A. M. (1988) *Polymer* **29**, 1539.
Buckley, C. P. and Kovacs, A. J. (1976) *Coll. Polym. Sci.* **254**, 695.
Conde Braña, M. and Gedde, U. W. (1992) *Polymer* **33**, 3123.
Cormier, C. M. and Wunderlich, B. (1966) *J. Polym. Sci.* **4**, 666.
Dosière, M., Colet, M. C. and Point, J. J. (1986) *J. Polym. Sci., Polym. Phys. Ed.* **24**, 345.
Eder, G., Janeschitz-Kriegl, H. and Liedauer, S. (1992) *Progr. Coll. Polym. Sci.* **87**, 129.
Ergoz, E., Fatou, J. G. and Mandelkern, L. (1972) *Macromolecules* **5**, 147.
Frank, F. C. (1974) *J. Cryst. Growth* **22**, 233.
Gedde, U. W. (1992) *Progr. Coll. Polym. Sci.* **87**, 8.
Gedde, U. W., Eklund, S. and Jansson, J.-F. (1983) *Polymer* **24**, 1532.
Goldbeck-Wood, G. and Sadler, D. M. (1989) *Molec. Sim.* **4**, 15.
Haudin, J. M. and Billon, N. (1992) *Progr. Coll. Polym. Sci.* **87**, 132.
Hoffman, J. D. and Miller, R. L. (1989) *Macromolecules* **22**, 3502.
Hoffman, J. D., Frolen, L. J., Ross, G. S. and Lauritzen, J. I. Jr (1975) *J. Res. Nat. Bur. Std. – A. Phys. Chem.* **79A**, 671.
Hoffman, J. D., Guttman, C. M. and DiMarzio, E. A. (1979) *Disc. Faraday Soc.* **68**, 177.
Janeschitz-Kriegl, H. (1992) *Progr. Coll. Polym. Sci,* **87**, 117.
Keller, A. (1957) *Phil. Mag.* **2**, 1171.
Keller, A., Hikosaka, M., Rastogi, S., Toda, A., Barham, P. J. and Goldbeck-Wood, G. (1994) *J. Mater. Sci.* **29**, 2579.
Kovacs, A. J. (1955) *Ric. Sci. (Suppl.)* **25**, 668.
Lauritzen, J. I. Jr (1973) *J. Appl. Phys.* **44**, 4353.
Lauritzen, J. I. Jr and Hoffman, J. D. (1960) *J. Res. Nat. Bur. Std.* **64A**, 73.
Mansfield, M. L. (1988) *Polymer* **29**, 1755.
Mansfield, M. L. (1990) *Polymer Commun.* **31**, 285.
Mehta, A. and Wunderlich, B. (1975) *Coll. Polym. Sci.* **253**, 193.
Norton, D. R. and Keller, A. (1984) *J. Mater. Sci.* **19**, 447.
Point, J. J., Colet, M. C. and Dosière, M. (1986) *J. Polym. Sci., Polym. Phys. Ed.* **24**, 357.
Point, J. J. and Dosière, M. (1989) *Polymer* **30**, 2292.
Point, J.-J. and Villars, D. (1992) *Polymer* **33**, 2263.
Rastogi, S. and Ungar, G. (1992) *Macromolecules* **25**, 1445.
Rego Lopez, J. M., Conde Braña, M. T., Terselius, B. and Gedde, U. W. (1988) *Polymer* **29**, 1045.
Sadler, D. M. (1983) *Polymer* **24**, 1401.
Sadler, D. M. and Gilmer, G. H. (1984) *Polymer* **25**, 1446.
Sanchez, I. C. and DiMarzio, E. A. (1971) *J. Chem. Phys.* **55**, 893.
Wunderlich, B. (1978) *Macromolecular Physics: 2. Crystal Nucleation, Growth, Annealing.* Academic Press, New York and London.
Wunderlich, B. (1980) *Macromolecular Physics: 3. Crystal Melting.* Academic Press, New York and London.
Wunderlich, B. and Mehta, A. (1974) *J. Polym. Sci., Polym. Phys. Ed.* **12**, 255.
Yoon, D. Y. and Flory, P. J. (1977) *Polymer* **19**, 509.
Zachmann, H. G. (1967) *Koll. Z. Z. Polym.* **216–217**, 180.

8.10 SUGGESTED FURTHER READING

Bassett, D. C. (1981) *Principles of Polymer Morphology.* Cambridge University Press, Cambridge.
Hoffman, J. D. and Miller, R. L. (1988) *Macromolecules* **21**, 3038.
Ross, G. S. and Frolen, L. J. (1980) Nucleation and crystallization, in *Methods of Experimental Physics*, Vol. 16, Part B: *Polymers: Crystal Structure and Morphology* (R. A. Fava, ed.). Academic Press, New York.

CHAIN ORIENTATION

9.1 INTRODUCTION

Chain orientation is a phenomenon unique to polymers. The one-dimensional nature of the linear polymer chain makes it possible to obtain strongly anisotropic properties. The anisotropy arises when molecules are aligned along a common director (Fig. 9.1). The intrinsic properties of a polymer chain are strongly directionally dependent. The strong covalent bonds along the chain axis and the much weaker secondary bonds in the transverse directions cause a significant anisotropy of any given tensor property (**x**) (Fig. 9.1). The concept of orientation would be meaningless if the chain-intrinsic properties were isotropic.

Chain orientation always refers to a particular structural unit (Fig. 9.2). It may be a whole sample or a macroscopic part of a specimen. The core of an injection-moulded specimen is typically unoriented, whereas the material of the same specimen closer to the surface is oriented. The director varies as a function of location in both types of sample, which points to the fact that it is possible to find smaller 'domains' of higher degree of orientation (Fig. 9.2). In semicrystalline polymers, the crystalline and amorphous components show different degrees of orientation.

Quiescent polymeric melts with a size of millimetres or larger essentially never show global orientation. It is well known that liquid-crystalline polymers show local orientation (Chapter 6). However, if a molten polymer is subjected to external forces of mechanical, electric or magnetic origin, the molecules may align along a common director. The oriented state is, however, only temporary and prevails only under the influence of the external field. If the orienting field is removed, the molecules relax to adopt a random orientation. The rate at which the oriented system approaches the unoriented state depends on the molecular architecture (primarily molar mass), the order of liquid (e.g. the presence of liquid crystallinity) and the monomeric frictional coefficient (see Chapter 6 for a detailed discussion). Very rapid solidification of an oriented melt makes it possible to make permanent the chain orientation. Rapid solidification can be achieved by rapid cooling at constant pressure or by rapid elevation of the hydrostatic pressure at constant temperature.

9.2 DEFINITION OF CHAIN ORIENTATION

9.2.1 GENERAL BACKGROUND

Chain orientation should not be confused with mechanical strains or frozen-in deformations. Orientation is the result of deformation but a given strain may result in very different degrees of orientation, as the following example illustrates.

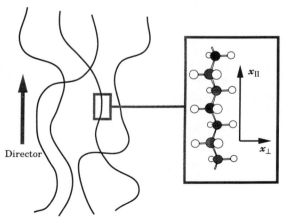

Figure 9.1 Schematic illustration of the concept of chain orientation. The average preferential direction (director) is shown. The insert illustrates intrinsic anisotropy of a polymer chain.

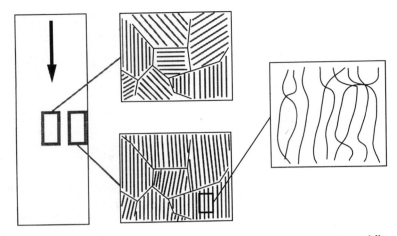

Figure 9.2 Chain orientation in relation to sample size showing the orientation at two different locations in the cross-section of an injection-moulded sample (the arrow on the left indicates flow direction). The middle sample is unoriented, whereas the surface sample is highly oriented along the flow direction.

The average end-to-end distance $\langle r \rangle_0$ of a random coil molecule is given by:

$$\langle r \rangle_0 = Cl\sqrt{n} \qquad (9.1)$$

where C is a constant which depends on the segmental flexibility, l is the bond length and n is the number of bonds. A fully oriented molecule has the end-to-end distance:

$$r_\infty = ln \qquad (9.2)$$

The strain (λ) necessary to reach the completely aligned state is given by the expression:

$$\lambda = \frac{r_\infty}{\langle r \rangle_0} = \frac{\sqrt{n}}{C} \qquad (9.3)$$

The strain to reach full extension, i.e. to attain complete orientation, thus increases with the square root of the molar mass. There is thus no unique relationship between degree of orientation and strain.

Figure 9.3 illustrates this with another example. The orientation of the segments of a molecule in the single crystal is very high indeed. More than 90% of the molecule is perfectly aligned provided that the crystal thickness is greater than 10 nm. The end-to-end vector is not, however, very long. The Gaussian chain shows a similar end-to-end group separation but the segmental orientation is completely random. Segmental orientation is revealed by measurement of optical birefringence, wide-angle X-ray diffraction and infrared spectroscopy. The orientation of the end-to-end vector has only more recently been assessed by neutron scattering. Furthermore, it has been noticed that the stiffness of ultra-oriented fibres does not correlate with the Hermans orientation function but instead with the (molecular) draw ratio

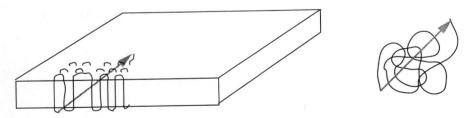

Figure 9.3 A chain in a single crystal and a Gaussian chain.

which more sensitively reflects the extension of the end-to-end vector. So, in conclusion, many properties depend strongly on segmental orientation, while other properties are more controlled by end-to-end vector orientation.

9.2.2 THE HERMANS ORIENTATION FUNCTION

The Hermans orientation function (f) is probably the quantity most frequently used to characterize the orientation. This orientation function was derived by P. H. Hermans in 1946 and is part of an equation which relates optical birefringence to chain (segmental) orientation. A brief derivation is presented here. The first assumption made is that the polarizability associated with each chain segment can be described by a component parallel to the chain axis (p_1) and a component perpendicular (p_2) to the same chain axis (Fig. 9.4). The orientation of the chain segment with respect to a coordinate system is schematically shown in Fig. 9.4. It is assumed that the electrical vector of the propagating light is along the z axis. The electrical field parallel to the chain axis is $E_z \cos \phi$ and the polarization along the chain segment is given by $p_1 E_z \cos \phi$. The contribution of this polarization to the polarization along the z axis amounts to $p_1 E_z \cos^2 \phi$.

The contribution to polarization along z from the transverse polarization of the segment (i.e. along '2') is by analogy equal to $E_z p_2 \sin^2 \phi$. The total polarization along the z axis caused by the electrical vector in the z direction (P_{zz}) is equal to the sum of these two contributions:

$$P_{zz} = E_z(p_1 \cos^2 \phi + p_2 \sin^2 \phi) \qquad (9.4)$$

The polarizability in the z direction (p_{zz}) is given by:

$$p_{zz} = \frac{P_{zz}}{E_z} = p_1 \cos^2 \phi + p_2 \sin^2 \phi \qquad (9.5)$$

Let us now consider the case where the electrical vector is oriented along the x axis. The polarizability in the x direction (p_{xx}) becomes:

$$p_{xx} = p_1 \cos^2 \alpha + p_2 \sin^2 \alpha \qquad (9.6)$$

which, since $\cos \alpha = \cos v \sin \phi$, may be modified to:

$$p_{xx} = p_2 + (p_1 - p_2)\sin^2 \phi \cos^2 v \qquad (9.7)$$

The polarizability tensor (p_{ij}) can be converted to the corresponding refractive index tensor (n_{ij}) according to the Lorentz–Lorenz equation:

$$\frac{n_{ij}^2 - 1}{n_{ij}^2 + 2} = \frac{4\pi}{3} p_{ij} \qquad (9.8)$$

Insertion of eqs (9.6) and (9.7) into eq. (9.8) yields the expressions:

$$\frac{n_{zz}^2 - 1}{n_{zz}^2 + 2} = \frac{4\pi}{3}(p_1 \cos^2 \phi + p_2 \sin^2 \phi) \qquad (9.9)$$

$$\frac{n_{xx}^2 - 1}{n_{xx}^2 + 2} = \frac{4\pi}{3}(p_2 + (p_1 - p_2)\sin^2 \phi \cos^2 v) \qquad (9.10)$$

The birefringence (Δn) and the average refractive index $\langle n \rangle$ are defined as follows:

$$\Delta n = n_{zz} - n_{xx} \qquad (9.11)$$

$$\langle n \rangle = \tfrac{1}{2}(n_{xx} + n_{zz}) \qquad (9.12)$$

Combination of eqs (9.11) and (9.12) yields the expressions:

$$n_{xx} = \langle n \rangle - \frac{\Delta n}{2} \qquad (9.13)$$

$$n_{zz} = \langle n \rangle + \frac{\Delta n}{2} \qquad (9.14)$$

Subtraction of eq. (9.10) from eq. (9.9) yields:

$$\frac{n_{zz}^2 - 1}{n_{zz}^2 + 2} - \frac{n_{xx}^2 - 1}{n_{xx}^2 + 2} = \frac{6\langle n \rangle \Delta n}{(n_{zz}^2 + 2)(n_{xx}^2 + 2)}$$

$$\approx \frac{6\langle n \rangle \Delta n}{(\langle n \rangle^2 + 2)^2} \qquad (9.15)$$

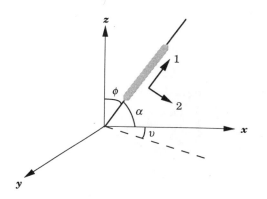

Figure 9.4 Chain segment and coordinate system.

Insertion of eqs (9.9) and (9.10) into eq. (9.15) yields, after algebraic simplification, the following expression:

$$\Delta n = \frac{(\langle n \rangle^2 + 2)^2}{6\langle n \rangle} \cdot \frac{4\pi}{3} (p_1 - p_2)$$
$$\times (1 - \sin^2 \phi - \cos^2 \upsilon \sin^2 \phi) \quad (9.16)$$

A great number of segments are considered and the assumption made at this stage is that orientation is **uniaxial**. All angles υ are equally probable. The average square of the cosine of υ may be calculated as follows:

$$\langle \cos^2 \upsilon \rangle = \frac{1}{2\pi} \int_0^{2\pi} \cos^2 \upsilon \, d\upsilon = \frac{1}{2} \quad (9.17)$$

Insertion of eq. (9.17) in eq. (9.16) gives:

$$\Delta n = \frac{(\langle n \rangle^2 - 2)^2}{6\langle n \rangle} \cdot \frac{4\pi}{3} \cdot (p_1 - p_2) \cdot \left(1 - \frac{3\langle \sin^2 \phi \rangle}{2}\right) \quad (9.18)$$

Hermans orientation function (f) is defined as

$$f = \frac{\Delta n}{\Delta n_0} = 1 - \frac{3}{2} \langle \sin^2 \phi \rangle$$
$$= \frac{3\langle \cos^2 \phi \rangle - 1}{2} \quad (9.19)$$

where the maximum (intrinsic) birefringence Δn_0 is given by

$$\Delta n_0 = \frac{(\langle n \rangle^2 - 2)^2}{6\langle n \rangle} \cdot \frac{4\pi}{3} \cdot (p_1 - p_2) \quad (9.20)$$

Hermans orientation function f takes value 1 for a system with complete orientation parallel to the director and $-\frac{1}{2}$ for the very same sample with the

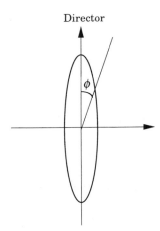

Figure 9.6 Illustration of orientation function $f(\phi)$. The orientation is a density function of only one parameter.

director perpendicular to the chain axis (Fig. 9.5). Non-oriented samples have an f value equal to zero.

It is possible to obtain estimates of the orientation function (f) by a number of methods other than birefringence measurements, e.g. X-ray diffraction and infrared spectroscopy.

The orientation can be viewed in more general terms (Fig. 9.6). Let us represent the orientation by an orientation probability function $f(\phi)$, where the angle ϕ is the angle between the director and the molecular segment. It is implicit in this statement that the orientation in the plane perpendicular to the director is random, i.e. orientation is uniaxial. It is possible to represent $f(\phi)$ by a series of spherical harmonics (Fourier series), and it can be shown that the following equation holds:

$$f(\phi) = \sum_{n=0}^{\infty} \left(n + \frac{1}{2}\right) \langle f_n \rangle f_n(\phi) \quad (9.21)$$

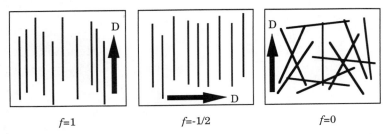

Figure 9.5 Values for the Hermans orientation function for three simple cases. The director (D) is shown for each case.

where the odd components are all zero and the first three even components are given by:

$$f_2(\phi) = \tfrac{1}{2}(3\cos^2\phi - 1) \quad (9.22)$$
$$f_4(\phi) = \tfrac{1}{8}(35\cos^4\phi - 30\cos^2\phi + 3) \quad (9.23)$$
$$f_6(\phi) = \tfrac{1}{6}(231\cos^6\phi - 15\cos^4\phi + 105\cos^2\phi - 5) \quad (9.24)$$

The parameters $\langle f_n \rangle$ are the average values (amplitudes). Note that f_2 is the Hermans orientation function. The full description of uniaxial orientation $f(\phi)$ cannot thus be attained by a single measurement of birefringence. The Hermans orientation function can be given a simple interpretation. A sample with orientation f may be considered to consist of perfectly aligned molecules of the mass fraction f and randomly oriented molecules of the mass fraction $1 - f$. Liquid-crystalline polymers are often characterized by their order parameter, denoted S (Chapter 6). This quantity is equivalent to the Hermans orientation function.

Polymers may exhibit a biaxial orientation. The segmental orientation function is in this case a function of two angular variables, i.e. $f(\phi, v)$. The measurement of the biaxial case is discussed in section 9.4.

9.3 METHODS FOR ASSESSMENT OF UNIAXIAL CHAIN ORIENTATION

There are several methods which are commonly used for the determination of chain orientation: e.g. measurement of in-plane birefringence, wide-angle X-ray diffraction, small-angle X-ray diffraction, infrared spectroscopy and sonic modulus measurements. The first four are briefly presented in this section. Recommended reading for a more in-depth understanding is given at the end of the chapter.

We deal in this section only with the simplest case, namely uniaxial orientation (Fig. 9.7). There is only one unique direction, the $z(3)$ direction. Orientation 'exists' only in the (x, z) and (y, z) planes. The uniaxial system appears to be isotropic in the (x, y) plane.

9.3.1 BIREFRINGENCE

Birefringence measurements are classical. The Hermans orientation function (f) is proportional to the birefringence (Δn):

$$f = \frac{\Delta n}{\Delta n_0} \quad (9.25)$$

where Δn_0 is the maximum birefringence, which can take negative or positive values. Polymers with polarizable units in the main chain, e.g. polyethylene and poly(ethylene terephthalate) have positive Δn_0 values, whereas polymers with strongly polarizable pendant groups, e.g. polystyrene, show negative Δn_0 values. The measured Δn can be the result of contributions from several components (phases), from internal stresses (deformations) and from interfacial effects (so-called form birefringence)

$$\Delta n = \Delta n_f + \Delta n_d + \Delta n_c + \Delta n_a \quad (9.26)$$

where Δn_f is form birefringence, Δn_d is deformation birefringence, and Δn_c and Δn_a are the orientation-induced birefringences originating from the crystalline and the amorphous components. Form birefringence occurs only in multiphase systems. The general

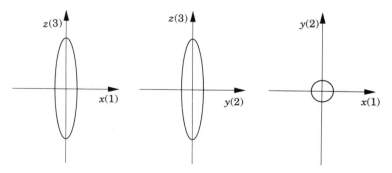

Figure 9.7 Uniaxial orientation.

conditions for it to occur are that the phases have different refractive indices, that at least one of the dimensions of the dispersion is of the order of the wavelength of light, and that the shape of the multiphase structure is anisotropic. Folkes and Keller (1971) showed that an unoriented block copolymer of polystyrene and polybutadiene exhibited a pronounced form birefringence. The polystyrene-rich phase consisted of thin (30 nm) cylinders dispersed in the matrix of the polybutadiene-rich phase. Semi-crystalline polymers may also exhibit form birefringence. Bettelheim and Stein (1958) indicated that the form effects may constitute 5–10% of the total birefringence in polyethylene. Other researchers, e.g. Samuels (1974), have claimed that the form birefringence in oriented isotactic polypropylene is insignificant. Deformation may cause stresses in bonds which also lead to birefringence. This deformation birefringence is usually insignificant and is mentioned only for completeness.

If orientation is uniaxial and both form and deformation birefringence can be neglected, eq. (9.26) can be rewritten:

$$\Delta n = \Delta n_c + \Delta n_a = \Delta n_0 [w_c f_c + (1 - w_c) f_a] \quad (9.27)$$

where f_c and f_a are the Hermans orientation functions of the crystalline and the amorphous components and w_c is the crystallinity.

Birefringence measurements can be made in a polarized light microscope using the configuration described in Fig. 9.9. Analyser and polarizer should be crossed and should be oriented at an angle of 45° to the main optical axes of the sample. A compensator which is a birefringent component is introduced between the sample and the analyser. The compensator makes it possible to change the optical retardation of the vertically and horizontally polarized light components so that the optical

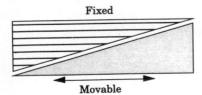

Figure 9.8 Babinet compensator which consists of two quartz wedges with crossed optical axes.

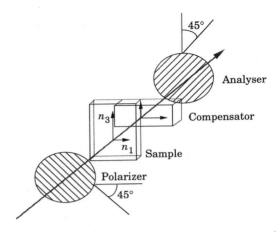

Figure 9.9 Optical system for the measurement of orientation in the plane.

retardation introduced by the sample is compensated for. There are different types of compensator. The Babinet and the tilt compensators are probably the most common.

The Babinet compensator consists of two quartz wedges cut so that their optical axes are mutually perpendicular (Fig. 9.8). A longitudinal shift of the lower wedge changes the optical retardation (R_{31}). A series of fringes is observed around the centre provided that monochromatic light has been used. The longitudinal shift of the lower wedge which shifts the zero-order fringe to the centre gives, after suitable calibration, R_{31} of the sample. With a white light source, containing a continuous spectrum of energy in wavelengths from 400 nm to 750 nm, coloured fringes are observed on both sides of the central black fringe with an optically isotropic sample. The optical retardation is wavelength-dependent and the complex subtraction of light of different wavelengths means that each optical retardation is 'characterized' by a specific colour. This colour scale, i.e. colour as a function of optical retardation, is given in the Michel–Levy chart.

The tilt compensators consist of either a single birefringent calcite plate (Berek type) or two plates of either calcite or quartz cemented together (Ehringhaus compensator). The change in optical retardation is achieved by rotation of the plate which causes a change in both absolute travelling length and refractive index.

The in-plane birefringence ($\Delta n = n_3 - n_1$) is calculated from the measured optical retardation (R_{31}) as follows:

$$\Delta n = \frac{R_{31} \lambda}{d} \quad (9.28)$$

where λ is the wavelength of the light and d is the sample thickness. The compensator method is ideal for static measurements. For dynamic measurements, it is possible to remove the compensator and to measure the intensity of the transmitted light (I) from which the optical retardation (R_{31}) can be calculated:

$$I = I_0 \sin^2(\pi R_{31}) \quad (9.29)$$

where I_0 is the intensity of the incoming light. This equation is only applicable to samples showing an optical retardation less than the wavelength of the light. The in-plane birefringence can then be calculated using eq. (9.28).

Optically clear amorphous polymers are readily studied also using relatively thick samples. However, semicrystalline polymers scatter light and the method of using visible light is only applicable to the study of thinner samples (less than 100 μm). Thicker samples (of millimetre thickness) of polyethylene have been studied with far-infrared interferometry.

9.3.2 WIDE-ANGLE X-RAY DIFFRACTION

The in-plane orientation of (hkl) planes in semicrystalline polymers is revealed by wide-angle X-ray diffraction. Figure 9.10 shows the diffraction patterns of two samples of the same polymer. The unoriented polymer shows concentric rings for the intrachain (00l) and interchain (hk0) reflections. The oriented polymer shows diffractions of both (00l) and (hk0), which are strongly dependent on the azimuthal angle (ϕ). The intrachain reflection is concentrated to the meridian and the interchain reflection to the equator.

The orientation expressed by the average square of the cosine of angle ϕ of the diffracting planes (hk0) and (00l) can be calculated from the equation:

$$\langle \cos^2 \phi_{hkl} \rangle = \frac{\int_0^\pi I_{hkl}(\phi) \cos^2 \phi \sin \phi \, d\phi}{\int_0^\pi I_{hkl}(\phi) \sin \phi \, d\phi} \quad (9.30)$$

where $I_{hkl}(\phi)$ is the scattered intensity of the (hkl) reflection at the azimuthal angle ϕ. It is here assumed that orientation is uniaxial. The $\sin \phi$ factor originates from the fact that the number of populating crystals increases with the radius of the circle, i.e. with $\sin \phi$ (Fig. 9.11).

We can relate the orientation of the different crystal planes (100), (010) and (001) of an orthorhombic cell to a common director as follows:

$$\langle \cos^2 \phi_{100} \rangle + \langle \cos^2 \phi_{010} \rangle + \langle \cos^2 \phi_{001} \rangle = 1 \quad (9.31)$$

Equation (9.31) may be applied to the simple case

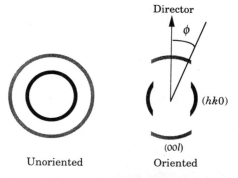

Figure 9.10 Schematic X-ray diffraction patterns from an unoriented and an oriented crystalline polymer. Definition of the azimuthal angle ϕ.

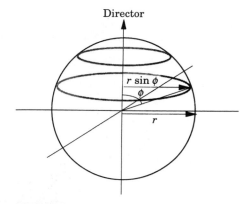

Figure 9.11 Illustration of the dependence of the number of diffracting planes on the angle ϕ.

shown in Fig. 9.10 with only one interchain reflection:

$$2\langle \cos^2 \phi_{hk0}\rangle + \langle \cos^2 \phi_{00l}\rangle = 1 \quad (9.32)$$

Equations (9.31) and (9.32) are readily transformed into the corresponding Hermans orientation functions (f_{hkl}):

$$f_{100} + f_{010} + f_{001} = 0 \quad (9.33)$$

and

$$2f_{hk0} + f_{00l} = 0 \quad (9.34)$$

One of the main attractions of the X-ray method is that the orientation function of crystalline planes, $f(\phi)$ in the case of uniaxial orientation and $f(\phi, v)$ in the case of biaxial orientation, can be revealed. The orientation can also be expressed in the Hermans orientation function. X-ray diffraction primarily assesses crystalline orientation. However, the amorphous halo may show azimuthal angle dependence and, the amorphous orientation can be estimated provided that a certain amorphous peak can be assigned to either an interchain or an intrachain spacing.

9.3.3 SMALL-ANGLE X-RAY DIFFRACTION

Small-angle X-ray diffraction provides information about the period of the lamella stacking in semicrystalline polymers and about the layer thickness of smectic liquid-crystalline polymers. The azimuthal angle dependence of the small-angle pattern provides information about the orientation of these 'superstructures' (Fig. 9.12).

The two-point pattern indicates that the lamella planes are perpendicular to the director. The orientation of these lamellar structures can be calculated from eq. (9.30). The four-point pattern indicates that the lamella normal scatters around an angle β to the director.

9.3.4 INFRARED (IR) SPECTROSCOPY

IR spectroscopy is very useful for the assessment of chain orientation. The measurement of IR dichroism requires the use of IR radiation with parallel and perpendicular polarization to a selected reference direction. Figure 9.13 shows the fundamental principle underlying the dichroism.

The theory developed by Beer predicts that the absorbance (A) depends very strongly on the angle (κ) between the electric vector and the transition moment vector:

$$A \cong [|\bar{\mathbf{E}}| \cdot |\bar{\mathbf{M}}| \cdot \cos \kappa]^2 \quad (9.35)$$

The IR radiation is strongly absorbed when the electric vector and the transition moment vector are parallel (Fig. 9.13). In the case of perpendicular orientation, no absorption occurs (Fig. 9.13). The dichroic ratio (R) is defined as:

$$R = \frac{A_\parallel}{A_\perp} \quad (9.36)$$

where $A_\parallel$ is the absorbance of polarized light parallel to the director and $A_\perp$ is the absorbance of polarized light perpendicular to the director. Hermans orientation function (f) is given by:

$$f = \frac{(R-1)\cdot(R_0+2)}{(R+2)\cdot(R_0-1)} \quad (9.37)$$

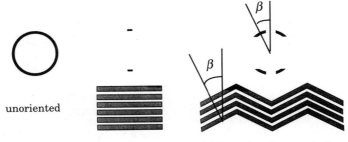

Figure 9.12 Small-angle X-ray patterns of different lamella organizations.

Methods for assessment of uniaxial chain orientation

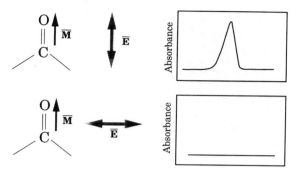

Figure 9.13 Absorption of infrared light of different polarization ($\bar{E}$) with respect to the transition moment vector ($\bar{M}$).

where R_0 is the dichroic ratio for a sample with perfect uniaxial orientation and is dependent on the angle between the transition moment vector and the chain axis (ψ, Fig. 9.14):

$$R_0 = 2 \cot^2 \psi \quad (9.38)$$

Absorption bands with $\psi = 54.76°$ show no dichroism ($R_0 = 1$). Absorption bands associated with a perpendicular transition moment vector have $R_0 = 0$ and the Hermans orientation function is given by:

$$f = 2 \cdot \frac{(1 - R)}{(R + 2)} \quad (9.39)$$

One of the attractions of the IR technique is that the dichroism of the absorption bands assigned to different groups, and the orientation of different groups of the repeating unit can be determined. The orientation of a side-chain smectic poly(vinyl ether) (Fig. 9.15) was measured by Gedde et al. (1993). The cyano-stretching band at 2230 cm^{-1} with a transition moment vector parallel to the long axis of the mesogen showed a dichroic ratio corresponding to

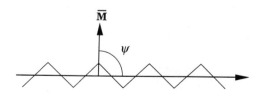

Figure 9.14 Definition of angle between transition moment vector and chain axis.

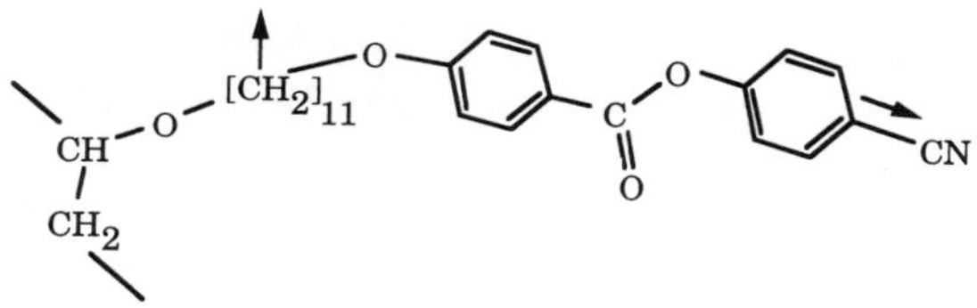

Figure 9.15 Structure of repeating unit of side-chain poly(vinyl ether). Approximate directions of the transition moment vectors are indicated.

$f = 0.7$–0.8, whereas the symmetric and asymmetric CH-stretching bands at 2854 cm^{-1} and 2927 cm^{-1} associated essentially with the spacer showed that the space group was considerably less oriented, $f = 0.3$–0.4.

The dichroic ratio can also be measured using ultraviolet or visible radiation (UV-visible spectroscopy). Polymers with carbonyl groups, phenyl groups and other conjugated systems may be studied.

9.3.5 SONIC MODULUS

The velocity of sound depends on the modulus, and the velocity of sound is greater when it propagates along than transverse to the chain axis. In a semicrystalline polymer, both the crystals and the amorphous phase contribute in proportion to their relative contents. In that sense, the sonic modulus is similar to birefringence. The following expression relating sonic modulus (E) and Hermans orientation factor for a semicrystalline polymer was derived by Samuels (1974):

$$\frac{3}{2}\left(\frac{1}{E} - \frac{1}{E_u}\right) = \frac{v_c f_c}{E_{t,c}} + \frac{(1 - v_c)f_a}{E_{t,a}} \quad (9.40)$$

where E_u is the modulus of the unoriented polymer, v_c is the volume crystallinity, f_c and f_a are the Hermans orientation functions of the crystalline and amorphous components, and $E_{t,c}$ and $E_{t,a}$ are the moduli for the propagation of the sound perpendicular to the chain in the crystalline and the amorphous fractions.

9.3.6 AMORPHOUS AND CRYSTALLINE ORIENTATION

Equation (9.27) states that the overall orientation of a semicrystalline polymer is the sum of contributions from the amorphous and crystalline components. The overall orientation (f) can be obtained by measurement of birefringence or the sonic modulus, and the crystalline orientation (f_c) can be obtained by wide-angle X-ray diffraction. The amorphous orientation (f_a) can then be obtained indirectly from the equation:

$$f_a = \frac{f - f_c w_c}{1 - w_c} \quad (9.41)$$

It is relatively difficult to make a direct determination of the amorphous orientation function of a semicrystalline polymer. Determination by X-ray diffraction recording the azimuthal angle dependence of the amorphous halo and infrared dichroism of an amorphous absorption band may be useful in addition to the indirect method suggested above.

9.4 METHODS FOR ASSESSMENT OF BIAXIAL CHAIN ORIENTATION

Biaxiality is illustrated in Fig. 9.16. In this case, it is possible to find two orthogonal planes in both of which the chains have a preferential direction. The orientation function is a function of two angular variables.

The biaxial orientation can be assessed by optical methods measuring the in-plane birefringence in the three orthogonal directions. The three Δn values may be derived by tensor summation of eq. (9.16):

$$n_3 - n_1 = \Delta n_0 \left(f - \frac{g}{2} \right) \quad (9.42)$$

$$n_3 - n_2 = \Delta n_0 \left(f + \frac{g}{2} \right) \quad (9.43)$$

$$n_1 - n_2 = \Delta n_0 g \quad (9.44)$$

where $f = \frac{1}{2}(3\langle \cos^2 \phi \rangle - 1)$ and $g = \langle \sin^2 \phi \cos 2v \rangle$.

X-ray diffraction can also be used, the X-ray diffraction patterns being recorded at different angles. It is in principle possible to obtain the complete crystalline orientation function ($= f(\phi, v)$). Biaxiality may not only refer to the chain axis. In a single crystal of polyethylene, for example, the chain orientation is indeed very high (f typically greater than 0.95), but

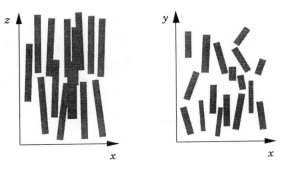

Figure 9.16 Schematic drawing showing the orientation of uniaxial moieties in two orthogonal planes.

the single crystal also shows equally significant a-axis and b-axis orientations. These 'secondary' bond orientations are obviously less important for properties because the largest difference found is that between the covalent bond and the secondary bond. However, certain variations in properties arise from a-axis and b-axis orientation, particularly in cases involving very strong secondary bonds, e.g. hydrogen bonds.

9.5 HOW CHAIN ORIENTATION IS CREATED

The methods by which chain orientation is created may be divided into solid-state and liquid processes. Solid-state processes involve a plastic deformation of an isotropic or weakly anisotropic solid. Glassy amorphous polymers can only be drawn to moderate strains. The polymers which can be substantially oriented are semicrystalline with deformable amorphous ($T > T_g$) and crystalline phases. The remaining discussion in this chapter is confined to this group of polymers. Deformation can be achieved by cold-drawing, extrusion or rolling (Fig. 9.17).

Cold-drawing leads to necking, the formation of a localized zone in which the 'unoriented' structure is transformed into a fibrous structure. The neck zone, i.e. the shoulder, travels through the specimen until the entire sample is drawn to a fibrous structure. A commonly used parameter which characterizes the extent of drawing is the draw ratio (λ):

$$\lambda = \frac{L}{L_0} \approx \frac{A_0}{A} \quad (9.45)$$

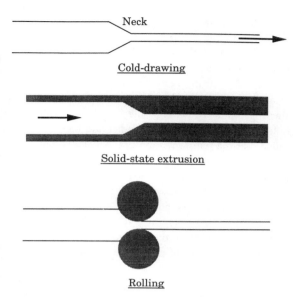

Figure 9.17 Solid-state processes to orient a polymer.

where L is the specimen length after drawing, L_0 is the original specimen length, A_0 is the original cross-sectional area and A is the cross-sectional area after drawing. The necking also causes the formation of a great many voids. Due to these voids the cold-drawn fibrous material is thus often opaque. The draw ratio reached after necking is denoted the natural draw ratio and takes values between 4 and 10. Polymers of suitable morphology can be drawn after necking to very high draw ratios, e.g. to $\lambda = 40$ for linear polyethylene. In many cases, these very high draw ratios can only be attained by drawing at elevated temperatures, typically 10–40°C below the melting point.

During solid-state extrusion, the solid polymer is pushed through a narrow hole and a very significant pressure is built up. A plug of solid polymer is pushed through the hole in the cold extrusion process, whereas the solid plug is surrounded by pressure-transmitting oil in the case of hydrostatic extrusion. Solid-state extrusion to high draw ratios is only possible at temperatures above the onset temperature for the crystalline α process (see Chapter 7). Linear polyethylene is extruded at temperatures between 80°C and 110°C. The entrance angle is important. Small angles lead to slow deformation and a possibility for the material to return to the unoriented state by relaxation. Higher entrance angles cause more rapid deformation and a higher efficiency of the orientation process. Fracture may occur when very high entrance angles are used. Zachariades, Mead and Porter (1979) developed an extrusion process which was intermediate between the solid-like and liquid-like processes. The hydrostatic extrusion of polyethylene was performed at 132–136°C and above a certain critical deformation rate. The pressure increased rapidly, which promoted oriented crystallization in the capillary entrance. The output of this process was a highly oriented and optically clear fibre.

The ability of a polymer to be drawn or extruded to an ultra-oriented fibre is dependent on several material factors:

- The presence of a crystalline α relaxation is a necessary condition for the possibility to extrude a polymer to a high draw ratio. The α process involves slippage of the chains through the crystals, and polymers having this ability also tend to be able to deform plastically into a highly oriented morphology. Polyamides exhibit no crystalline α process and solid-state extrusion of polyamide 6 is only possible at small draw ratios. Drawing or extrusion is preferably carried out at a temperature between the temperature of onset of the α process and the melting temperature.
- Chain entanglements make the plastic deformation difficult and high molar mass polymers crystallized under normal conditions cannot be drawn to very high draw ratios. It is also known that crystallization conditions leading to the formation of only very few chain entanglements favour extensibility and the attainment of an ultra-oriented fibre.
- The number of interlamellar tie chains must be sufficiently high to prevent early brittle fracture.

Cold-drawing/solid-state extrusion of semicrystalline polymers involves initially the deformation of the spherulitic structure, the subsequent transformation of the spherulitic structure to a fibrillar structure and, finally, the plastic deformation of the fibrillar structure.

The most intriguing part is the transformation stage. It was suggested by Peterlin (1979), one of the pioneers in the field, that the crystal lamellae twist (rotate) and break up into smaller crystallites which are pulled into long and thin microfibrils (a sandwich consisting of crystal blocks 10 nm thick and 10 nm wide, connected with many taut tie chains). This mechanism seems not to account for the observed change in the crystal thickness accompanying the transformation process. Peterlin showed that the crystal thickness (L_c) of the fibrous polymer depends on the degree of supercooling (ΔT) prevailing during the transformation of the spherulitic to a fibrillar structure, according to an expression valid for any crystallization:

$$L_c = \frac{C_1}{\Delta T} + C_2 \quad (9.46)$$

where C_1 and C_2 are constants. The validity of eq. (9.46) indicated that the transformation was accompanied by both melting and recrystallization. In a recent study by small-angle neutron scattering, Sadler and Barham (1990) showed that cold-drawing of linear polyethylene at temperatures greater than 70–90 °C indeed involved melting and recrystallization. However, at temperatures lower than 70–90 °C, no melting occurred and the ΔT-dependence of the crystal thickness must be explained differently. Sadler and Barham suggested that laterally small crystalline blocks are broken out from the crystal lamellae so that the regular stacking is partially lost. The small-angle X-ray diffraction pattern is smeared out. Furthermore, Sadler and Barham suggested that the thinner crystals are not disintegrated to the same extent but instead rotate to adapt to the fibrillar orientation, leaving the regularity of the lamella stacking and providing the small-angle X-ray diffraction patterns recorded. Sadler and Barham also found that the transformation of the melt-crystallized spherulitic structure to a fibrillar structure was accomplished by affine deformation of the molecules. The molecular draw ratio was thus the same as the measured macroscopic draw ratio. Drawing of single crystal mats of polyethylene led, however, to non-affine deformation during necking.

The phenomenon of necking is closely related to that of yielding. Figure 9.18 shows that the yield stress is strictly proportional to the crystal thickness and that the regression line intersects the axes at the origin. The yield stress increases with increasing mass crystallinity (at constant crystal thickness). It may be concluded that the yield as a precursor to necking is controlled by deformation within the crystals. The exact mechanism for the yielding of semicrystalline polymers is not known but several proposals have been made: intralamellar slip along the chain axis involving different ($hk0$) planes, twinning, and thermal activations of screw dislocations.

Sadler and Barham (1990) also recorded the change in molecular draw ratio during the post-necking

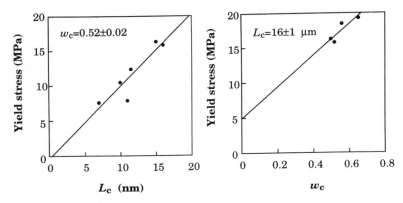

Figure 9.18 Yield stress of polyethylene as a function of crystal thickness (L_c) at constant crystallinity ($w_c = 0.52$) and as a function of mass crystallinity (w_c) at constant crystal thickness ($L_c = 16$ nm). Drawn after data from Young (1988).

deformation and found that it was equal to the change in the measured macroscopic draw ratio. Post-necking deformation is thus affine. Earlier data by electron microscopy and Raman spectroscopy of ultra-oriented cold-drawn or solid-state extruded polyethylene indicated the presence of extended-chain fibrillar crystals together with oriented folded-chain crystals. It is believed that this longitudinal continuity of the molecules has a profound positive effect on the stiffness of the fibre.

The liquid processes include melt-spinning and solution-spinning. The melt or the solution is strained and the molecules are extended from their equilibrium isotropic shapes. High molar mass species show the longest relaxation times and are the most likely to crystallize while being oriented and hence to form extended-chain crystals. Later crystallizing species form epitaxial folded-chain lamellae. The success of a 'liquid process' is strongly dependent on the oriented state being frozen in with a minimum of relaxation. The mobile oriented liquid may be 'quenched' by a rapid elevation of the hydrostatic pressure or by rapid cooling. The resulting morphology is referred to as 'interlocked shish-kebab' (Chapter 7). The melt-drawing process was developed for high molar mass polyethylene by Mackley and Keller (1973) and an axial modulus approaching 100 GPa was achieved in the best cases. Kevlar (polyterephthalamide) is spun from a solution in concentrated sulphuric acid. The solution shows nematic mesomorphism and is readily oriented.

Injection moulding causes a more moderate orientation of the material near the surfaces of the mould. The chain orientation is generally along the major flow direction. Figure 9.19 shows the structural layering of a liquid-crystalline polymer. The outermost layer is formed by elongational flow. The orientation within the next layer is due to shear flow. Conventional polymers show principally a similar orientation pattern, although the orientation is weaker and the thickness of the oriented layers is considerably smaller than that of liquid-crystalline polymers.

Finally, a few words about other methods used to achieve orientation: Electrical and magnetic fields have been used to align both monomers, particularly liquid-crystalline, and polymers. The field strength needed to orient polymers is very high, several orders

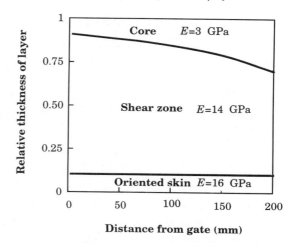

Figure 9.19 Relative thickness of microstructural layers of a 3 mm thick injection-moulded ruler of liquid-crystalline poly(p-hydroxybenzoic acid-co-ethylene terephthalate). The axial moduli (E) of the materials in the different layers are shown. Drawn after data from Hedmark et al. (1988).

of magnitude greater than that required to orient small molecules.

9.6 PROPERTIES OF ORIENTED POLYMERS

The first question asked in this section is concerned with the properties of a perfectly aligned polymer. Large samples showing perfect orientation have not been prepared. The properties are either measured on a very small piece, e.g. a single crystal (crystallite), or are calculated from other measurements. The highest possible axial elastic modulus of a polymer can be experimentally determined from X-ray diffraction data by measuring the change in the distance between crystal planes in accurately stressed crystallites. It can also be calculated from data obtained by vibrational spectroscopy. The latter provides data on the elastic constants of the individual bonds deformed by stretching, and the bond angle deformation, and a theoretical modulus can be calculated. The axial and transverse thermal expansivities are readily obtained from X-ray diffraction data. Table 9.1 presents axial elastic modulus data (along the chain axis) for a few selected polymers.

Table 9.1 The maximum elastic modulus at room temperature of a few selected polymers

Polymer	Elastic modulus (GPa)
Polyethylene	240–360
Isotactic polypropylene	42
Polyoxymethylene	54
Poly(ethylene terephthalate)	140
Polyamide 6	250
Diamond	800

Source: Holliday and White (1971).

A simple but very useful principle is to consider that the elastic deformation of a single molecule occurs by three possible mechanisms (Fig. 9.20): bond stretching, bond-angle deformation and torsion about a σ bond. The elastic constants for these three deformation mechanisms are very different. The following comparative values may be used: 100 (bond stretching), 10 (bond angle deformation) and 1 (torsion about a σ bond).

The high axial elastic modulus of polyethylene and polyamide 6 is due to the fact that these polymers have a preferred conformation that is fully extended, i.e. all-trans. The elastic deformation is caused by the deformation of bond angles and by bond stretching, both showing high elastic constants. Isotactic polypropylene and polyoxymethylene crystallize in helical conformations and therefore exhibit a maximum stiffness which is only 20% of the maximum stiffness of the all-trans polymers. The elastic deformation of a helical chain involves, in addition to the deformation of bond angles and bond stretching, deformation by torsion about the σ bonds. The latter shows a very low elastic constant. The relatively low maximum stiffness of the fully extended (all-trans) poly(ethylene terephthalate) is due the low molecular packing of this polymer. The three-dimensional covalent-bond structure of diamond leads to its extraordinarily high stiffness by largely inhibiting bond-angle deformation.

The transverse modulus is determined by the weak secondary bonds and is for this reason only a small fraction of the axial (chain) elastic modulus. X-ray diffraction indicates the following values for linear polyethylene: $E_c \approx 300$ GPa and $E_a \approx E_b \approx 3$ GPa.

The sonic modulus and thermal expansivity are both closely related to the elastic properties of the molecules. The thermal expansivity parallel to the chain axis (α_2) is negative for many polymers (Chapter 7). Thermal vibration leads however to thermal expansion in the perpendicular directions (α_1). The optical properties belong to a different group. The refractive index (tensor) is related to the polarizability (tensor) according to the Lorentz–Lorenz equation (eq. 9.8). In one group of polymers, polarizability is larger in the chain axis than in the transverse direction. This leads to a polymer with an intrinsic birefringence (Δn_0) greater than zero. The other group of polymers have strongly polarizable pendant groups and a negative intrinsic birefringence. It is difficult to make direct measurement of perfectly aligned samples. Highly oriented samples are measured and a low limit of Δn_0 can be obtained. In some cases, it is possible to determine the Hermans orientation function independently and, from a measurement of the in-plane birefringence of this particular sample, Δn_0 can be calculated using eq. (9.19). Theoretical calculation

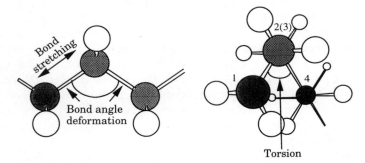

Figure 9.20 Deformation of a polyethylene molecule by bond stretching, bond angle deformation (both left-hand side) and torsion about the σ bond linking carbons 2 and 3 (right-hand side).

Table 9.2 Summary of relationships between chain orientation and a few selected properties

Property	Relation	Other relations
Birefringence (Δn)	$\Delta n = \Delta n_0 f$ Uniaxial orientation[a]	$n_3 - n_1 = \Delta n_0 \left(f - \dfrac{g}{2}\right)$ $n_3 - n_2 = \Delta n_0 \left(f + \dfrac{g}{2}\right)$ $n_1 - n_2 = \Delta n_0 g$ Biaxial orientation[b]
Thermal expansivity (α)	Linear expansivities:[c] $\alpha_\parallel = \alpha_0 - \tfrac{2}{3}(\alpha_2 - \alpha_1)f$ $\alpha_\perp = \alpha_0 + \tfrac{1}{3}(\alpha_2 - \alpha_1)f$ Uniaxial orientation	Volume expansivity: $\alpha_v = \alpha_\parallel + 2\alpha_\perp =$ const.
Thermal conductivity (λ_c)	$\dfrac{1}{\lambda_{c\parallel}} = \dfrac{1}{\lambda_{c0}} - \dfrac{2}{3}\left(\dfrac{\frac{1}{\lambda_{c2}} - \frac{1}{\lambda_{c1}}}{}\right)f$ Uniaxial orientation[d]	$\dfrac{1}{\lambda_{c\perp}} = \dfrac{1}{\lambda_{c0}} + \dfrac{1}{3}\left(\dfrac{\frac{1}{\lambda_{c2}} - \frac{1}{\lambda_{c1}}}{}\right)f$ Uniaxial orientation
Elastic compliance (J)	$\dfrac{J_\parallel}{J_0} = 1 - f$ Uniaxial orientation simplified formula; small/medium f values	For ultra-oriented polymers: good correlation only with draw ratio

[a] f is the Hermans orientation function and Δn_0 is the maximum birefringence obtained at perfect uniaxial orientation.
[b] f is the Hermans orientation function (3 is the reference direction); $g = \langle \sin^2 \phi \cdot \cos 2v \rangle$.
[c] $\alpha_1 =$ thermal expansivity along chain axis; $\alpha_2 =$ thermal expansivity perpendicular to chain axis; $\alpha_0 =$ thermal expansivity of unoriented polymer.
[d] $\lambda_{c1} =$ thermal conductivity along chain axis; $\lambda_{c2} =$ thermal conductivity perpendicular to chain axis; $\lambda_{c0} =$ thermal conductivity of unoriented polymer.

based on polarizability data is another possible method.

What is the relationship between degree of orientation and a given property? The answer obviously depends on what property is asked for and also what polymer is considered. Table 9.2 presents a very brief answer involving only a few properties.

Let us discuss in more detail the elastic properties of highly oriented polymers. It is known that the elastic modulus shows a good correlation with the draw ratio. For a given polymer, this relationship is fairly uncommon. The correlation between elastic modulus and the Hermans orientation functions f, f_c or f_a is not unique in that sense. It may be argued on the basis of the data of Sadler and Barham (1990) that the relevant correlation is between the Young's modulus and molecular draw ratio (Fig. 9.21). A high c-axis orientation is achieved already at relatively low draw ratios and it increases only moderately on further drawing. The molecular draw ratio assesses the extension of the end-to-end vectors and the axial molecular continuity of the microfibrils. It is believed that long and thin molecular connections, amorphous or crystalline, constitute a 'third' strongly reinforcing component in the oriented semicrystalline polymer.

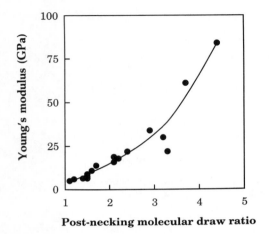

Figure 9.21 Young's modulus of cold-drawn polyethylene as a function of the post-necking molecular draw ratio. The molecular draw ratio precisely after necking is close to 8. Drawn after data from Sadler and Barham (1990).

9.7 SUMMARY

Chain orientation arises from the one-dimensional character of polymers. Most properties are direction-dependent with respect to the polymer chain. The origin of this intrinsic anisotropy lies in the presence of two different types of bonds, namely the covalent bonds and the family of weak, secondary bonds. Polarizability also shows a very strong directional dependence. The elastic modulus of polymers is always highest parallel to the chain axis. Polymers with a fully extended (all-trans) conformation show room-temperature (intrinsic) values between 150 and 350 GPa. The modulus of polymers with a helical conformation is only about 50 GPa. The transverse moduli which are controlled by the secondary bonds are equal to only a few gigapascals. The directional dependence of the polarizability leads to birefringence. The intrinsic (maximum) birefringence can, when strong polarizable groups are present in the backbone chain, amount to 0.2–0.3.

Chain orientation is a purely geometrical quantity of a system. If the polymer segments have a preferential direction, the system is said to be oriented. The 'system' can be a whole specimen, a small part of that specimen, the crystalline component of the specimen, the amorphous component of the specimen, etc. The preferential direction is called the 'director' and the angle between the chain segments and the director is denoted ϕ. Hermans orientation function (f), the most frequently used quantity for the characterization of uniaxial orientation, is defined as follows:

$$f = \frac{3\langle\cos^2\phi\rangle - 1}{2} \quad (9.47)$$

P. H. Hermans showed in the 1940s that the in-plane birefringence ($\Delta n = n_3 - n_1 = n_3 - n_2$, where '3' denotes the director) is proportional to the orientation function:

$$\Delta n = \Delta n_0 f \quad (9.48)$$

where Δn_0 is the maximum (intrinsic) birefringence. The Hermans orientation function takes value 1 for a system with perfect orientation parallel to the director, and takes value $-\frac{1}{2}$ for the very same sample but with the director perpendicular to the chain axis. The Hermans orientation function is zero for a unoriented sample. Liquid-crystalline polymers are often characterized by their order parameter (denoted S). This quantity is equivalent to the Hermans orientation function.

The orientation can be viewed in more general terms by an orientation probability function $f(\phi)$. It is implicit in this statement that the orientation in the plane perpendicular to the director is random, i.e. orientation is uniaxial. It is possible to represent $f(\phi)$ by a series of spherical harmonics (Fourier series):

$$f(\phi) = \sum_{n=0}^{\infty}\left(n + \frac{1}{2}\right)\langle f_n\rangle f_n(\phi) \quad (9.49)$$

where the odd components are all zero and the first three even components are given by:

$$f_2(\phi) = \tfrac{1}{2}(3\cos^2\phi - 1) \quad (9.50)$$
$$f_4(\phi) = \tfrac{1}{8}(35\cos^4\phi - 30\cos^2\phi + 3) \quad (9.51)$$
$$f_6(\phi) = \tfrac{1}{6}(231\cos^6\phi - 15\cos^4\phi + 105\cos^2\phi - 5) \quad (9.52)$$

The parameters $\langle f_n\rangle$ are the average values. Note that f_2 is the Hermans orientation function. The full description of uniaxial orientation $f(\phi)$ can thus not be given by a single measurement of birefringence.

Polymers may also show biaxial orientation. The segmental orientation function is in this case a function of two angular variables, i.e. $f(\phi, v)$. Biaxial orientation can be assessed by optical methods measuring the in-plane birefringence in the three orthogonal directions:

$$n_3 - n_1 = \Delta n_0 \left(f - \frac{g}{2} \right) \quad (9.53)$$

$$n_3 - n_2 = \Delta n_0 \left(f + \frac{g}{2} \right) \quad (9.54)$$

$$n_1 - n_2 = \Delta n_0 g \quad (9.55)$$

where $f = \frac{1}{2}(3\langle \cos^2 \phi \rangle - 1)$ and $g = \langle \sin^2 \phi \cos 2v \rangle$.

Chain (segmental) orientation can be determined by a number of methods, e.g. optical methods measuring birefringence, wide-angle X-ray diffraction (crystalline orientation), infrared spectroscopy and measurement of sonic modulus. Determination of birefringence and sonic modulus provides only data yielding the Hermans orientation function whereas the X-ray diffraction method provides information about the full orientation function ($f(\phi)$ or $f(\phi, v)$). The orientation of the end-to-end vector has only more recently been assessed by neutron scattering.

Orientation is the result of deformation. 'Mobile' molecules are extended by the application of an external force field (mechanical, electric or magnetic). The oriented state is frozen in by 'quenching' of the sample, e.g. by rapid cooling or by a pressure increase. The methods by which chain orientation is obtained may be divided into solid-state and liquid-state processes. The solid-state processes, i.e. cold-drawing, extrusion or rolling, involve a plastic deformation of an isotropic or weakly anisotropic solid. The polymers which can be substantially oriented are semicrystalline, with deformable amorphous and crystalline phases. Cold-drawing/solid-state extrusion of a semicrystalline polymer involves initially the deformation of the spherulitic structure, the subsequent transformation of the spherulitic structure to a fibrillar structure and finally the plastic deformation of the fibrillar structure. A polymer can only be cold-drawn or solid-state extruded to an ultra-oriented fibre if the following conditions are fulfilled: a crystalline α relaxation must be present; chain entanglements must be largely absent; and the number of interlamellar tie chains must be sufficiently high to prevent early brittle fracture.

The liquid processes include melt-spinning and solution-spinning of crystalline or liquid-crystalline polymers. The melt or the solution is strained and the molecules are stretched from their equilibrium isotropic Gaussian states (flexible chains) or simply aligned with the flow field (rigid rods). It is important that the oriented liquid-like state is rapidly transformed to the solid state with a minimum of relaxation. This is more readily achieved with high molar mass polymers. The oriented liquid may be 'quenched' by a rapid elevation of the hydrostatic pressure or by rapid cooling producing an 'interlocked shish-kebab' morphology in the case of polyethylene.

Most properties are strongly influenced by chain orientation. Birefringence, thermal expansivity, thermal conductivity and the elastic modulus depend on the Hermans orientation function according to relatively simple formulae. However, the elastic modulus of ultra-oriented polymers depends more directly on the macroscopic and molecular draw ratio. The latter reflects the extension of the end-to-end vector and the axial chain continuity.

9.8 EXERCISES

9.1. Calculate the chain orientation (Hermans orientation function) of the schematic molecules shown in Fig. 9.22.

9.2. Struik (1990, p. 302) gives the following data on the maximum birefringence ($\Delta n_0 = n_c - n_t$, where n_c

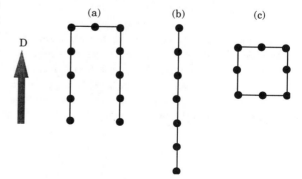

Figure 9.22 Three molecules oriented differently in the plane with a common director (D).

is the refractive along the chain axis and n_t is the refractive index in the transverse direction) of the four polymers:

Polymer	Δn_0
poly(ethylene terephthalate)	0.25
polycarbonate	0.24
poly(vinyl chloride)	0.01
polystyrene	−0.16

Source: Struik (1990).

Discuss the molecular reasons for the large differences in Δn_0 among the different polymers.

9.3. Calculate the crystalline chain orientation from the azimuthal angle dependence of the $(hk0)$ X-ray reflection as given below:

ϕ (deg.)	0	20	60	65	70	75	80	85	90
I	0	0	0	2	4	8	16	40	60

Note that the director is set parallel to the average chain-axis direction.

9.4. Calculate the amorphous chain-axis orientation for the same polymer sample. The overall chain orientation was determined by optical methods to be 0.8. The crystallinity was determined by DSC to be 50%.

9.5. Draw schematic X-ray diffraction patterns of $(00l)$ in a sample with uniaxial orientation along the z-axis in the (x, z), (y, z) and (x, y) planes.

9.6. Draw the same X-ray diffraction patterns for a sample showing biaxial orientation.

9.7. Draw schematically the infrared absorbance at 1700 cm^{-1} (carbonyl stretching band) as a function of the angle (ϕ) between the polarization direction of the infrared light and the fibre axis of a polyester fibre. Indicate also how the chain-axis orientation can be obtained from the IR absorbance data.

9.8. Define molecular draw ratio. How can it be measured?

9.9. Why is it impossible to extrude PA 6 through a very narrow die at temperatures well below the melting point? Why is the same action possible in the case of polyethylene?

9.9 REFERENCES

Bettelheim, F. A. and Stein, R. S. (1958) *J. Polym. Sci.* **27**, 567.
Folkes, M. J. and Keller, A. (1971) *Polymer* **12**, 222.
Gedde, U. W., Andersson, H., Hellermark, C., Johnsson, H., Sahlén, F. and Hult, A. (1993) *Progr. Coll. Polym. Sci.* **92**, 129.
Hedmark, P. G., Rego Lopez, J. M., Westdahl, M., Werner, P.-E. and Gedde, U. W., (1988) *Polym. Eng. Sci.* **28**, 1248.
Holliday, L. and White, J. W. (1971) *Pure & Appl. Chemistry* **26**, 245.
Mackley, M. R. and Keller, A. (1973) *Polymer* **14**, 16.
Peterlin, A. (1979) Mechanical properties of fibrons polymers, in *Ultrahigh Modulus Polymers* (A. Ciferri and I. M. Ward, eds), p. 279. Applied Science Publishers, London.
Sadler, D. M. and Barham, P. J. (1990) *Polymer* **31**, 46.
Samuels, R. J. (1974) *Structured Polymer Properties: the Identification, Interpretation and Application of Crystalline Polymer Structure.* Wiley, New York.
Struik, L. C. E. (1990) *Internal Stresses, Dimensional Instabilities and Molecular Orientation in Plastics.* Wiley, Chichester and New York.
Young, R. J. (1988) *Materials Forum* **210**.
Zachariades, A. E., Mead, W. T. and Porter, R. S. (1979) Recent developments in ultramolecular orientation of polyethylene by solid state extrusion, in *Ultrahigh Modulus Polymers* (A. Ciferri and I. M. Ward, eds), p. 77. Applied Science Publishers, London.

9.10 SUGGESTED FURTHER READING

Hay, I. L. (1980) Production and measurement of orientation, in *Methods of Experimental Physics*, Vol. 16, Part C (R. A. Fava, ed.). Academic Press, New York.
Read, B. E., Duncan, J. C. and Meyer, D. E. (1984) *Polymer Testing* **4**, 143–164.
Ward, I. M. (ed.) (1975) *Structure and Properties of Oriented Polymers.* Applied Science Publishers, London.

THERMAL ANALYSIS OF POLYMERS

10.1 INTRODUCTION

According to the definition originally proposed in 1969 by the Nomenclature Committee of the International Confederation for Thermal Analysis (ICTA) and later reaffirmed in 1978, thermal analysis includes a group of analytical methods by which a physical property of a substance is measured as a function of temperature while the substance is subjected to a controlled temperature regime. Thus, thermal analysis involves a physical measurement and not, strictly speaking, a chemical analysis. Figure 10.1 presents a summary of the different thermal analytical methods available. In this chapter, calorimetry, i.e. differential scanning calorimetry (DSC), and differential thermal analysis (DTA), thermogravimetry (TG), thermal mechanical analysis (TMA, DMTA), thermal optical analysis (TOA) and dielectric thermal analysis (DETA) are discussed. In the first part, the methods are briefly described. The use of thermal analysis on polymers requires special attention, as is discussed in the final section of the chapter. Examples from the melting and crystallization of flexible-chain polymers, the glass transition of amorphous polymers, phase transitions in liquid-crystalline polymers and chemical reactions including the degradation of polymers are presented to illustrate the non-equilibrium effects which are typical of polymers.

Thermo-analytical methods are powerful tools in the hands of the polymer scientist. Thermometry is the simplest and oldest method in thermal analysis. A sample is heated by a constant heat flow rate. Any phase transition is recorded as an invariance in temperature.

The number of phenomena which can be directly studied by thermal analysis (DSC (DTA), TG, TMA, DMTA, TOA and DETA) is impressive. Typical of these methods is that only small amounts of sample (a few milligrams) are required for the analysis. Calorimetric methods record exo- and endothermic processes, e.g. melting, crystallization, liquid-crystalline phase transitions, and chemical reactions, e.g. polymerization, curing, depolymerization and degradation. Second-order transitions, e.g. glass transitions, are readily revealed by the calorimetric methods. Thermodynamic quantities, e.g. specific heat, are sensitively determined. TG is a valuable tool for the determination of the content of volatile species and fillers in polymeric materials and also for studies of polymer degradation. The majority of the aforementioned physical transitions can also be monitored by TMA (dilatometry). DMTA and DETA

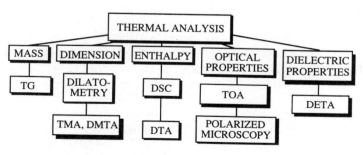

Figure 10.1 Thermo-analytical methods.

provide information about relaxation processes, both the glass transition and the secondary transitions (subglass processes). Additional information about the nature of the phase transitions in both crystalline and liquid-crystalline polymers may be obtained by TOA.

10.2 THERMO-ANALYTICAL METHODS

It is not the purpose of this chapter to give full details of the thermoanalytical methods. A very brief survey is, however, necessary in order to provide a basis for the understanding of the later discussion of the phenomena in relation to polymers.

10.2.1 DIFFERENTIAL THERMAL ANALYSIS AND CALORIMETRY

Accurate temperature-measuring devices – thermocouple, resistance thermometer and optical pyrometer – were developed in the late 1800s. These instruments were used by Le Chatelier in the late nineteenth century on chemical systems to study curves of change in the heating rate of clay (see Le Chatelier 1887). The first DTA method was conceived by the English metallurgist Roberts-Austen (1889). The modern DTA instrumentation was introduced by Stone (1951). This apparatus permitted the flow of gas or vapour through the sample during the temperature scans. The undesirable feature of the classical DTA (Fig. 10.2(b)) in sensor–sample interaction was overcome by Boersma (1955). This type of technique has since been referred to as 'Boersma DTA' (Fig. 10.2(c)). The temperature sensors are placed outside the sample and the reference.

The first differential scanning calorimeter was introduced by Watson et al. (1964) (Fig. 10.2(a)). A number of new developments in the instrumentation have since been made. The temperature scan is controlled, and data are collected and analysed by computers in today's instruments. Simultaneous measurements of differential temperature (ΔT) and sample weight, i.e. combined DTA and TG as well as combined DTA and TOA, are now commercially available. High-pressure DTA instruments have been in use since the early 1970s. In 1966, Cohen and co-workers constructed a DTA cell which could be

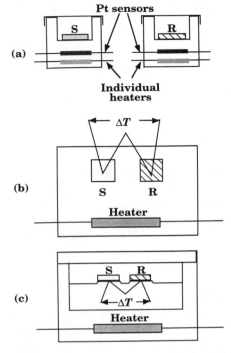

Figure 10.2 Schematic representation of (a) DSC; (b) 'classical' DTA and (c) 'Boersma' DTA. S and R denote sample and reference, respectively.

used at pressures up to 5 MPa. DSC instruments equipped with a UV cell were developed during the 1980s.

A new class of highly sensitive (100 nW) isothermal or slowly scanning calorimeters to be used at temperatures lower than 120°C were developed by Suurkuusk and Wadsö starting in the late 1960s (see Suurkuusk and Wadsö 1982). The first instrument in the series was a batch reaction calorimeter. Similar systems had been built a few years earlier by Calvet and Prat (1963) and by Benzinger and Kitzinger (1963). These thermopile heat conduction calorimeters consist of a calorimetric vessel surrounded by thermopiles, often Peltier elements through which the heat is conducted to or from the surrounding heat sink.

The calorimetric methods, DSC and DTA, are schematically presented in Fig. 10.2. DSC relies on the so-called 'null-balance' principle (Fig. 10.3). The temperature of the sample holder is kept the same as that of the reference holder by

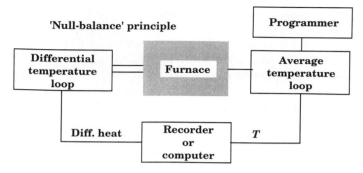

Figure 10.3 Principle of Perkin-Elmer DSC apparatus.

continuous and automatic adjustment of the heater power. The sample and reference holders are individually heated. A signal proportional to the difference between the heat power input to the sample and to the reference, dH/dt, is recorded. In the classical and Boersma DTA systems, the sample and reference are heated by a single heat source. Temperatures are measured by sensors embedded in the sample and reference material (classical) or attached to the pans which contain the material (Boersma).

Strictly speaking, DTA measures the difference in temperature (ΔT) between sample and reference, but it is possible to convert ΔT into absorbed or evolved heat via a mathematical procedure. The conversion factor is temperature-dependent. However, a DTA which accurately measures calorimetric properties is referred to as a differential scanning calorimeter. A DSC is thus a DTA that provides calorimetric information. The DSC instrument made by Perkin-Elmer is more a 'true' calorimeter since it directly measures differences in heat between an inert reference and the sample.

Figure 10.4, showing the DSC thermogram of an undercooled, potentially semicrystalline polymer, illustrates the measurement principle. At low temperatures, the sample and the reference are at the same temperature (balance). When the glass transition is reached, an increase in the (endothermal) heat flow to the sample is required in order to maintain the two at the same temperature. The change in level of the scanning curve is thus proportional to Δc_p. The polymer crystallizes at a higher temperature and exothermal energy is evolved. The heat flow to the sample should in this temperature region be less than the heat flow to the reference. The integrated difference between the two, i.e. the area under the exothermal peak, is thus equal to the crystallization

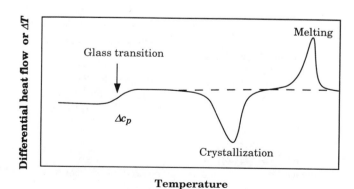

Figure 10.4 Schematic DSC traces showing three transition types.

enthalpy. At further higher temperatures, melting, which is an endothermal process, occurs. The heat flow to the sample is higher than that to the reference, and the peak points upwards. The area under the endothermal peak is thus proportional to the melting enthalpy.

The following equation is valid for the evolved heat flow (dH/dt) in the DTA apparatus:

$$R\frac{dH}{dt} = (T_s - T_r) + R(C_s - C_r)\frac{dT_r}{dt} + RC_s\frac{d(T_s - T_r)}{dt} \quad (10.1)$$

where the subscripts s and r refer respectively to the sample and the reference, T_s and T_r are the temperatures, C_s and C_r are the heat capacities, and R is the thermal resistance between the block and the sample cell. The heat (ΔH) involved in an endothermal (or exothermal) process can be derived from eq. (10.1) under the assumption that $C_s \approx C_r$:

$$dH = \frac{1}{R}\Delta T \, dt + C_s d(\Delta T)$$

$$\Delta H = \frac{1}{R}\int_0^\infty \Delta T \, dt = \frac{1}{R}S \quad (10.2)$$

where $\Delta T = T_s - T_r$, and S is the area under the DTA (ΔT) peak. The calibration factor ($1/R$) is temperature-dependent and the DTA apparatus has to be calibrated by several standards in the different temperature regions of interest.

An equation analogous to eq. (10.1) can be obtained for the DSC system:

$$\frac{dH}{dt} = -\frac{dq}{dt} + (C_s - C_r)\frac{dT_r}{dt} - RC_s\frac{d^2q}{dt^2} \quad (10.3)$$

where dq/dt is the differential heat flow. Equation (10.3) can be integrated to yield the ΔH of a first-order transition:

$$q = -\Delta H \quad (10.4)$$

The calibration factor transforming peak surface area to enthalpy is independent of the thermal resistance ($1/R$) and is simply an 'electrical' conversion factor with no temperature dependence.

There are a few significant differences between a DSC (Perkin-Elmer type) and a DTA. The mass of the sample and reference holders in the DSC apparatus is very low and the maximum cooling rate is greater in the typical (Perkin-Elmer) DSC apparatus than in DTA. The maximum measurement temperature for a DSC is only about 725°C. The upper temperature limit for a number of DTA instruments is significantly greater. This is not, however, a crucial question for organic polymers. One obstacle to calorimetric operation remains unsolved for the DTA method: the factor converting the observed peak area to energy is temperature-dependent (eq. (10.2)). This is particularly relevant for polymers which typically melt over a wide temperature range. The melting curve is asymmetric and requires the use of a complex conversion factor. The calibration constant in DSC is independent of temperature and quantitative operation is inherently simpler than with DTA.

The accuracy in the determination of transition temperatures by DSC/DTA is dependent on several factors:

- Standardized sample geometry and mass. The sample should be flat and have good thermal contact with the sample pan. Heat-conductive, 'thermally inert' liquid media may be used to improve the thermal contact.
- The purge gas and the sample pan material should be 'inert'.
- Thermal lag (difference) between sample and thermometer may be corrected for by using the slope of the leading edge of the melting of highly pure indium (or similar metal).
- Parallel processes should be inhibited. Melting of polymer crystals is accompanied by crystal thickening (parallel process). Further details about this particular case are given in section 10.3.1.

The accuracy in the determination of transition enthalpies depends on the level of the enthalpy change, on the temperature region of the transition and on the linearity (or control) of the base line.

10.2.2 THERMOGRAVIMETRY (TG)

The fundamental components of TG have existed for thousands of years. Mastabas or tombs in ancient Egypt (2500 BC) have wall carvings and paintings displaying both the balance and the fire. The two components were, however, first coupled in the fourteenth century AD for studies of gold refining. Honda pioneered the modern TG analytical technique (thermobalance) in 1915. The automatic thermobalance was introduced by Cahn and Schultz (1963).

TG is carried out in a so-called thermobalance which is an instrument permitting the continuous measurement of sample weight as a function of temperature/time. The following components are included in a typical TG instrument: recording balance; furnace; furnace temperature controller; and computer (Fig. 10.5). Figure 10.6 shows a schematic TG trace for a filled polymer. The furnaces can be run at temperatures up to 2400°C or more in a great variety of atmospheres, including corrosive gases. The sensitivity of the best modern recording balances is extremely high, in the microgram range.

It should be noted that the sample size and form affect the shape of the TG curve. A large sample may develop thermal gradients within the sample, a temperature deviation from the set temperature due to endo- or exothermal reactions and a delay in mass loss due to diffusion obstacles. Finely ground samples are preferred in quantitative analysis for the aforementioned reasons.

TG is often combined with various techniques to analyse the evolved gas. Infrared (IR) spectroscopy and gas chromatography (GC), the latter often combined with mass spectrometry (MS) or IR, are used for identification of the volatile products.

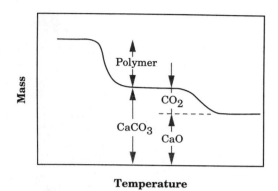

Figure 10.6 Sample mass as a function of temperature. Schematic curve of a polymer filled with $CaCO_3$.

10.2.3 DILATOMETRY/THERMAL MECHANICAL ANALYSIS (TMA)

The classical way of measuring sample volume as a function of temperature is by a Bekkedahl dilatometer. More recent types of instrumentation, e.g. the thermal mechanical analysers (Fig. 10.7), have been developed by several companies. These instruments not only measure volume and linear thermal expansion coefficients but also modulus as a function of temperature.

When thermal expansion or penetration (modulus) is being measured, the sample is placed on a platform of a quartz sample tube. The thermal expansion coefficient of quartz is small (about $0.6 \times 10^{-6}\,K^{-1}$) compared to the polymer materials. The quartz tube is connected to the armature of a linear variable

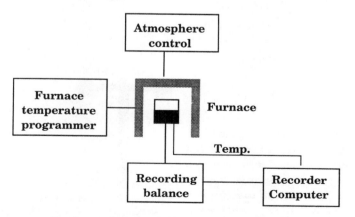

Figure 10.5 Schematic presentation of a TG apparatus.

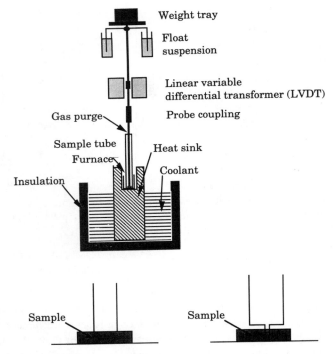

Figure 10.7 Schematic presentation of the Perkin-Elmer thermal mechanical analyser (top). Bottom left: expansion probe for measurement of linear thermal expansion coefficient. Bottom right: penetration probe for measurement of stiffness.

differential transformer (LVDT) and any change in the position of the core of the LVDT, which floats frictionless within the transformer coil, results in a linear change in the output voltage. The upper temperature limit for the currently available commercial instruments is about 725 °C. Gillen (1978) showed that the tensile compliance as obtained by TMA using the penetration probe and suitable compressive loads is comparable with the data measured by conventional techniques on considerably larger samples.

10.2.4 DYNAMIC MECHANICAL THERMAL ANALYSIS (DMTA)

DMTA measures stress and strain in a periodically deformed sample at different loading frequencies and temperatures. It provides information about relaxation processes in polymers, specifically the glass transition and subglass processes. The theoretical background is presented in section 5.5.1. It should be noted that the strains involved in the measurements should be small (less than 0.5%) to avoid a nonlinear response, i.e. nonlinear visco-elasticity. There are essentially three main types of instrument: the torsion pendulum, an apparatus based on the resonance method, and forced oscillation instruments (Fig. 10.8).

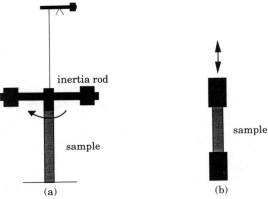

Figure 10.8 Schematic representation of (a) torsion pendulum and (b) reversed uniaxial tension (forced oscillation).

A highly schematic representation of the torsion pendulum is given in Fig. 10.8(a). The sample, which may be a cylindrical rod, is rigidly held at one end, and at the other end supports an inertia rod. The inertia rod is set into oscillation and the polymer sample is subjected to a sinusoidal torsion which, depending on the relative size of the viscous component, gradually dampens. The frequency range of operation is 0.01–50 Hz The real (G') and imaginary (G'') parts of the shear modulus are given by:

$$G' = \frac{2lM\omega^2}{\pi r^4} \quad (10.5)$$

$$G'' = \frac{2l^2 M}{\pi r^4} \omega^2 \frac{\Lambda}{\pi} \quad (10.6)$$

where l is the length of the sample rod, M is the moment of inertia of the inertia rod, ω is the angular frequency, r is the radius of the sample rod and Λ is the logarithmic decrement, defined as:

$$\Lambda = \frac{1}{k} \ln\left(\frac{A_n}{A_{n+k}}\right) \quad (10.7)$$

where k is the number of swings and A_i is the amplitude of the ith swing. It is possible to vary the angular frequency by changing the moment of inertia (M) of the inertia rod.

The forced oscillation technique can be used over a wider frequency range than the torsion pendulum ranging from 10^{-4} to 10^4 Hz. Uniaxial extension (Fig. 10.8(b)), bending, torsion and shear are used in different commercial instruments. A sinusoidal strain is applied to the specimen and the stress (force) is accurately measured using a strain gauge transducer as a function of time. Knowing both the strain and stress as functions of time enables the complex modulus (E^* or G^*) to be determined.

10.2.5 THERMAL OPTICAL ANALYSIS (TOA)

TOA is normally carried out using a polarized light microscope (crossed polarizers) equipped with a hot-stage by which the temperature of the sample can be controlled. The microscopic image of the typically 10 μm thick sample can be viewed directly in binoculars or the transmitted light intensity can be

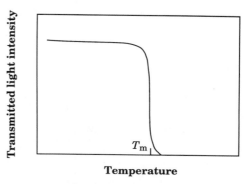

Figure 10.9 Melting of a crystalline polymer as recorded by TOA.

recorded with a photodiode. Melting (crystallization) and mesomorphic phase transitions in liquid-crystalline polymers can be directly studied by TOA (Fig. 10.9). Apparatus for TOA of sheared polymer melts was developed during the 1980s.

10.2.6 DIELECTRIC THERMAL ANALYSIS (DETA)

DETA, which is also referred to as dielectric spectroscopy, provides information about the segmental mobility of a polymer. Chemical bonds between unlike atoms possess permanent electrical dipole moments. Many polymers with significant bond dipole moments show no molecular dipole moments due to symmetry, i.e. the bond moments of a given central atom (or group of atoms) counteract each other and the vector sum of the dipole moments is zero. Other polymers, however, have repeating unit structures such that the dipole moments can vectorially accumulate into a repeating unit moment in the possible conformational states. This group of polymers can be studied by DETA. Polarizability (α) is the sum of dipole moments per unit volume. Figure 10.10 shows the variation of α as a function of the frequency of the alternating electric field. The electric field induces a distortion of the electronic clouds at a frequency of about 10^{15} Hz, which corresponds to the optical ultraviolet range. Electronic polarizability (α_e) is closely related to the refractive index (n) of visible light according to the Clausius–Mesotti equation:

$$\frac{n^2 - 1}{n^2 + 2} = \frac{4\pi \rho N_A}{3M} \cdot \alpha_e \quad (10.8)$$

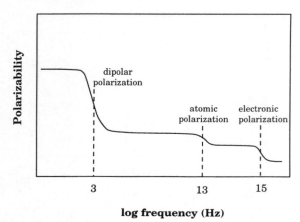

Figure 10.10 Schematic representation of polarizability as a function of frequency.

where ρ is the density, N_A is Avogadro's number and M is the molar mass.

So-called atomic polarization arises from small displacements of atoms under the influence of the electric field at a frequency of approximately 10^{13} Hz (optical infrared range). The atomic polarizability cannot be determined directly but is normally small compared to the electronic polarizability. Electronic and atomic polarization occur in all types of polymer, even polymers with no permanent dipole moments. Polymers with permanent dipoles show no macroscopic polarization in the absence of an external electric field. If an alternative electric field is applied and if the electric field frequency is sufficiently low with reference to the jump frequency of segments of the polymer, the dipoles orient in the field and the sample shows not only electronic and atomic polarization but also a dipolar polarization. DETA, which operates in a frequency range from 10^{-4} to 10^8 Hz, is used to monitor dipole reorientation induced by conformational changes. These are referred to as dielectric relaxation processes.

Let us consider two parallel plates with opposite surface charges ($\pm\sigma$) and a dielectric material inserted between the plates. The electric field (**E**) causes a displacement field (**D**) which for a configuration of charged parallel plates is related to the polarization (**P**) by the expression:

$$\mathbf{D} = \kappa_0 \mathbf{E} + \mathbf{P} \quad (10.9)$$

where κ_0 is the permittivity of free space. The polarization is the induced dipole moment per unit volume and can be expressed as:

$$\mathbf{P} = \chi\kappa_0\mathbf{E} \quad (10.10)$$

where χ is called the susceptibility. Insertion of eq. (10.10) into eq. (10.9) yields:

$$\mathbf{D} = (1 + \chi)\kappa_0\mathbf{E} = \varepsilon\kappa_0\mathbf{E} \quad (10.11)$$

where $\varepsilon = 1 + \chi$ is the dielectric constant. The capacitance (C) of a capacitor is defined as the amount of charge that can be stored per unit voltage, and it can be shown that the dielectric constant of the medium inserted between the plates is given by:

$$\varepsilon = \frac{C}{C_0} \quad (10.12)$$

where C_0 is the capacitance of the capacitor *in vacuo*. Most dielectric work is concerned with a periodic **E** field with an angular frequency ω, and the dielectric constant is conveniently represented as a complex quantity:

$$\varepsilon^* = \varepsilon' - i\varepsilon'' \quad (10.13)$$

where ε' is referred to as the dielectric constant and ε'' as the dielectric loss. It can be shown that the complex displacement is given by:

$$\mathbf{D}^* = (\sqrt{\varepsilon'^2 + \varepsilon''^2})\kappa_0\mathbf{E}_0 \exp[i(\omega t - \delta)] \quad (10.14)$$

where δ is the phase difference between the applied **E** field and the displacement field. The phase angle δ is given by:

$$\tan\delta = \frac{\varepsilon''}{\varepsilon'} \quad (10.15)$$

A simple assumption may be made that the polarization that originates from a reorientation of dipoles is proportional to its displacement from equilibrium (χ_0):

$$\frac{d\chi}{du} = -k(\chi - \chi_0) \quad (10.16)$$

The angular frequency (ω) dependence of the complex dielectric constant is given by:

$$\varepsilon^* = \varepsilon_u + \frac{(\varepsilon_r - \varepsilon_u)}{(1 + i\omega\tau)} \quad (10.17)$$

where τ is the (single) relaxation time, ε_u is the unrelaxed dielectric constant (the dielectric constant obtained at such a high frequency that no dipole relaxation takes place) and ε_r is the relaxed dielectric constant which is the value that the dielectric constant takes at such a low frequency that all dipoles move with the field. The difference between the two limiting dielectric constants ($\varepsilon_r - \varepsilon_u$) is called the relaxation strength. The real and imaginary parts of the complex dielectric constant are obtained as follows:

$$\varepsilon' - i\varepsilon'' = \varepsilon_u + \frac{(\varepsilon_r - \varepsilon_u)(1 - i\omega\tau)}{(1 + i\omega\tau)(1 - i\omega\tau)}$$

$$= \varepsilon_u + \frac{(\varepsilon_r - \varepsilon_u) - i\omega\tau(\varepsilon_r - \varepsilon_u)}{(1 + \omega^2\tau^2)}$$

so that

$$\varepsilon' = \varepsilon_u + \frac{(\varepsilon_r - \varepsilon_u)}{(1 + \omega^2\tau^2)} \quad (10.18)$$

$$\varepsilon'' = \frac{(\varepsilon_r - \varepsilon_u)\omega\tau}{(1 + \omega^2\tau^2)} \quad (10.19)$$

The dielectric constant and loss according to eqs (10.18) and (10.19) are plotted as a function of log $\omega\tau$ in Fig. 10.11. A maximum in ε'' and an inflection point

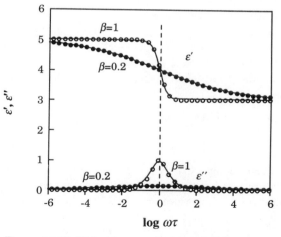

Figure 10.11 Dielectric constant (ε') and dielectric loss (ε'') as a function of angular frequency (ω) assuming a single relaxation time (τ), $\beta = 1$; and symmetric broadening according to the Cole–Cole equation with a symmetric broadening factor (β) of 0.2. In both cases, the following limiting values for the dielectric constant have been used: $\varepsilon_u = 3$ and $\varepsilon_r = 5$.

in ε' appear at the angular frequency $\omega = \tau^{-1}$. Note that the dielectric constant takes the relaxed value ($\varepsilon_r = 5$) at low frequencies ($\omega < 0.01\tau^{-1}$) and the unrelaxed value ($\varepsilon_u = 3$) at frequencies greater than $100\tau^{-1}$.

It is very uncommon that the relaxation processes in polymers can be described by a single relaxation time model, and eq. (10.17) has been modified empirically. The Cole–Cole equation includes a symmetric broadening factor β which takes values between 0 and 1 (single relaxation time):

$$\varepsilon^* = \varepsilon_u + \frac{(\varepsilon_r - \varepsilon_u)}{(1 + (i\omega\tau_0)^\beta)} \quad (10.20)$$

where τ_0 is the central relaxation time. The real and imaginary parts of the complex dielectric constant are in this case given by:

$$\varepsilon' = \varepsilon_u + \frac{(\varepsilon_r - \varepsilon_u)\left(1 + (\omega\tau_0)^\beta \cos\left(\frac{\beta\pi}{2}\right)\right)}{\left(1 + 2(\omega\tau_0)^\beta \cos\left(\frac{\beta\pi}{2}\right) + (\omega\tau_0)^{2\beta}\right)}$$

(10.21)

$$\varepsilon'' = \frac{(\varepsilon_r - \varepsilon_u)(\omega\tau_0)^\beta \sin\left(\frac{\beta\pi}{2}\right)}{\left(1 + 2(\omega\tau_0)^\beta \cos\left(\frac{\beta\pi}{2}\right) + (\omega\tau_0)^{2\beta}\right)}$$

(10.22)

The effect of a lowering of the broadening factor from 1 (single relaxation time) to 0.2 is shown in Fig. 10.11. An appreciable change in the dielectric constant occurs over a frequency range which is greater than 12 orders of magnitude. The dielectric loss curve shows only a weak maximum at $\omega\tau_0 = 1$.

The Argand plot is a very useful diagram by which several important parameters of the relaxation process can be obtained (Fig. 10.12). This type of diagram is also commonly called a Cole–Cole plot. The single relaxation time ($\beta = 1$) data form a half circle with the x-axis intercepts at ε_u and ε_r. The symmetrically broadened relaxation ($\beta = 0.2$) also results in a circular arc with the x-axis intercepts at ε_u and ε_r, but the radius of the circle is considerably larger than in the case of a single relaxation time. The equation for the

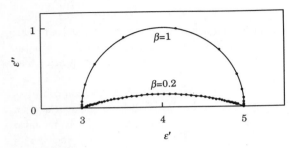

Figure 10.12 Argand plot showing the circular arcs of the data presented in Fig. 10.11.

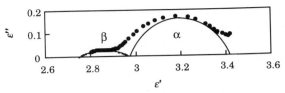

Figure 10.13 Argand plot showing two dielectric relaxation processes α and β for a hyper-branched polyester measured at 260 K. Experimental data are indicated by the points and the lines are obtained by fitting the Havrilak–Negami equation to experimental data. After data from Malmström et al. (1994).

circle is given by:

$$\left[\varepsilon' - \frac{\varepsilon_r - \varepsilon_u}{2}\right]^2 + \left[\varepsilon'' + \left(\frac{\varepsilon_r - \varepsilon_u}{2}\right)\cot\left(\frac{\beta\pi}{2}\right)\right]^2$$
$$= \left[\frac{1}{2}(\varepsilon_r - \varepsilon_u)\mathrm{cosec}\left(\frac{\beta\pi}{2}\right)\right]^2 \quad (10.23)$$

which is a circle with its centre at the coordinates $(\frac{1}{2}(\varepsilon_r - \varepsilon_u), \frac{1}{2}(\varepsilon_r - \varepsilon_u)\cot(\beta\pi/2))$ and a radius of $\frac{1}{2}(\varepsilon_r - \varepsilon_u)\mathrm{cosec}(\beta\pi/2)$.

A number of other related empirical equations have been proposed. The Cole–Davidson equation includes an asymmetric broadening factor (γ):

$$\varepsilon^* = \varepsilon_u + \frac{(\varepsilon_r - \varepsilon_u)}{(1 + i\omega\tau_0)^\gamma} \quad (10.24)$$

The combination of the Cole–Cole equation (eq. (10.20)) and the Cole–Davidson equation (eq. (10.24)) is, after the inventors, referred to as the Havriliak–Negami equation:

$$\varepsilon^* = \varepsilon_u + \frac{(\varepsilon_r - \varepsilon_u)}{(1 + (i\omega\tau_0))^\beta)^\gamma} \quad (10.25)$$

This equation shows a great flexibility and can be fitted to most dielectric data. Figure 10.13 shows dielectric data presented in an Argand plot. Two circular arcs indicating symmetric broadening ($\gamma = 1$) of the two relaxation processes α and β are fitted to the experimental data. From the positions of the centres of the circular arcs, it is evident that the symmetric broadening factor is larger for the α process than for the β process.

The temperature dependence of a relaxation process is conveniently represented in an Arrhenius diagram (Fig. 10.14). Subglass processes display an Arrhenius temperature dependence,

$$f_{max} \propto \frac{1}{\tau_0} = A_1 \exp[-(\Delta E/RT)]$$

where ΔE is the activation energy, R is the gas constant and A_1 is a constant, whereas the glass transition is characterized by the WLF behaviour,

$$f_{max} \propto \frac{1}{\tau_0} = A_2 \exp[-(A_3/T - T_0)]$$

where A_2 and A_3 are constants, and T_0 is a temperature.

The instruments used for the determination of dielectric constants are based on Wheatstone bridges (10–10^7 Hz), Schering bridges (10^{-4} – 10 Hz), distributed circuits and time-domain spectrometers. A detailed presentation is given by Boyd (1980).

10.3 THERMAL BEHAVIOUR OF POLYMERS

10.3.1 SEMICRYSTALLINE POLYMERS

The melting of polymer crystals exhibits many instructive features of non-equilibrium behaviour. It has been known since the 1950s that the crystals of flexible-chain polymers, e.g. polyethylene (PE), are lamella-shaped with the chain axis almost parallel to the normal of the lamella. The lamellar thickness (L_c) is of the order of 10 nm, corresponding to approximately 100 main chain atoms, which is considerably less than the total length of the typical polymer chain. This fact led to the postulate that the macroconformation of the chains must be **folded**. The

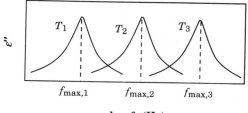

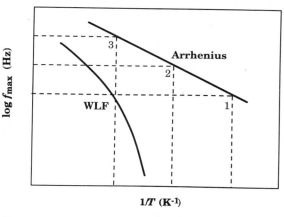

Figure 10.14 Arrhenius diagram showing WLF and Arrhenius temperature dependence. The lines are obtained from isothermal dielectric loss data as a function of log frequency.

transverse dimensions (W) are two to three orders of magnitude greater than the lamellar thickness. The melting point (T_m) of thin lamellar crystals is related to the lamellar thickness (L_c) according to the Thompson–Gibbs equation:

$$T_m = T_m^0 \left[1 - \frac{2\sigma}{\Delta h^0 L_c \rho_c} \right] \quad (10.26)$$

where T_m^0 is the melting point of the infinitely thick crystal, σ is the fold surface free energy, Δh^0 is the heat of fusion and ρ_c is the density of the crystalline component. The Thompson–Gibbs equation is explored in more detail in Chapter 7. Thus, factors such as molar mass, degree of chain branching and cooling rate may influence crystal thickness and hence the melting point.

The lamellar shape ($L_c/W \approx 0.01$–0.001, where W is the width of the crystal) of the polymer crystals is the reason for the rearrangement occurring at

Thermal behaviour of polymers 227

temperatures well below the melting point. The equilibrium shape of the crystals can be calculated on the basis of the surface free energies of the fold surface (σ) and the lateral surface (σ_L):

$$\frac{L_c}{W} = \frac{\sigma}{\sigma_L} \quad (10.27)$$

For linear polyethylene (LPE), the surface free energies are equal respectively to $\sigma = 93$ mJ m^{-2} and $\sigma_L = 14$ mJ m^{-2}, resulting in an equilibrium L_c/B value of 6.6 which is three orders of magnitude greater than the experimental value. The crystal thickening of polymer crystals expected to occur on the basis of these thermodynamic arguments has been verified by X-ray diffraction experiments. The recorded dependence of melting point on heating rate of single crystals of linear polyethylene presented in Fig. 10.15 is consonant with this view. At low heating rates, the crystal thickening occurs to a much greater extent than at high heating rates, which in turn leads to a greater 'final' crystal thickness and a higher melting point after slow heating. The melting point value obtained at the higher heating rates is thus more in agreement with that of the original crystals.

Extended-chain crystals of polyethylene are produced by high-pressure crystallization at elevated temperatures, typically at 0.5 MPa and 245°C. These micrometre-thick crystals display a distinctly different melting behaviour from that of the thin folded-chain single crystals grown from solution. The recorded

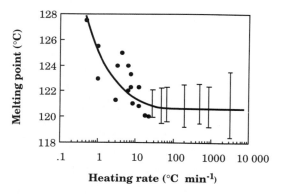

Figure 10.15 Melting point of solution crystals of LPE (0.05% (w/w) in toluene at 81°C) as a function of heating rate. Drawn after data from Hellmuth and Wunderlich (1965).

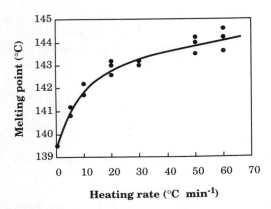

Figure 10.16 Melting point of extended-chain crystals of linear polyethylene (0.48 MPa; 227°C). Drawn after data from Hellmuth and Wunderlich (1965).

increase in melting point with increasing heating rate shown in Fig. 10.16 is due to **superheating**.

Crystals of intermediate thickness display approximately zero-entropy-production melting, i.e. the melting point is almost independent of heating rate. Interestingly, as is shown in Fig. 10.17, linear polyethylene samples with a broad molar mass distribution contain crystals with a great variety of thicknesses, displaying in a single sample reorganization, zero-entropy-production melting and superheating in order of increasing crystal thickness, i.e. of increasing melting point.

Polymers in general melt over a wide temperature range, typically covering more than 30°C. This is due, first, to their multicomponent nature. Polymers always exhibit a distribution in molar mass and occasionally also in monomer sequence (copolymers). The different molecular species crystallize at different temperatures and this leads to a significant variation in crystal thickness and melting point. Polymer samples which have first crystallized under isothermal conditions and have later been rapidly cooled to lower temperatures frequently exhibit bimodal melting for these reasons. The high-temperature peak is associated with the high molar mass species which have crystallized under the isothermal conditions and the low-temperature peak is due to the low molar mass component able to crystallize only during the subsequent cooling phase.

Second, reorganization of crystals may occur on heating, involving either partial melting followed by crystallization forming thicker and more perfect crystals of essentially the same unit cell of the original crystals or the transformation of one crystal structure to another. Samples which undergo these premelting transitional phenomena display bimodal or multi-modal melting provided that the original crystals are of about the same lamella thickness.

Figure 10.18 presents four cases of bimodal melting with different reasons for the bimodality. The melting of the once-folded orthorhombic crystals of n-$C_{294}H_{590}$ leads to recrystallization into extended-chain crystals which melt at a higher temperature in the high-temperature melting peak. Isothermal crystallization (at T_c, Fig. 10.18) almost always leads to two melting peaks with a minimum in between at a temperature in the vicinity of T_c. High-temperature annealing leads to similar effects. Immiscible, not cocrystallizing binary blends of polymers with different melting points show two melting peaks. Numerous polymers have different possible crystalline structures, i.e. they have different polymorphs, each with a certain melting point.

The content of the crystalline component, the crystallinity, in semicrystalline polymers is a major factor affecting their material properties, e.g. modulus, permeability and density. The crystallinity of a sample

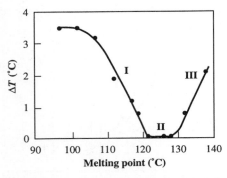

Figure 10.17 Difference (ΔT) in melting point as recorded at $10°C\ min^{-1}$ from that recorded at high heating rates (region I: crystal thickening) or zero heating rate (region III: superheating) plotted against melting point as recorded at $10°C\ min^{-1}$. Region II refers to zero-entropy-production melting. Drawn after data from Gedde and Jansson (1983) on linear polyethylene.

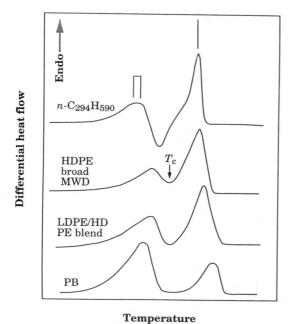

Figure 10.18 Heating thermograms showing different bimodal melting traces. Schematic curves.

can be determined by several techniques, e.g. X-ray diffraction, density measurements and DSC/DTA.

The mass crystallinity (w_c) obtained from the heat of fusion is based on the measurement of the area under the DSC melting peak. The choice of base line is crucial, particularly for polymers of low crystallinity, e.g. poly(ethylene terephthalate) (PETP). Another problem arises from the fact that the heat of fusion is temperature-dependent. What temperature should

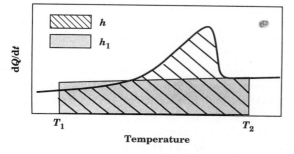

Figure 10.19 Schematic diagram of the DSC melting trace. Application of the total enthalpy method.

Thermal behaviour of polymers 229

be selected? Two rigorous methods have been proposed by Gray (1970) and Richardson (1976): the total enthalpy method and the peak area method. Excellent agreement is obtained for crystallinity data obtained for samples of linear, branched and chlorinated polyethylene and PETP by the DSC and total enthalpy methods with those obtained by X-ray diffraction method. Referring to Fig. 10.19, the mass crystallinity at temperature T_1 is given by:

$$w_c(T_1) = \frac{h - h_1}{\Delta h^0(T_1)} \quad (10.28)$$

where $\Delta h^0(T_1)$ is the heat of fusion at temperature T_1 given by:

$$\Delta h^0(T_1) = \Delta h^0(T_m^0) - \int_{T_1}^{T_m^0} (c_{pa} - c_{pc}) dT \quad (10.29)$$

where c_{pa} and c_{pc} are the specific heats of the amorphous and crystalline components. Numerous specific heat data on different polymers were collected in Wunderlich and Baur (1970).

The greatest problem is to determine the crystallinity of polymers which degrade at low temperatures in the melting range. Polyvinylchloride belongs to this group. The base-line definition is always a problem due to early thermal degradation and the low overall crystallinity which leads to melting over a very broad temperature range.

The kinetics of crystallization, which is of interest for both academic and industrial reasons, is preferably studied under isothermal conditions by DSC, dilatometry or TOA. These methods reveal the overall crystallinity, volume (v_c) or mass (w_c) crystallinity as a function of time (t), and the general Avrami equation can be applied:

$$1 - \frac{v_c}{v_{c\infty}} = \exp(-Kt^n) \quad (10.30)$$

where n and K are constants and $v_{c\infty}$ is the maximum crystallinity attained.

The Avrami exponent (n) depends on nucleation type, the geometry of crystal growth and the kinetics of crystal growth (see Chapter 8). The kinetics at low degrees of conversion usually follows the Avrami equation but deviates from the linear trend in the plot of $\ln(-\ln(1 - v_c/v_{c\infty}))$ against $\ln t$ at higher degrees of conversion.

Most of the kinetic work has dealt with the temperature dependence of the growth rate in accordance with the kinetic theory of Lauritzen and Hoffman (see Hoffman et al. 1975). The experimental data, the linear growth rate (G) of spherulites (axialites), are obtained by hot-stage polarized light microscopy at different constant temperatures and the data are adapted to the equation:

$$G = G_0 \exp(-U^*/R(T - T_\infty))\exp(-K_g/T\Delta Tf) \quad (10.31)$$

where G_0 is a constant, U^* is a constant (measured in joules per mole) for short-distance diffusion of the crystallizing segments, R is the gas constant, T is the crystallization temperature, T_∞ is a temperature which is related to the glass temperature, K_g is the 'nucleation kinetic constant' which depends on the surface energies of the crystals formed and the mechanisms of crystallization, ΔT is the degree of supercooling ($= T_m^0 - T_c$, where T_m^0 is the equilibrium melting point and T_c is the crystallization temperature), and f is a correction factor close to unity which takes into account changes in the heat of fusion with temperature.

A DSC analogue of eq. (10.31) may be written as follows:

$$\frac{1}{t_{0.5}} = C \exp(-U^*/R(T - T_\infty))\exp(-K_g/T\Delta Tf) \quad (10.32)$$

where $t_{0.5}$ is the time required to attain 50% conversion and C is a constant.

10.3.2 AMORPHOUS POLYMERS

The glass transition temperature is possibly the most prominent temperature for an amorphous polymer. The stiffness of a typical amorphous polymer changes by three orders of magnitude from a few gigapascals in the glassy state (low-temperature side) to a few megapascals in the rubbery state (high-temperature side). The glass transition appears at first sight to be a second-order phase transition, i.e. volume (V) and enthalpy (H) are continuous functions through the

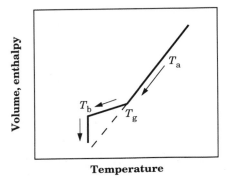

Figure 10.20 Schematic representation of the glass transition in amorphous polymers. The decrease in volume and enthalpy at T_b is referred to as physical ageing.

transition temperature interval (Fig. 10.20). However, if the amorphous polymer is cooled from temperature T_a, a break in the rectilinear H, $V = f(T)$ curve appears at the glass transition temperature (T_g). If the cooling is stopped at a temperature near T_g, at T_b in Fig. 10.20, and the sample is kept at this temperature, the enthalpy and volume of the sample decrease as a function of time. This so-called physical ageing clearly shows that glassy amorphous polymers are not in equilibrium and that the measured glass transition temperature is a 'kinetic' temperature rather than a temperature associated with a true thermodynamic transition. Physical ageing thus leads to more densely packed material of higher stiffness and lower impact strength. The kinetics of physical ageing can be followed by DSC/DTA and dilatometry. The fundamental aspects and practical implications of physical ageing have been the subject of many papers. Physical ageing was reviewed by Struik (1978).

The non-equilibrium phenomenon described above also leads to a significant dependence on cooling rate of the recorded T_g. Atactic polystyrene, for example, shows a change in T_g from 365 K at a cooling rate of 1 K h^{-1} to 380 K at a cooling rate of 1 K s^{-1}. Even with a cooling rate of 1 K yr^{-1} the observed T_g does not decrease below 351 K. It is important to note that these data refer to the T_g recorded on cooling. On heating, glassy amorphous polymers exhibit superheating effects. If a slowly cooled amorphous polymer is rapidly heated, the T_g is shifted to significantly higher temperatures and an endothermic hysteresis peak is observed above the glass transition. In conclusion, the most reliable and precise way of measuring T_g is by cooling the melt at a specified low cooling rate and recording the step in the specific heat curve.

The morphology of amorphous polymer blends is indeed of great practical importance. The impact strength of stiff and brittle glassy polymers (e.g. atactic polystyrene) may be greatly improved by including a few per cent of an immiscible elastomer such as polybutadiene. Provided that the polymers have different glass transition temperatures, DSC(DTA) or TMA are valuable tools for seeking the answer to the crucial question of whether the polymers are miscible or whether they exist in different phases.

Figure 10.21 shows the thermogram obtained from an immiscible mixture of styrene-acrylonitrile (SAN) copolymer and polybutadiene. This blend is a high impact strength material referred to as acrylonitrile-butadiene-styrene (ABS) plastic. Two glass transitions are observed, the low-temperature transition associated with the polybutadiene and the high-temperature transition associated with the SAN copolymer.

Miscible blends are less commonly found. One of the most studied blends is that between polystyrene and poly(phenylene oxide) (PPO). This blend is miscible in all proportions and the films of the blends are optically clear. Only one glass transition intermediate in temperature between the T_g's of polystyrene and PPO has been reported. The compositional dependence of the T_g of a compatible binary blend follows in some cases, e.g. the Fox equation for blends of polyvinylchloride and ethylene-vinyl acetate copolymers:

$$\frac{1}{T_g} = \frac{w_1}{T_{g1}} + \frac{w_2}{T_{g2}} \qquad (10.33)$$

where w_i and T_{gi} are the weight fractions and glass transition temperatures of the pure polymers. Several other equations relating T_g to the composition have also been proposed (Chapters 4 and 5).

10.3.3 LIQUID-CRYSTALLINE POLYMERS

Liquid-crystalline polymers are a relatively new group of polymers which have aroused considerable interest during the last decade. A detailed presentation of liquid-crystalline polymers is given in Chapter 6. **Main-chain** polymers, with the mesogenic stiff units implemented in the main chain, exhibit a unique combination of good processing and good mechanical properties and hence are now used as engineering plastics. **Side-chain** polymers, with the mesogens in the side chains, are mainly used in speciality polymers with potential application in electronics and optronics. Liquid-crystalline polymers exhibit a number of thermal transitions, as shown in Fig. 10.22.

Thermal analysis, DSC/DTA, TMA and TOA and hot-stage microscopy are, together with X-ray diffraction, the popular methods for structural assessment at different temperatures. A combination of these methods is commonly used. X-ray diffraction of aligned samples is the most reliable method for structural assessment (Chapter 6).

When the isotropic melt cools from very high temperatures, it is ultimately transformed into a liquid-crystalline phase, i.e. a so-called **mesophase**. A great number of different liquid-crystalline structures have been reported. Two major groups, differing in degree of order, exist: **nematics** and **smectics**. The high-temperature transition is readily revealed by DSC (DTA) as an exothermic first-order transition (see Fig. 10.22) and by TOA and hot-stage polarized microscopy from the formation of birefringent structures. The enthalpy involved in an isotropic–nematic transition is significantly smaller than that involved in an isotropic–smectic transition. The phase assignment of the mesophase can be achieved by

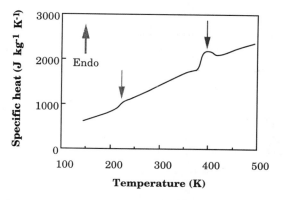

Figure 10.21 DSC thermogram of ABS showing two glass transitions. Drawn after data from Bair (1970).

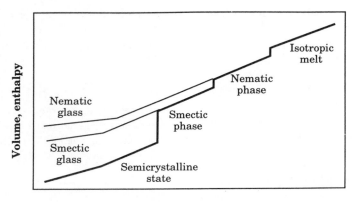

Figure 10.22 Typical phase transitions in thermotropic liquid-crystalline polymers.

polarized light microscopy according to the scheme by Demus and Richter (1978) or preferably by X-ray diffraction (Chapter 6).

At lower temperatures, a number of liquid-crystalline transitions may occur which again can be recorded by DSC/DTA as exothermic first-order transitions (Fig. 10.22). Hot-stage microscopy and X-ray diffraction are used to determine the nature of these transitions. At lower temperatures, solid crystals may be formed. The latter are revealed by DSC/DTA as an exothermic first-order transition, by TMA as an increase in sample stiffness and by X-ray diffraction as sharp Bragg reflections. Some liquid-crystalline polymers, e.g. copolyesters, are supercooled to a glassy state without crystallizing (Fig. 10.22).

10.3.4 POLYMER DEGRADATION

Polymers are, with few exceptions, very sensitive to degradation reactions occurring both during the melt-processing and during use. Reactions with oxygen, thermal oxidation and photo-oxidation, are for many polymers the dominating degradation reactions. Stabilizers, e.g. antioxidants, increase the stability of polymers and extend the life of many products considerably. The exposure of a plastic material to heat and oxygen leads to consumption of the antioxidant. It has therefore been important to develop efficient methods for the determination of antioxidant content. Thermal analyses, DSC/DTA and TG, are among the most frequently used methods for this purpose.

The oxidative induction time (OIT) is measured by first heating the polymer sample while keeping it in a nitrogen atmosphere to a high temperature, typically 200°C for polyethylene. After the establishment of constant temperature, the atmosphere is switched to oxygen and the time (OIT) to the start of an exothermic (oxidation) process is measured (Fig. 10.23). It has been shown that the OIT exhibits an Arrhenius temperature dependence and that there is a linear relationship between OIT and the content of efficient antioxidant. This follows from the fact that

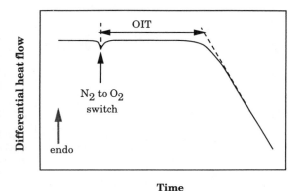

Figure 10.23 Typical thermogram from an OIT measurement.

the consumption of antioxidant at these conditions follow zero-order kinetics:

$$\frac{dc}{dt} = k_0 \exp(-\Delta E/RT)\sqrt{p_{O_2}}$$

$$\frac{dc}{dt} = k_0' \exp(-\Delta E/RT) \quad (p_{O_2} = \text{const.}) \quad (10.34)$$

where c is the concentration of efficient antioxidant, t is the time, k_0 and k_0' are constants, ΔE is the activation energy, R is the gas constant, T is the temperature (in kelvin) and p_{O_2} is partial pressure of O_2. Equation (10.34) can be integrated under isothermal conditions to yield:

$$c_0 = k_0' \left[\exp(-\Delta E/(RT))\right] \cdot \text{OIT} \quad (10.35)$$

where c_0 is the initial concentration of efficient antioxidant. It is important to note that, in order to use OIT data as an absolute method for the determination of the antioxidant content, the system must be calibrated to determine the constants k_0' and ΔE. Similar measurements can be made by TG. The first indication of oxidation is a small increase in sample mass. At a later stage, when degradation leads to the formation of volatile products, the sample mass decreases strongly.

When the constants k_0' and ΔE have been established, the antioxidant content can be indirectly measured by recording the thermogram at a constant heating rate in oxygen atmosphere and determining the oxidation temperature (T_{ox}) as the temperature of the exothermic deviation from the scanning base line. The relationship between T_{ox} and the antioxidant concentration (c_0) is obtained by integrating eq. (10.34) as follows:

$$c_0 = \frac{k_0'}{\frac{dT}{dt}} \int_0^{T_{ox}} \exp(-\Delta E/(RT))dT \quad (10.36)$$

The integral can only be solved numerically. An example is presented in Fig. 10.24. Since OIT is proportional to c_0, it is evident that the dynamic method is less suitable for determination of antioxidant content in highly stabilized systems (high c_0), but it is sufficiently sensitive for small variations in c_0 in systems with small c_0 values.

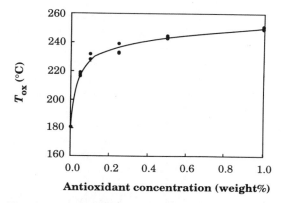

Figure 10.24 Oxidation temperature as a function of antioxidant concentration (Irganox 1010 in a medium-density polyethylene). With permission from Elsevier (Karlsson, Assargren and Gedde 1990).

OIT/T_{ox} measurements provide rapid and reliable results. Many materials contain a system of antioxidants, often a combination of a primary antioxidant (hindered phenol or amine) and a secondary antioxidant (phosphite, thioester, etc.). Thermal analysis provides no information about the concentration of the different antioxidants separately but rather an overall assessment of the stability. Extraction followed by chromatography (HPLC) is one of the main techniques for the determination of antioxidant concentration, but it is no doubt much more time-consuming than DSC/DTA.

10.3.5 FURTHER APPLICATIONS OF THERMAL ANALYSIS IN POLYMER SCIENCE AND TECHNOLOGY

TG is a very suitable method for determining the concentration of volatile species. A typical analysis involves heating the sample to a temperature at which the vapour pressure of the sorbed component is significant, and holding of the sample at that particular temperature until the sample mass reaches a constant value. The relative decrease in sample mass is then equal to the mass fraction of the volatile component. Sorbed water in polyamide 6, sorbed propane in polyethylene and plasticizing dioctylphthalate (DOP) in polyvinylchloride can be determined in this way.

TG is also very suitable for the determination of the overall content of fillers in polymers. The fraction of carbon black can be selectively determined, since it oxidizes completely at elevated temperatures, forming volatile species.

Chemical reactions in polymer technology, polymerization and vulcanization, both typically strongly exothermic reactions, are conveniently recorded by DSC/DTA. However, it is difficult in many cases to follow the kinetics under isothermal conditions. The reaction is initiated very rapidly at elevated temperatures, and this sets the upper temperature limit. The lower temperature limit is set by the activation energy (shift in rate constant) and the sensitivity of the calorimeter. Some reactions can thus only be studied at constant heating-rate conditions. The method of Barton (1973) can be applied to such cases and kinetic parameters, e.g. the activation energy, can be obtained. Reactions with unknown kinetic expressions can be treated in a simplified manner by measuring the time (t_x) to reach a certain degree of conversion (x) and by plotting t_x against temperature in an Arrhenius diagram as in Fig. 10.25. The activation energy is obtained from the slope.

Another useful application of thermal analysis is in the determination of unreacted monomers in thermosets, e.g. unreacted styrene in crosslinked polyesters. Figure 10.26 shows schematic thermograms of samples with different concentrations of unreacted styrene.

It is possible to determine the molar fraction (x) of unreacted species simply from the evolved exothermal enthalpy (ΔH_{exo}), and the molar energy of

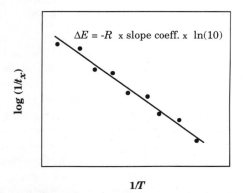

Figure 10.25 Arrhenius diagram correlating chemical reaction rate, the reciprocal time to reach a certain degree of conversion, with temperature.

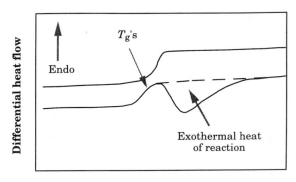

Figure 10.26 Heating thermogram of polyester (unreacted styrene) polymers (lower thermogram) and after a heating and cooling scan of this sample (upper curve).

polymerization (crosslinking), ΔH_{cross}:

$$x \propto \frac{\Delta H_{exo}}{\Delta H_{cross}} \qquad (10.37)$$

10.4 SUMMARY

Thermal analysis involves a set of analytical methods by which a physical property of a sample is measured as a function of temperature (time). The properties that are studied are typically enthalpy (DSC/DTA), dimensions (TMA, dilatometry), visco-elastic properties (DMTA), mass (TG), dielectric properties (DETA) and optical properties (TOA). Reproducible and accurate results are currently available and allow a great number of materials and phenomena involving both physical and chemical aspects to be studied. Thermal transitions of polymers, involving both crystallization/melting and glass formation, are irreversible processes and great care must be taken in the interpretation of the data.

10.5 EXERCISES

10.1. Determine the mass crystallinity from the following DSC data: $\Delta h_{21} = 140$ J g^{-1}, Δh^0 ($T_m^0 \approx 418$ K) = 293 J g^{-1}; $T_1 = 370$ K. Use the specific data of Wunderlich and Baur (1970) and calculate mass crystallinity according to the total enthalpy method.

Temperature (K)	c_{pc} (cal mol^{-1} K^{-1})[a]	c_{pa} (cal mol^{-1} K^{-1})[a]
310	5.40	7.68
330	5.89	7.84
350	6.38	7.99
370	7.02	8.15
380	7.37	8.23
390	7.72	8.31
400	8.14	8.38
410	8.63	8.46
420	–	8.54

[a] Per mole of repeating units (14 g mol^{-1}).

10.2. Calculate the molar entropy of fusion for linear polyethylene from the following data: equilibrium enthalpy of fusion $\Delta h^0 = 293$ J g^{-1} and the equilibrium melting point $T_m^0 = 415$–418 K (different values have been reported).

10.3. The entropy of fusion of polyamide 6 is similar in relation to that of polyethylene counted per mole of main-chain atoms. However, the equilibrium melting point is 533 K for polyamide 6 compared with 415–418 K for polyethylene. Explain.

10.4. A medium-density polyethylene exhibits two melting peaks, appearing at 385 K and 408 K respectively, after a constant-rate cooling from the melt. Give possible reasons.

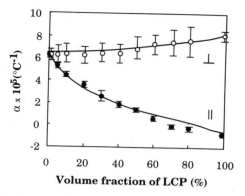

Figure 10.27 Linear thermal expansion coefficient parallel ($\parallel$) and perpendicular ($\perp$) to the moulding directions at room temperature. Samples: blends of Vectra and PES. Drawn after data from Engberg et al. (1994).

10.5. Describe a suitable thermal analysis experiment by which the moisture content in polyamide 6 can be determined.

10.6. It is known that a thermoplastic material consists of two fillers: $CaCO_3$ and carbon black. Design a suitable thermo-analytical experiment by which the filler contents can be determined.

10.7. The thermal expansion data given in Fig. 10.27 were obtained for injection-moulded samples of binary

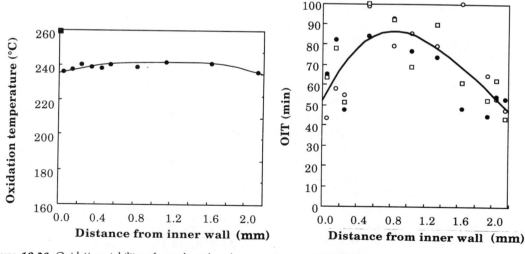

Figure 10.28 Oxidative stability of samples taken from an unexposed MDPE pipe as a function of distance from the inner wall. Oxidation temperature measurements were obtained at a 10°C min^{-1} scan rate and OIT at 190°C. With permission from Society of Plastic Engineers (Karlsson, Smith and Gedde 1992).

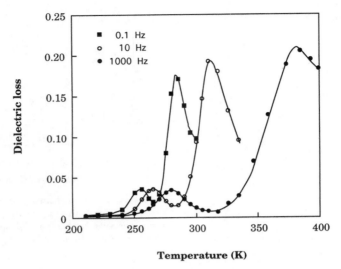

Figure 10.29 Dielectric loss as a function of temperature. Isochronal curves of a hyperbranched polyester. Drawn after data from Malmström et al. (1994).

blends of a liquid-crystalline polymer (Vectra) and polyethersulphone (PES). What do the data say about the structure of the blends?

10.8. Calculate the volume expansion coefficients for blends of 0%, 40% and 80% of Vectra. Use the data shown in Fig. 10.27.

10.9. Draw schematic DSC curves for a fully amorphous polymer cooled at two different rates, 1 and 100 K min^{-1}. The glass–rubber transition is within the temperature range of the cooling scan. Draw schematically the heating thermograms for the sample originally cooled at 1 K min^{-1} when it is heated at 0.1 and 100 K min^{-1}.

10.10. Suggest a series of DSC experiments by which the kinetics of physical ageing can be recorded.

10.11. The DSC trace of a polymer shows a large low-temperature first-order transition and a very small first-order transition at higher temperatures. TOA shows that the major change in light intensity is associated with the high-temperature transition. Present an explanation for these data.

10.12. The oxidation temperature data shown in Fig. 10.28 show only moderate variations, 236–242 K, whereas OIT data show considerable variation, from 50 to 100 min. Are the two sets of data mutually consistent?

10.13. The activation energy for the consumption of antioxidant in the molten state was determined to be 176 kJ mol^{-1}. Calculate the OIT variation (with distance from inner wall) at 220 and 230°C.

10.14. Calculate the activation energies of the relaxation processes shown in the isochronal scans given in Fig. 10.29.

10.6 REFERENCES

Bair, H. E. (1970) *Polym. Eng. Sci.* **10**, 247.
Barton, J. M. (1973) *Makromol. Chem.* **171**, 247.
Benzinger, T. H. and Kitzinger, C. (1963) *Temperature – its Measurement and Control in Science and Industry*, Vol. 3, Part 3 (J. D. Hardy, ed.). Reinhold, New York.
Boersma, S. L. (1955) *J. Am. Ceramic Soc.* **38**, 281.
Boyd, R. H. (1980) Dielectric constant and loss, in *Methods of Experimental Physics*, Vol. 16, Part C (R. A. Fava, ed.). Academic Press, New York.
Cahn, L. and Schulz, H. (1963) *Anal. Chem.* **35**, 1729.
Calvet, E. and Prat, H. (1963) *Recent Progress in Microcalorimetry* (H. A. Skinner, ed. and trans.). Pergamon Press, London.
Demus, D. and Richter, L. (1978) *Textures of Liquid Crystals*. VEB Deutscher Verlag für Grundstoffindustrie, Leipzig.
Engberg, K., Strömberg, O., Martinsson, J. and Gedde, U. W. (1994) *Polym. Eng. Sci.* **34**, 1336.

Gedde, U. W. and Jansson, J.-F. (1983) *Polymer* **24**, 1521.

Gillen, K. T. (1978) *J. Appl. Polym. Sci.* **22**, 1291.

Gray, A. P. (1970) *Thermochimica Acta* **1**, 563.

Hellmuth, E. and Wunderlich, B. (1965) *J. Appl. Phys.* **36**, 3039.

Hoffman, J. D., Frolen, L. J., Ross, G. S. and Lauritzen, J. I. Jr (1975) *J. Res. Natl. Bur. Std. A* **6**, 671.

Karlsson, K., Assargren, C. and Gedde, U. W. (1990) *Polymer Testing* **9**, 421.

Karlsson, K., Smith, G. D. and Gedde, U. W. (1992) *Polym. Eng. Sci.* **32**, 649.

Le Chatelier, H. (1887) *Bull. Soc. Fr. Mineral Crystallogr.* **10**, 204.

Malmström, E., Liu, F., Boyd, R. H., Hult, A. and Gedde, U. W. (1994) *Polymer Bulletin* **32**, 679.

Richardson, M. J. (1976) *Plastics and Rubber: Materials and Applications* **1**, 162.

Roberts-Austen, W. C. (1889) *Metallographist* **2**, 186.

Stone, R. L. (1951) *Bull. Ohio State Univ., Eng. Exp. Stn.* **146**, 1.

Struik, L. C. E. (1978) *Physical Aging in Amorphous Polymers and Other Materials*. Elsevier, Amsterdam, Oxford and New York.

Suurkuusk, J. and Wadsö, I. (1982) *Chimica Scripta* **20**, 155.

Watson, E. S., O'Neill, M. J., Justin, J. and Brenner, N. (1964) *Anal. Chem.* **36**, 1233.

Wunderlich, B. and Baur, H. (1970) *Adv. Polym. Sci.* **7**, 151.

10.7 SUGGESTED FURTHER READING

Billingham, N. C., Bott, D. C. and Manke, A. S. (1981) Application of thermal analysis methods to oxidation and stabilisation of polymers, in *Developments in Polymer Degradation*, Vol. 3 (N. Grassie, ed.). Applied Science Publishers, London.

Chiu, J. (1987) Thermogravimetry for chemical analysis of polymers, in *Applied Polymer Analysis and Characterization: Recent Developments in Techniques, Instrumentation, Problem Solving* (J. Mitchell Jr, ed.). Hanser, Munich.

McCrum, N. G., Read, B. E. and Williams, G. (1967) *Anelastic and Dielectric Effects in Polymeric Solids*. Wiley, New York.

Turi, E. A. (ed.) (1981) *Thermal Characterization of Polymeric Materials*. Academic Press, New York.

Wendtland, W. W. (1974) *Thermal Methods of Analysis*, 2nd edn. Wiley, New York.

Wunderlich, B. (1990) *Thermal Analysis*. Academic Press, Boston.

MICROSCOPY OF POLYMERS

11.1 INTRODUCTION

Microscopy is the name given to a group of experimental methods which permit magnification of morphological structures to make details visible.

The discovery of the optical microscope was made in 1590 by the Dutch optician, Zacharias Jensen, whose microscope consisted of a 45 cm long tube with two lenses (objective and eyepiece). The discovery of the optical microscope led to a rapid development in both physical and biological sciences. The development of instruments proceeded apace during the succeeding centuries. E. Abbé gave optical microscopy a theoretical foundation. He explained the limitation in resolution (minimum separation of two adjacent points that can be seen as two different items) by diffraction effects. The first transmission electron microscope was constructed by M. Knoll and E. Ruska in 1931. The resolution of that instrument was several tens of nanometres. This development was preceded by the realization that particles such as electrons have an associated wave nature (de Broglie) and by the discovery by H. Busch in 1926 that axially symmetric magnetic fields can focus electrons. Several commercial instruments with claimed resolutions of 0.2 nm were developed in Europe, America and Japan on the basis of the principles outlined by Knoll and Ruska. The first scanning electron microscope (SEM) was built by M. von Ardenne in 1938. It took many years, however, until 1965, before SEM became commercially available thanks to the pioneering work of Oatley and collaborators. The scanning transmission electron microscope was developed at the beginning of the 1970s.

The morphology of polymer blends, block copolymers, semicrystalline polymers and liquid-crystalline polymers can be assessed by microscopy. The morphology can be directly observed and the structure is in most cases assessed without the need for any sophisticated model. Small-angle X-ray scattering provides information about the long period (thickness of crystal and amorphous layer) in semicrystalline polymers (Chapters 7 and 12). The structural information gained from the scattering pattern requires, however, the use of a model. Transmission electron microscopy of samples treated with etchants or with staining compounds makes direct observation of the crystal lamellae possible (Fig. 11.1).

What must be questioned is whether the image obtained by microscopy is free from distortions and artefacts. It is quite possible that the dimensions of the structural features will be affected by the radiation. Radiation effects are very rare in optical microscopy but they are common in electron microscopy. Some polymers show a great sensitivity to the electron beam and rapidly degrade, leaving a hole in the specimen. More moderate damage, e.g. shrinkage of the specimen, may also occur.

The purpose of the sample preparation is to obtain a contrast-rich representation of the 'true' structure. If contrast is not naturally present in the specimen, it may be created by staining, i.e. adding a substance with a characteristic colour or high (electron) density selectively to a specific phase. The contrast may also be achieved by selective etching, leaving a topography indicative of the phase structure. It is possible, however, that the preparation method used may produce artificial structures.

A third problem originates from the fact that only a small part of the whole object is studied. Is that part representative of the object? The amount of information obtained from the observations is substantial and the microscopist has to select only a few features which are typical of the specimen. It is, however, a clear possibility that non-typical features are presented.

Microscopy may be divided into the following main groups: **optical microscopy** (OM), **scanning electron microscopy** (SEM) and **transmission electron microscopy** (TEM). These main groups

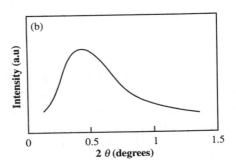

Figure 11.1 (a) Transmission electron micrograph of chlorosulphonated polyethylene. (b) Small-angle X-ray scattering pattern of the same polyethylene sample. Intensity is plotted on an arbitrary units scale. Unpublished data of M. Hedenqvist, Dept of Polymer Technology, Royal Institute of Technology, Stockholm, Sweden.

should be considered as families, each family consisting of several individual techniques. Several new techniques have recently been developed which may briefly be described here. Infrared microscopy provides the infrared spectrum of fine morphological details. Ultraviolet microscopy enables the assessment of local variations in UV absorption, and has been used for the morphological assessment of the content of UV-absorbing additives. Atomic force microscopy (AFM) is a high-resolution method which sensitively records the surface topography, and which has great potential for non-conductive materials such as polymers. Acoustic microscopy provides information about local variations in the mechanical properties in a sample and may thus reveal local variations in chain orientation and composition. These special techniques are not further discussed in the text.

It is possible to use both the optical and electron microscopes for scattering/diffraction studies. The observation of the 'scattering pattern' in an optical microscope is referred to as **conoscopy**. It is also possible by closing the field iris and the aperture iris to obtain a small-angle light scattering pattern (Chapter 7). Electron diffraction, which is carried out in an electron microscope, is a classical method. The scattering methods are presented in Chapter 12.

Table 11.1 presents a summary of the features of OM, SEM and TEM.

Table 11.1 Comparison between optical microscopy (OM), scanning electron microscopy (SEM) and transmission electron microscopy (TEM)

	OM	SEM	TEM
Size of studied objects	0.2–200 μm	0.004–4000 μm	0.0002–20 μm
Depth of field	Small	Large	Large
Objects	Surface or bulk structure	Surface structure	Bulk structure; surface structure (replicate)
Specimen environment	Ambient	High vacuum	High vacuum
Radiation damage	None	Some	Severe
Specimen preparation	Easy	Easy	Difficult
Chemical analysis	No	Yes	Yes
Detection of chain orientation	Yes	No	Yes

Source: Sawyer and Grubb (1987).

11.2 OPTICAL MICROSCOPY (OM)

11.2.1 FUNDAMENTALS

The fundamental principles of optical microscopy are here briefly recapitulated. Detailed presentations of the fundamentals of optical microscopy are found in most textbooks on optics. The magnification of the modern type of optical microscope (compound microscope) is obtained by two lens systems, referred to as the **objective** and the **eyepiece**. The objective generates a magnified real image of the specimen. The real image is further magnified by the eyepiece and a magnified real image is formed at the retina of the eye. The maximum magnification obtained by OM is about 2000×.

A modern research optical microscope can be modified in different ways. The surface topography of a specimen is studied in reflected light mode, whereas the bulk structure is studied with the light transmitted through the specimen. **Bright-field microscopy** is microscopy in which the specimen is illuminated essentially parallel to the main optical axis (either in reflected or in transmission mode). In **dark-field microscopy**, the illumination is at a very large angle to the main optical axis. It is used both in reflected and transmission modes. The term 'dark-field' is used because the phase boundaries appear light against a dark background, which is the inverse of the bright-field image.

The eye can detect variations in light intensity and in colour. Light of different states of polarization cannot be distinguished with the unaided eye. There are several techniques by which a variation in optical path (OP) is converted to a corresponding variation in light intensity (Fig. 11.2). The variation in optical path can be due to a variation in refractive index, in specimen thickness and/or in surface topography (reflected light). The techniques using this principle are polarized microscopy, phase-contrast microscopy and differential interference-contrast microscopy. These techniques are presented in section 11.2.2.

The resolution is defined as the least distance between two 'points' that can be recognized as two separate items. Rayleigh showed that two points can be resolved if the light intensity between the points is 85% or less of the maximum value at the centres of the points. The resolution (d) that is controlled by diffraction is given by:

$$d = \frac{\lambda}{NA_{obj} + NA_{cond}} \quad (11.1)$$

where λ is the wavelength of the light and NA_{obj} and NA_{cond} are the numerical apertures of the objective and the condenser, defined as:

$$NA = n \sin \alpha \quad (11.2)$$

where n is the refractive index and α is half the acceptance angle (Fig. 11.3). Equation (11.1) only holds under the condition that $NA_{obj} \geq NA_{cond}$. High resolution is obtained when components of high numerical aperture are used. The resolution increases with decreasing wavelength of the light.

All optical components are imperfect. Lenses suffer from different so-called aberrations: chromatic (light of different wavelengths is refracted differently), spherical, coma, astigmatism, field curvature (the sharp image falls on a curved surface) and distortion. Modern lenses correct for some of the existing aberrations: achromats and apochromats correct for chromatic aberration. Lenses corrected for field

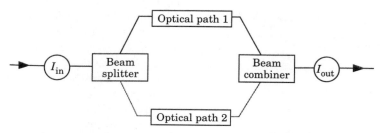

Figure 11.2 The principle behind the techniques that reveal variations in optical path. A variation in optical path (OP) between the two beams (OP1 ≠ OP2) leads to a lowering of the intensity of the outgoing light ($I_{out} < I_{in}$).

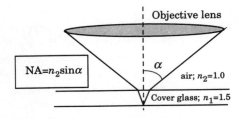

Figure 11.3 Definition of numerical aperture (NA) of objective lens.

curvature are denoted 'plano'. However, the aberrations reduce the resolution of the microscope and the d value calculated according to eq. (11.1) should be considered to be the maximum resolution.

The components in a transmission optical microscope are shown schematically in Fig. 11.4. The

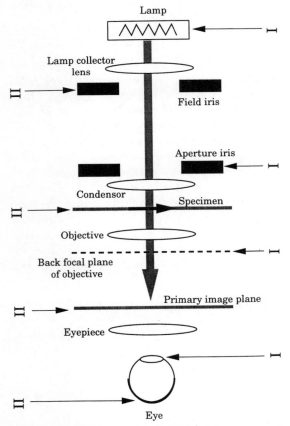

Figure 11.4 Schematic representation of the compound microscope, including the conjugate planes.

light source in a modern microscope is of the filament type and the commonest type, made of tungsten, produces a continuous spectrum in the range 300–1500 nm. Colour-correction filters are used to obtain a narrower spectrum. The intensity of the light should not be adjusted by changing the voltage but rather by inserting a neutral filter. The collector lens is placed so that the focal point of the lens is at the filament of the bulb. The field iris adjusts the size of the viewed field. The aperture iris adjusts the part of the lamp filament that illuminates the specimen.

The condenser concentrates the light on the specimen, provides a uniform illumination and matches the aperture of the illuminating cone with the numerical aperture of the objective. A common bright-field condenser consists of a top lens that can be swung into the light beam to work with higher-magnification objectives whereas lower-magnification objectives are used with the lower condenser lens alone. The condenser may provide special illumination. Dark-field condensers illuminate the specimen at a high angle to the main optical axis. Special condensers are used for phase-contrast and differential interference-contrast microscopy. More details of these techniques are presented in section 11.2.2. The condensers are characterized by their numerical aperture and optical correction.

The objective provides the major part of the magnification and also controls the aperture which in turn controls the resolution. The maximum objective magnification is about $100\times$. The objective is characterized by magnification, numerical aperture, optical correction (achromate, fluorite, apochromate, plano, etc.). objectives marked 'oel' are so-called oil-immersion objectives and should always be used with immersion oil between the cover glass and the objective lens. The working distance, i.e. the distance between the objective lens and the specimen, is very short (≤ 1 mm) for high-magnification objectives. Special objectives ($25-35\times$ magnification) with high working distances (c. 10 mm) have been constructed for hot-stage work. Special objectives are used for polarized microscopy (using stress-free glass; denoted POL), phase-contrast and differential interference-contrast microscopy. The eyepiece is used to magnify the primary image further and to introduce scales and pointers in the primary image plane.

'Conjugate planes' is the term given to planes in the optical system which are equivalent. An object placed in one of them will appear as a sharp image in the other subsequent planes of the series. There are two separate series of conjugate planes in the compound microscope (Fig. 11.4):

- Series I consists of the lamp filament, the front focal plane of the condenser (which is located at the aperture iris), the back focal plane of the objective and the exit pupil of the eye.
- Series II consists of the lamp iris diaphragm (field iris), the specimen plane, the primary image plane and the retina of the eye.

Series II includes the specimen plane. The field iris controls the diameter of the area that is viewed. Pointers or scales must be inserted in the primary image plane. Series I is related to the filament. An image of the filament is found at the back focal plane of the objective. The principle behind the uniform illumination of the specimen is the so-called Köhler illumination. Every part of the filament illuminates every part of the specimen. The correct illumination is obtained by adjusting the collector lens and the condenser. Detailed instructions are provided in microscopy manuals.

11.2.2 POLARIZED MICROSCOPY AND RELATED TECHNIQUES

A polarized microscope is equipped with two crossed polarizers. The first, the polarizer, is located in front of the specimen stage and the second, the analyser, is located behind the objective. The light entering the specimen is linearly polarized. The light passes through an optically isotropic sample without any change of polarization and the crossed analyser extinguishes the light. However, the outgoing light is elliptically polarized after passage through a birefringent sample and the analyser. The analyser reduces the intensity of the outgoing light but an appreciable fraction of the incoming light is transmitted.

It is customary to define a refractive index ellipsoid which becomes an ellipse in the plane. The lengths of the long axes of the refractive index ellipse are proportional to the maximum and the minimum

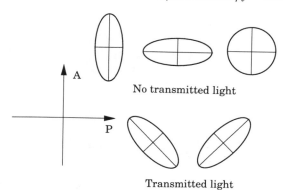

Figure 11.5 The influence of the orientation of refractive index ellipses with respect to the polarizer (P)/analyser (A) pair on the transmitted light through the analyser.

refractive indices in the plane. The polarization of the light remains unchanged if the refractive index ellipse has its long axes parallel to the polarizer/analyser pair. All other orientations of the ellipse will lead to a situation in which the light is partially transmitted through the analyser (Fig. 11.5).

These statements may be explained by the following qualitative discussion: the polarization of the light may be represented by an electrical field vector $\bar{E}$. The incoming light is linearly polarized and $\bar{E}$ may be divided into two components, E_1 and E_2, where '1' denotes the direction of maximum refractive index and '2' the direction of minimum refractive index. The two components E_1 and E_2 are initially in phase. However, the refractive index difference causes a retardation of the '1' component with respect to the '2' component and the outgoing light is the combination of the two linearly polarized components that are out of phase, i.e. an elliptically polarized light. If the phase difference is a multiple of the wavelength, then the two components combine to give a linearly polarized light. If '1' is parallel to either P or A, the partitioning of $\bar{E}$ results in only one of E_1 and E_2, the other being perpendicular and hence zero and the outgoing light will be linearly polarized and blocked by the analyser (Fig. 11.5).

The lambda plate and other compensators are accessories commonly used in polarized microscopes. A detailed presentation of compensators and their use for the assessment of in-plane birefringence is given in Chapter 9. It is impossible to distinguish the two

244 *Microscopy of polymers*

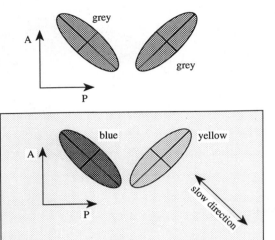

Figure 11.6 Schematic illustration of the images of birefringent structures of different orientation with respect to the polarizer (P)/analyser (A) pair. Top: crossed polarizers only. Bottom: crossed polarizers with an inserted lambda plate.

differently birefringent objects shown in Fig. 11.6 with conventional polarized microscopy. The two birefringent objects can be distinguished by inserting a lambda plate at 45° to the A/P pair (Fig. 11.6, bottom). The reason for the presence of colours (presented in a so-called Michel–Levy chart) is given in Chapter 9. Superstructures present in semicrystalline polymers are conveniently studied with polarized light microscopy. The optical features of spherulites are discussed in Chapter 7.

Phase-contrast microscopy converts differences in refractive index into variations in light intensity. Light which is scattered by the specimen travels in a different direction than undeflected light. The scattered light is phase-shifted and destructive interference takes place when the two are recombined to form the image. Contrast is obtained when the relative intensities of the scattered and the unscattered (transmitted) light vary from one location in the sample to another. Zernike phase contrast uses a special condenser with an opaque central plate, which blocks the light, and a matching phase ring in the back focal plane of the objective. Only the unscattered light will travel through the phase ring and its phase is changed compared to that of the scattered light. The combination of scattered and unscattered light leads to a change in light intensity. The phase contrast image typically shows bright halos around the fine structure.

Differential interference-contrast microscopy (DICM) involves the splitting of the illuminating beam into two components. The two beams are displaced and then combine, and constructive or destructive interference may take place. A region of constant refractive index and constant specimen thickness appears without any contrast. However, any sudden change in refractive index gives a strong contrast. DICM is mostly used in reflective mode.

11.3 ELECTRON MICROSCOPY

Fine details of the microstructure cannot be resolved by optical microscopy, but electron microscopes can resolve details smaller than 1 nm. Electro-optical imaging can be achieved in different ways. The electron beam is controlled by lenses consisting of magnetic fields. Rotational symmetric electromagnets focus the electron beam in the same way as convex lenses do in optical microscopes. The transmission electron microscope is built according to the same principle as the optical microscope, with a condenser lens, an objective lens and a projector lens (the analogue of the eyepiece), as shown in Fig. 11.7. A magnified image is obtained on a fluorescent screen or on a photographic film. The electron source is commonly a hairpin tungsten filament or a lanthanum boride (LaB_6) filament heated with a low-voltage source. The potential of the filament is highly negative and the electrons are accelerated towards an anode held at a small positive potential. A typical voltage used in transmission electron microscopes (TEM) is 100 kV. The acceleration voltage used in scanning electron microscopes (SEM) is considerably lower, typically about 15 kV.

The resolution of the electron microscope depends on the voltage. The wavelength of the accelerated electrons is, according to de Broglie, equal to:

$$\lambda = \frac{h}{mv} \qquad (11.3)$$

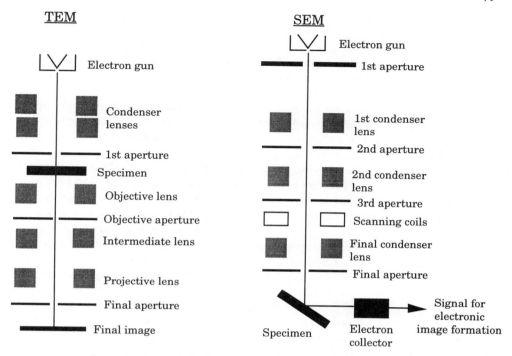

Figure 11.7 Schematic description of conventional TEM and SEM.

where h is Planck's constant, m is the mass of the electron and v is the velocity of the electron. For an electron with charge e subjected to a voltage U, the following expression holds provided that relativistic effects are negligible:

$$eU = \frac{mv^2}{2} \quad (11.4)$$

Insertion of eq. (11.4) into eq. (11.3) yields:

$$\lambda = \sqrt{\frac{h^2}{2meU}} \quad (11.5)$$

The high resolution (d) of the electron microscope is due to the short wavelength of the electrons, i.e.:

$$d = \sqrt{\left(\frac{0.61\lambda}{\alpha}\right)^2 + (C_s\alpha^3)^2} \quad (11.6)$$

where α is half the angular aperture (in radians) and C_s is the spherical aberration coefficient. The first term is the resolution given by diffraction and the second term is that due to chromatic aberration. The wavelength of the accelerated elecrons in a 100 kV TEM is, according to eq. (11.5), equal to 0.0037 nm. Insertion of this value and of $\alpha = 6 \times 10^{-3}$ rad and $C_s = 1.6$ mm into eq. (11.6) yields a resolution of $d = 0.5$ nm. The d value is in practice higher due to other aberrations of the magnetic lenses. The resolution increases with increasing voltage (cf. eqs (11.5) and (11.6)). The diffraction-dependent resolution of a 20 kV SEM is approximately 2 nm. However, most modern SEM instruments have a secondary electron image resolution of 10–30 nm. The depth of field is given by $2d/\alpha$. The aperture is normally very small (α is 10^{-2} to 10^{-3} rad) and the depth of field is of micrometre size for both SEM and TEM, which is significantly larger than for OM.

The electron microscope operates under high vacuum. The electron beam in a conventional TEM is stationary. The beam is, by appropriate adjustment of the condenser lenses, highly collimated (diameter of 5 nm) and is allowed to scan over the specimen surface under the control of scanning coils in both SEM and scanning TEM (Fig. 11.7). The advantage of scanning TEM over conventional TEM is that no image-forming lenses are used in the scanning TEM after the specimen. This eliminates chromatic aberration due to multiple scattering in the sample. Thicker samples and lower acceleration voltages for the same resolution can consequently be used in the scanning TEM. It is also possible to perform electron diffraction on areas as small as 10 nm² in a scanning TEM. There are nevertheless some drawbacks with scanning TEM compared with conventional TEM: space-charge effects from the small-diameter and high-current beam, and poorer diffraction contrast images.

SEM provides information about the topography of a specimen. The electrons that hit the surface of the specimen yield secondary electrons (used for the image), backscattered electrons (used for the image), Auger electrons (50–2000 eV) and X-ray radiation. The secondary electrons have relatively low energies, less than 50 eV, with an average value close to 4 eV. They cannot escape from depths greater than 10 nm in the specimen. The wavelength of the emitted X-ray is unique for a given element. Some SEMs are equipped with an X-ray microanalyser, and these instruments provide information about the elemental composition of the surface material. The characteristic X-rays emitted from the lighter elements such as hydrogen, carbon, oxygen and nitrogen cannot be measured with sufficient accuracy to permit an elemental composition analysis of polymers. However, additives and 'foreign' particles containing heavier elements (e.g. sulphur) can be detected by conventional instruments.

Figure 11.8 shows schematically the origin of diffraction contrast in TEM. The sample scatters (diffracts in the case of a crystal) the electrons, and it is possible to intercept all scattered/diffracted electrons using a small objective aperture. Only the undiffracted electrons form the final image in the bright-field case. A region of high density scatters more electrons than a region of low density. Regions of low density appear bright (high electron density) and regions with high density appear dark (low electron density). Dark-field images are obtained by adjusting the objective aperture so that only diffracted electrons are transmitted through the aperture. It is possible to select specified crystalline diffraction spots and to obtain dark-field images of crystals. The dark-field image is the complement of the bright-field image. Regions with prolific diffraction appear dark (low electron density) in bright-field and light (high electron density) in dark-field images. It should be noted that the naturally occurring variations in density are mostly insufficient. Staining involves selective addition of heavy elements to one of the phases, making it denser than other regions.

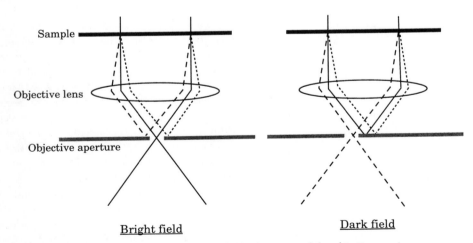

Figure 11.8 Bright-field and dark-field adjustment of the objective aperture.

Phase contrast also appears in electron microscopy. The passage of the wavefront through a specimen causes a phase shift which can be converted into image intensity by interference between the retarded wave with another wave. Components which cause phase contrast have been tried in the transmission electron microscope and do work, but they are impractical.

11.4 PREPARATION OF SPECIMENS FOR MICROSCOPY

A general rule in microscopy is that the preparation of a specimen takes a much longer time than the actual examination. The difficulty in preparing good samples is substantial.

11.4.1 PREPARATION OF SAMPLES FOR OPTICAL MICROSCOPY

Specimens for OM should be thin (1–20 μm) and without scratches or zones of plastic deformation on the surface. Thin films are readily obtained by compression moulding or by solution casting of 1% solutions. Small drops of the solution are placed on an object glass kept on a hot plate. The temperature of the plate should be 30–40°C above the melting point of the polymer. The solvent is evaporated and the sample is covered with a cover glass. Solution casting is particularly suited for preparing thin specimens. Samples can also be obtained by microtoming from thicker objects. This is a particularly difficult preparation method which requires both skill and experience. A correct combination of pre-specimen geometry, knife and temperature has to be used. Soft polymers should be sectioned at cryogenic temperatures. The size of the section specimen must be small. The specimen is inserted between object and cover glass. It may be helpful to use a refractive-index-matching immersion oil to avoid disturbing scattering from surface scratches.

11.4.2 PREPARATION OF SAMPLES FOR SCANNING ELECTRON MICROSCOPY

Polymer specimens to be examined by SEM are coated with a thin layer of electrically conductive material. The electrons which hit the specimen are partly scattered and subsequently recorded by the detector and partly conducted to the grounded electrode. The coating is achieved by vacuum evaporation or sputtering of a heavy metal (e.g. Au or Pd) or carbon. The specimen is often glued with an electrically conductive adhesive to a metal-base specimen holder to achieve good electrical contact with the grounded electrode. Samples prepared for elemental analysis using an X-ray microanalyser should be coated with carbon. SEM can be used for many different purposes. One is fractography, which requires no pretreatment of the sample other than metal or carbon coating. Another is the study of fibre orientation in fibre composites, which requires grinding and polishing using abrasive pastes. The phase morphology of polymer blends can be studied on samples prepared by different methods: solvent or degrading extraction of a polished sample to remove one of the components; or fractography of samples cracked at cryogenic temperatures. Spherulitic structure in semicrystalline polymers can be studied on solvent-etched samples. Degrading etchants can also be used.

11.4.3 PREPARATION OF SAMPLES FOR TRANSMISSION ELECTRON MICROSCOPY

Preparation of samples for TEM is difficult and time-consuming. The natural variation in density is seldom sufficient to achieve adequate contrast. Contrast is obtained by staining or by etching followed by replication. Staining involves a chemical reaction between one of the components and the staining reagent. Atoms of higher elements are added selectively to this phase and regions of different electron density are achieved in the image (Fig. 11.9).

Treatment of polydienes with osmium tetroxide (OsO_4) was introduced by Andrews (1964) and Andrews and Stubbs (1964). A later development by Kato (1965; 1966; 1967) showed that OsO_4 was able to stain the rubber component in polymer blends. Osmium tetroxide adds to the double bonds and the density of rubber phase is increased. Since it crosslinks the rubber polymer by reacting with two double bonds located in two different polymer chains (Fig. 11.10), the rubber is fixed and sectioning can be carried out without smearing. Osmium tetroxide

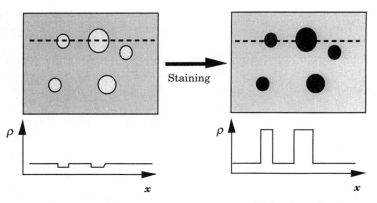

Figure 11.9 The effect of staining. The heavy element is added only to the discrete phase.

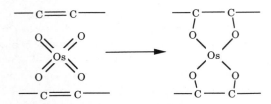

Figure 11.10 Simplified reaction scheme for how OsO_4 is added to unsaturated polymer chains.

staining has been used on ABS, high-impact polystyrene, other rubber-containing plastics, styrene-butadiene-styrene copolymers and polyethylene-terephthalate. The morphology of semicrystalline polyisoprenes has also been studied. The osmium tetroxide diffuses only into the amorphous component, leaving the crystals unaffected. Osmium tetroxide has a high vapour pressure, making vapour staining possible. Osmium tetroxide should be treated with great care since it reacts not only with double bonds in polymers but also with the membranes of the eye.

Ruthenium tetroxide is a stronger oxidizing staining agent than OsO_4, which enables RuO_4 to react also with phenyl groups, e.g. in polystyrene. Furthermore, ruthenium tetroxide has been used to stain polyethylene and polypropylene in addition to the list of polymers stained by OsO_4.

The staining ability of chlorosulphonic acid of the amorphous component of polyethylene was first reported by Kanig (1973). The strong acid diffuses and reacts almost exclusively with the amorphous component, leaving the crystals unaffected. Sulphur, chlorine and oxygen are introduced into the amorphous component and the density of the amorphous component is increased above the density of the crystalline component. Elemental analysis reported by Kalnins, Conde Braña and Gedde (1992) showed that chlorosulphonation causes a major reduction in hydrogen content from about 14% to only 4.3% and that the chlorosulphonated material contains 38.5% of C, 25.4% of O, 25.2% of S and 6.2% of Cl. The ratio of O to S is about twice that in the SO_2Cl group, which indicates that the polymer is oxidized by the chlorosulphonic acid treatment. The chlorosulphonic acid also crosslinks the polymer, thus fixing the stained material to be sectioned in a similar way to OsO_4 in unsaturated rubber. The contrast is further enhanced by treatment with uranyl acetate. Chlorosulphonation of polyethylene makes it possible to assess the lamellar structure, although only crystals with their fold surfaces parallel to the electron beam are clearly revealed. More details of the chlorosulphonation of polyethylene are presented in Chapter 7. A polyethylene sample treated with chlorosulphonic acid turns black and the thickness of the black skin layer increases approximately linearly with time (Fig. 11.11). The rate of staining decreases with increasing density (crystallinity) of the polymer. Polypropylene also reacts with chlorosulphonic acid, but this polymer disintegrates rather than crosslinks and the black layer remains very thin even after prolonged treatment (Fig. 11.11).

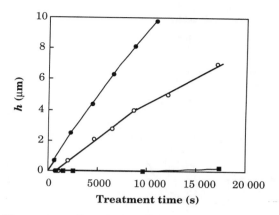

Figure 11.11 Penetration depth (h) as a function of chlorosulphonic acid treatment time at room temperature for LDPE (●), HDPE (○) and PP (■). After Kalnins, Conde Braña and Gedde (1992); with permission from Elsevier Applied Science, London.

Table 11.2 presents a summary of commonly used staining agents. The list is not complete, but the more commonly used staining agents are presented.

The typical specimen which is examined in a TEM consists of a series of thin (50–100 nm) sections of stained polymer on a microscopy grid. Thin sections are produced in an ultramicrotome. Sectioning requires great care, skill and experience. The following steps are included: specimen mounting, optional embedding (epoxy, polyesters and methacrylates are commonly used) and curing, trimming and finally sectioning. Glass knives and diamond knives are used far more than any other. Glass knives are cheap and easy to make but only remain sharp for less than 1 h. They can be used for polymers softer than glass. Diamond knives are very expensive but sharp (and can also be sharpened). Figure 11.12 shows the knife, the trough and how the sectioned samples float in a row in the trough. Soft polymers cannot be sectioned at room temperature without a fixing (staining) agent. They require sectioning at low temperatures (cryo-ultramicrotomy). Several problems arise when sectioning at lower temperatures, such as freezing of the trough fluid and frost build-up. Reid (1975) is recommended reading on this subject.

Etching involves the removal of a specific component from the structure. The topography of an etched sample reflects the distribution of the removed component in the structure. It is essential to confirm

Table 11.2 Staining agents used for TEM

Functional groups	Polymers	Staining agents
$-CH_2-CH_2-$	PE	chlorosulphonic acid
$-CH_2-CHCH_3-$	PP	ruthenium tetroxide phosphotungstic acid
$-CH=CH-$	polydienes etc.	osmium tetroxide ruthenium tetroxide
$-OH; -COH$	polyalcohols, polyaldehydes	osmium tetroxide ruthenium tetroxide
$-O-$	polyethers	osmium tetroxide ruthenium tetroxide
$-COOR$	polyesters	hydrazine followed by osmium tetroxide, phosphotungstic acid, silver sulphide
$-CONH-$	polyamides	phosphotungstic acid tin chloride
$-Ph; -Ph-$		ruthenium tetroxide silver sulphide mercury trifluoroacetate

Source: Sawyer and Grubb (1987, p. 108).

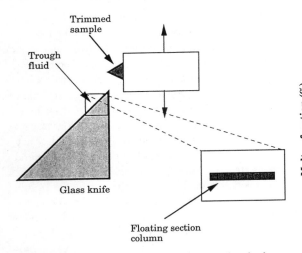

Figure 11.12 Ultramicrotomy using a glass knife.

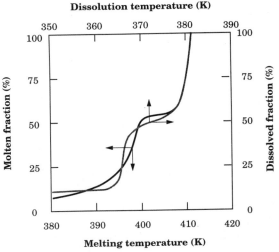

Figure 11.13 Cumulative melting and dissolution (in p-xylene) curves of a linear polyethylene crystallized at 401 K to completeness and then rapidly cooled to room temperature. Drawn after data from Gedde, Eklund and Jansson (1983).

the efficiency of the etching. There are two main types of etching: solvent etching and chemical etching.

Solvent etching removes whole molecules from the sample without any chemical reaction. It is impossible to remove one of the components from an immiscible polymer blend by this method. Low melting species in semicrystalline polymers can also be removed selectively. Solvent and treatment temperature are the two factors that are changed. Blends of amorphous polymers can be etched with good solvents of the minor component, leaving holes in the etched surface. Crystalline polymers require etching at elevated temperatures. Low melting species of polyethylene can be selectively removed by etching with p-xylene. The temperature chosen for the treatment should be approximately 30 K lower than the maximum melting point of the species dissolved and removed from the sample (Fig. 11.13). This relationship turns out to be applicable to linear polyethylene of low and intermediate molar mass. However, very high molar mass and slightly branched polyethylenes cannot be etched efficiently at lower temperatures.

Chemical etching is accomplished by degradation and removal of one of the components from the sample. Classical techniques are etching with plasma and ions. Both these techniques tend to produce artefacts. Acids, alkalis and n-alkyl amines are used to etch polyesters. Nitric acid degrades the amorphous component of polyethylene, leaving the crystals intact. However, degradation is so complete that the sample disintegrates into a powder.

The most important development in this area is due to Olley and Bassett (1982) who invented permanganic acid etching. The original recipe consisted of 7% of potassium permanganate in concentrated sulphuric acid. Later recipes also contained orthophosphoric acid and the amount of potassium permanganate was reduced to 1–2%. Etching using the original recipe led in some cases to artificial structures, particularly evident at low magnifications. The permanganic acid degrades the amorphous component more readily than the crystalline, and the lamellar structure of semicrystalline polymers is revealed. The lamellar structure of several semicrystalline polymers, including polyethylene, isotactic polypropylene, poly(butene-1), poly(4-methyl-pentene-1), isotactic polystyrene and polyetherether ketone has been studied. One advantage of the permanganic acid method over the staining methods (e.g. chlorosulphonation) is that all crystal lamellae are visible, regardless of their orientation. It is possible to assess their thickness from a 'side view' and their

Table 11.3 Etching techniques

Polymer	Etchant	Comments
Polyethylene	Hot p-xylene, toluene and CCl_4	Low melting species are removed
Polyethylene, isotactic polypropylene, isotactic polystyrene, poly(etherether ketone)	Permanganic acid	Lamellar morphology and superstructure are revealed. Artefacts have been reported in low-magnification images
Polyesters including PETP	n-alkylamines	Superstructures and lamellar morphology are revealed. Artefacts have been reported
ABS	10 M chromic acid Sulphuric acid, chromic acid and water	Degrades the rubber phase
PMMA in blends	Electron irradiation in electron microscope	PMMA is degraded and contrast develops
Polyamides	Aromatic and chlorinated hydrocarbons	Low melting species are removed

Source: Sawyer and Grubb (1987, p. 108).

lateral shape from a 'top view'. The staining methods reveal only crystallites viewed from the side.

Table 11.3 presents a summary of etching methods, again in this case not a complete list.

The surface topography of etched samples can be studied directly using SEM (after metal or carbon coating) or TEM (after replication). The replica technique has the advantage that the specimen examined contains no beam-sensitive polymer. One-stage or direct replicas show the best resolution and are fast, although difficult to make. The sample is coated with carbon, shadowed with a heavy metal (e.g. Pt) at an angle of 30–45°. The replica is stripped off the sample and transferred to the microscopy grid. The stripping is the difficult step. The preparation of the two-stage replica involves the formation of a replica using a polymer solution, stripping of the replica from the specimen, shadowing with a heavy metal, carbon coating, placing the shadowed replica on the grid and dissolution of the plastic film. More information is given by Sawyer and Grubb (1987).

A number of special preparation techniques have been developed for single-polymer crystals. The thickness of a single crystal is of the order of 10–20 nm. This thickness can be determined by shadowing, i.e. evaporation of a heavy metal at a known angle (Chapter 7). Gold decoration is a technique which provides information about the fold surface and the orientation of the steps in the fold surface. The sample is given a very thin gold layer and is later heated slightly. The gold migrates to the steps and orientates along the step boundaries.

11.4.4 ARTIFICIAL STRUCTURES

Artificial structures, or simply artefacts, are not true features of the structure of a material. Artefacts are created by the preparation method or during examination (radiation damage). A short list indicating just a few of the reported artefacts is presented below. The lesson to learn is that observed structures should never be taken for granted as long as the possibility of artefacts has not been considered:

- Etching with permanganic acid using the original recipe (7% $KMnO_4$ in concentrated H_2SO_4) leads

to the formation of large regular (hexagonal in some cases) structures that in some reports were mistaken for spherulites. It was later shown that the artefacts were formed by precipitation of a manganese salt.
- Solvent etching sometimes causes local swelling of the polymer and the formation of regular artificial topographical structures: CCl_4 vapour etching of crosslinked polyethylene led to large (10–100 μm) quilt-like patterns falsely reported to be spherulites. The 'true' spherulites in this sample had a diameter of 1–2 μm.
- Freeze-fracturing may lead to the formation of spherulite-shaped crack patterns with no resemblance to the true spherulitic structure.
- Chlorosulphonated sections of polyethylene may shrink considerably in the electron beam due to thermal effects. The shrinkage is more pronounced for samples treated with chlorosulphonic acid for a short period of time. The more numerous crosslinks present in samples subjected to prolonged acid treatment prevent shrinkage of the sections. The effect of shrinkage of the section is that the dimensions of morphological features apparently become smaller.

11.5 APPLICATIONS OF POLYMER MICROSCOPY

The list of applications of microscopy in polymer science and technology is almost infinite. Microscopy is useful in all fields and the understanding of many phenomena is founded on data generated by microscopy. However, at the same time, a number of reports have obviously presented data of doubtful character. The possibility of obtaining artefacts is not negligible. It is also possible that unrepresentative features have been reported. We shall here present a few areas in which microscopy has played a vital role. The list of good examples can, however, be made significantly longer.

11.5.1 SEMICRYSTALLINE POLYMERS

Spherulites as small as 1 μm in diameter have been reported; they cannot be resolved by conventional polarized microscopy. Small-angle light scattering is structures are conveniently assessed by polarized microscopy. The possible difficulties arise from problems in making the specimen sufficiently thin and also in the case of very small superstructures. Spherulites as small as 1 μm in diameter have been reported; they cannot be resolved by conventional polarized microscopy. Small-angle light scattering is the ideal way to analyse samples containing small spherulites (Chapters 7 and 12). It is also possible to obtain meaningful results from solvent-etched samples examined using SEM.

The lamellar structure is best studied using TEM. Let us take polyethylene. Two methods of preparation are commonly used: staining with chlorosulphonic acid/uranyl acetate and etching with permanganic acid.

Figure 11.14 shows a transmission electron micrograph of two chlorosulphonated samples. The crystal lamellae appear white and the amorphous component dark. Crystals with sharp contrast have their fold surfaces parallel to the direction of viewing, the latter being along [$uv0$]. It is typical to find 'blank' areas with no lamellar contrast on the micrographs. This is because tilted lamellae show no contrast. The crystal lamellae are only seen when they are (approximately) parallel to the electron beam. This is a clear disadvantage of this method. One of the advantages of the chlorosulphonation method is that the thickness of both crystal lamellae and the amorphous interlayer can be determined with high precision.

The location of the sections should be random with respect to the centres of the spherulites (axialites). From simple geometrical considerations, it can be deduced that the average distance between the section and the spherulite centre should be $R/\sqrt{3}$, where R is the radius of the spherulites. It is well established that [010] is parallel to the radius of a mature spherulite. Hence, crystal lamellae which appear sharp are predominantly viewed along [010]: 40% of the surface is within a 20° angle from [010] and 60% within a 30° angle from [010].

Linear polyethylene displays relatively straight and long crystal lamellae. The amorphous interlayers are generally very thin. Occasional roof-ridged lamellae are found. The number of lamellae per stack is high. Branched polyethylene

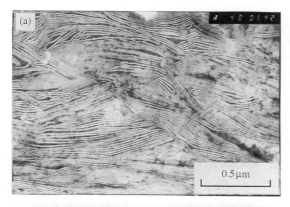

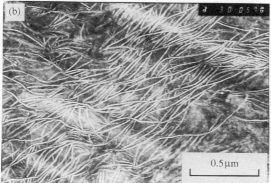

Figure 11.14 Transmission electron micrographs of chlorosulphonated samples: (a) linear polyethylene; (b) branched polyethylene. Micrographs of M. T. Conde Braña, Dept. of Polymer Technology, Royal Institute of Technology, Stockholm.

displays long S-shaped lamellae surrounding significantly shorter lamellae of the same or less thickness.

Voigt-Martin and Mandelkern (1989) compared crystal thickness data by TEM (chlorosulphonated thin sections), Raman spectroscopy and small-angle X-ray scattering (SAXS) on several linear polyethylenes. A sample with a narrow thickness distribution, an average crystal thickness of 21.5 nm and a distribution ranging from 10 to 30 nm, obtained by crystallization at relatively low temperatures, exhibited perfect agreement in crystal thickness data by TEM, Raman spectroscopy and SAXS. In samples exhibiting pronounced bimodal distributions or very broad distributions in crystal thickness, TEM clearly underestimated the thicker crystals. In a sample having a bimodal distribution with maximum values at 10

and 15 nm, TEM overemphasized the thinner crystals. Both populations appeared in the histograms but with incorrect weights. Another sample which was crystallized at 403 K, exhibited a very broad crystal thickness distribution with maxima at 10, 29, 36 and 40 nm. The 10, 29 and 36 nm crystals were revealed by TEM but with different weightings. The 40 nm crystals, on the other hand, were detected only by Raman spectroscopy and not by TEM.

It has been argued on several occasions that the crystal thickness values as obtained by TEM on chlorosulphonated specimens are too low. Hill, Bradshaw and Chevili (1992) showed that the sections may shrink when the sections are exposed to the electron beam even at low intensities, particularly in the case of samples which have been chlorosulphonated for only a short period of time.

Figure 11.15 shows transmission electron micrographs of replicas of samples etched with permanganic acid. One of the advantages is directly noted. There are no blank areas. All crystals, irrespective of orientation, are revealed. It is possible to find areas where the crystals are seen from the side with the chain axis in the sample plane (Fig. 11.15, top micrograph). It is also possible to find regions which reveal the lateral shape of the crystals, i.e. with the chain axis perpendicular to the sample plane (Fig. 11.15, bottom micrograph). Another advantage of the permanganic acid etching is that replica is not beam-sensitive. A disadvantage is the impossibility of measuring the thickness of the amorphous interlayer. By tilting the sample, it is possible, however, to make a fairly accurate assessment of the long period.

11.5.2 LIQUID-CRYSTALLINE POLYMERS

Polarized microscopy has been one of the most important methods for structural assessment of liquid-crystalline polymers. It goes back to Demus and Richter (1978) who presented a great number of photomicrographs of small-molecule liquid crystals. Liquid-crystalline polymers are presented in detail in Chapter 6. Figure 11.16 shows a series of photomicrographs of different liquid-crystalline and crystalline monomers and polymers. The schlieren

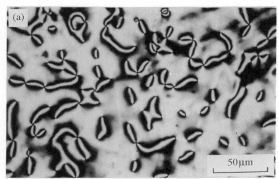

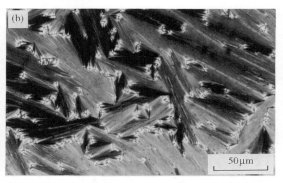

Figure 11.15 Transmission electron micrographs of replicas of polyethylene etched with permanganic acid. Micrographs of M. T. Conde Braña, Dept of Polymer Technology, Royal Institute of Technology, Stockholm.

Figure 11.16 Polarized photomicrographs of liquid-crystalline compounds: (a) nematic; (b) focal conic texture (s_A). Photomicrographs taken by F. Sahlén, Dept of Polymer Technology, Royal Institute of Technology, Stockholm.

texture (Fig. 11.16(a)) is typical of a nematic organization. The presence of disclinations of strength $\pm\frac{1}{2}$ is unique for the nematics. The focal conic texture (Fig. 11.16(b)) is typical of a smectic A phase.

Polarized microscopy can be used as one of several methods for the assessment of liquid-crystalline structure. Because of its simplicity it deserves to be used, but it is equally clear that the X-ray diffraction methods, particularly on aligned samples, provide more definite answers.

11.5.3 POLYMER BLENDS

The morphology of polymer blends is an area of significant scientific and technological importance. The issue is to assess the miscibility of the constituents. A detailed discussion of the thermodynamics of polymer–polymer blends and of the currently available techniques for the assessment of miscibility is presented in Chapter 5. In this section a few micrographs showing the result of different preparation methods are presented. Polyethylene and polystyrene are incompatible. By adding a few per cent of a styrene-b-ethylene-co-1-butene-b-styrene (SEBS) tri-block polymer it is possible to increase the interfacial adhesion and to reduce the size of the polystyrene domains in the polyethylene matrix. Figure 11.17 shows scanning electron micrographs of two tertiary blends of polyethylene-polystyrene-SEBS. Two different preparation techniques have been employed: solvent etching using chloroform (a good solvent for polystyrene) and fracturing.

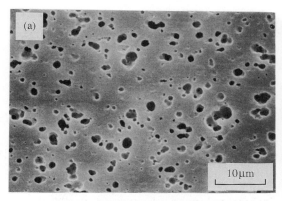

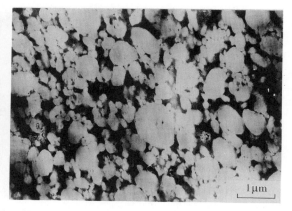

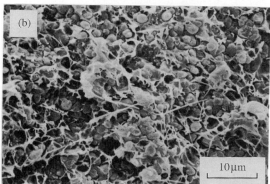

Figure 11.17 Scanning electron micrographs of blends of polyethylene/polystyrene/SEBS: (a) chloroform-etched sample (80/20/2); (b) fractured sample (50/50/5); at 23°C. With permission from Society of Plastic Engineers, USA (Gustafsson, Salot and Gedde 1993).

Figure 11.18 Transmission electron micrograph of an OsO_4-stained PVC/poly(ethylene-*co*-vinyl acetate (45% by weight)) (EVA-45) blend. The EVA phase was first hydrolysed by refluxe boiling in a solution of sodium hydroxide in methanol. The dark regions are the OsO_4 stained EVA-45 phase. Micrograph of B. Terselius, Dept of Polymer Technology, Royal Institute of Technology, Stockholm.

The solvent-etched specimen was first polished. The spherical holes indicate the presence of spherical polystyrene particles (Fig. 11.17(a)). The fracture surface contains many small spherical particles which consist of polystyrene. The polyethylene component had undergone some plastic deformation which is seen in the microfibrils appearing in the micrograph (Fig. 11.17(b)).

Finer details can be resolved by TEM either on stained samples (Fig. 11.18) or on replicas of etched samples. The classical osmium tetroxide treatment has proven very useful. Apart from staining of unsaturated rubber polymers, it effectively stains structures containing vicinal hydroxyl groups. Thus the vinyl-acetate groups of EVA-45 could be stained with OsO_4 after being hydrolysed to vinyl alcohol (Fig. 11.18).

11.5.4 SHORT-FIBRE COMPOSITES

Short-fibre composites consist of sub-millimetre- and millimetre-long fibres with different orientations. The modulus of the fibres is roughly two orders of magnitude greater than that of the polymeric matrix. The mechanical properties of these composites depend primarily on the fibre orientation distribution (FOD), the fibre length distribution and the volume fraction of fibres. Figure 11.19 depicts a scanning electron micrograph taken from a cross-section of an injection-moulded ruler. The specimen was polished to smoothness and coated with Au before examination by SEM. It is also possible to use OM but the accuracy of SEM measurements is higher. Contact microradiography can also be used for the assessment of FOD and fibre length.

A circular cross-section indicates that the fibre is parallel to the normal to the surface. Highly elongated fibre cross-sections appear for fibres with their long axes in the specimen plane. It is not, however, possible

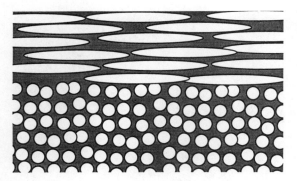

Figure 11.19 Drawing of a scanning electron micrograph of a polished short-fibre composite.

from a single micrograph to distinguish a fibre oriented from bottom left to top right from a fibre oriented from top left to bottom right.

11.6 SUMMARY

Optical microscopy (OM), transmission electron microscopy (TEM) and scanning electron microscopy (SEM) are major techniques for the assessment of morphology of semicrystalline polymers, liquid-crystalline polymers and polymer blends. The list of applications in other areas of polymer science and technology is extensive.

OM can resolve objects in a specimen as small as 1 μm. The depth of field is relatively small, making the assessment of topographical features difficult in many cases. Sample preparation is generally simple. Variations in absorption coefficient, sample thickness, refractive index and birefringence can be converted to contrast (light intensity) with the available OM techniques in the final image. Polarized microscopy, phase-contrast microscopy and differential interference-contrast microscopy convert differences in optical path (sample thickness, refractive index and birefringence) to variations in light intensity. Assessments of superstructure (spherulites, axialites, etc.) of semicrystalline polymers and mesomorphic structure (nematic, smectic, etc.) of liquid-crystalline polymers are important applications of OM.

SEM produces detailed topographical images. The image is obtained by recording the scattered secondary electrons. The resolution is typically between 10 and 30 nm. The depth of field is very large and sharp images can be obtained for specimens with large topographical variations. Sample preparation is relatively simple. Information about morphology is obtain from a topographical analysis of fracture surfaces and etched specimens. All polymeric samples have to be coated with a thin layer of conductive material prior to examination by SEM. Some scanning electron microscopes are equipped with an X-ray microanalyser and these instruments can detect additives and 'foreign' particles containing heavier elements.

TEM produces detailed images of the density variation in bulk samples and topographical variations. The resolution is below 1 nm and the depth of field is large. Sample preparation is difficult. The naturally occurring density variation in polymers is generally small and, in order to achieve contrast, it is necessary to add heavier elements selectively to one of the phases. This procedure is referred to as staining. Osmium tetroxide, reacting with unsaturated polymers, and chlorosulphonic acid, reacting selectively with the amorphous component of polyethylene, are two important staining agents. Another type of preparation technique for specimens for TEM is etching, which may further be divided into solvent etching and chemical etching. Permanganic acid (concentrated sulphuric acid, orthophosphoric acid and potassium permanganate) etching invented by Bassett and co-workers has proven to be a most useful etchant for a great variety of semicrystalline polymers. The replica technique makes it possible to assess the topography of the etched samples.

11.7 EXERCISES

11.1. Suggest suitable methods for determining the spherulite structure of two polyethylene samples. One is known to have large spherulites and the other to have very small spherulites.

11.2. Suggest suitable staining or etching methods for the following polymer blends: PVC/NR, ABS, PE/PS, PETP/LCP and PMMA/PVC.

11.3. Suggest another method (based on microscopy) for assessing miscibility in the polymer blends of question 11.2.

11.4. The lamellar structure of polyethylene can be studied after chlorosulphonation. What is the typical view on the micrographs?

11.5. How can chain-axis orientation be measured by TEM?

11.6 Explain the term 'dark-field' for optical microscopy and electron microscopy.

11.7. Fractography, to study a fracture surface, can be carried out using different microscopy techniques. Which is the preferred method? Explain.

11.8 REFERENCES

Andrews, E. H. (1964) *Proc. Roy. Soc. Lond. A* **227**, 562.

Andrews, E. H. and Stubbs, J. M. (1964) *J. R. Microsc. Soc.* **82**, 221.

Demus, D. and Richter, L. (1978) *Textures of Liquid Crystals*. VEB Deutscher Verlag für Grundstoffindustrie, Leipzig.

Gedde, U. W., Eklund, S. and Jansson, J.-F. (1983) *Polymer* **24**, 1532.

Gustafsson, A., Salot, R. and Gedde, U. W. (1993) *Polym. Comp.* **14**, 421.

Hill, M. J., Bradshaw, D. G. and Chevili, R. J. (1992) *Polymer* **33**, 874.

Kalnins, M., Conde Braña, M. T. and Gedde, U. W. (1992) *Polym. Test.* **11**, 139.

Kanig, G. (1973). *Kolloid Z. Z. Polym.* **251**, 782.

Kato, K. (1965) *J. Electron Microsc.* **14**, 20.

Kato, K. (1966) *J. Polym. Sci., Polym. Lett.* **4**, 35.

Kato, K. (1967) *Polym. Eng. Sci.* **7**, 38.

Olley, R. H. and Bassett, D. C. (1982) *Polymer* **23**, 1707.

Reid, N. (1975) Ultramicrotomy, in *Practical Methods in Electron Microscopy* (A. M. Glaubert, ed.). Elsevier, Amsterdam.

Sawyer, L. C. and Grubb, D. T. (1987) *Polymer Microscopy*. Chapman and Hall, London and New York.

Voigt-Martin, I. G. and Mandelkern, L. (1989) *J. Polym. Sci., Polym. Phys. Ed.* **27**, 967.

11.9 SUGGESTED FURTHER READING

Campbell, D. and White, J. R. (1989) *Polymer Characterization: Physical Techniques*. Chapman and Hall, London and New York.

Hemsley, D. A. (1989) *Applied Polymer Light Microscopy*. Elsevier Applied Science, London and New York.

Patel, G. N. (1980) Chemical methods in polymer physics, in *Methods of Experimental Physics*, Vol. 16, Part B: *Polymers: Crystal Structure and Morphology* (R. A. Fava, ed.). Academic Press, New York.

Vadimsky, R. G. (1980) Electron microscopy, in *Methods of Experimental Physics*, Vol. 16, Part B: *Polymers: Crystal Structure and Morphology* (R. A. Fava, ed.). Academic Press, New York.

Vaughan, A. S. (1993) Polymer microscopy, in *Polymer Characterisation* (B. J. Hunt and M. I. James, eds). Blackie Academic, London.

Woodward, A. E. (1989) *Atlas of Polymer Morphology*. Hanser, Munich, Vienna and New York.

SPECTROSCOPY AND SCATTERING OF POLYMERS

12.1 INTRODUCTION

This chapter deals with two rather separate topics: spectroscopic methods involving vibrational spectroscopy, i.e. infrared (IR) and Raman spectroscopy, nuclear magnetic resonance (NMR) spectroscopy and a few other spectroscopic techniques; and methods using scattering of various waves including X-rays, light and neutrons.

Spectroscopy is conventionally defined as the science of the interaction between radiation and matter, and deals with both experimental and theoretical aspects of these interactions. In this presentation we will restrict the field to electromagnetic radiation. Figure 12.1 shows the spectral range of various spectroscopic methods.

The polymer-related problems which can be solved by spectroscopy are many and varied. They may concern chemical aspects and chain structure, e.g. tacticity, 'mer' sequence distribution, chain branching or structure of radicals. They may concern physical aspects, e.g. chain orientation, crystallinity, crystal thickness, miscibility of polymers, chain conformation or chain dynamics.

Scattering is a phenomenon noted in everyday life. We can see a light beam in a black room from all angles. The light is scattered. The sky is blue due to scattering. The rainbow is another scattering phenomenon arising from 50 μm spheres. Scattering involves 'objects' which are of the same size as the wavelength of the radiation. Scattering is also very useful as an experimental tool for the assessment of the structure of matter, including polymers. Electromagnetic radiation is scattered by matter. We may think about the matter as a group of atoms, each atom scattering the incoming wave. A periodicity in the arrangement of the atoms (i.e. a crystal) causes constructive interference of the scattered waves which are collected in lines or points. This kind of scattering is referred to as **diffraction**. Diffraction thus occurs when the interacting 'objects' are regularly positioned and when they are of about the same size as the radiation wavelength. Scattering is more diffuse when the objects (e.g. atoms) possess only short-range order. X-rays have wavelengths of about 0.1 nm, making them suitable for studying interatomic distances, e.g. crystal unit cell structures. Visible light ($\lambda \approx 1$ μm) can be used to study larger objects, e.g. spherulites in semicrystalline polymers or polymer molecules in solutions. It is also possible to study much larger objects ('repeats') using a certain radiation by recording the scattered intensity at very small angles. Wide-angle X-ray scattering (WAXS) is used for crystallographic work assessing structure with

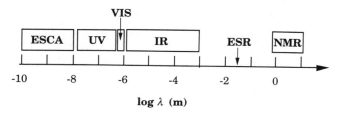

Figure 12.1 Spectral range (λ is the wavelength) of spectroscopic methods: NMR, electron spin resonance (ESR) spectroscopy, IR, visible (VIS) light spectroscopy, ultraviolet (UV) spectroscopy and electron spectroscopy for chemical analysis (ESCA).

repeating distances typically less than 1 nm, whereas small-angle X-ray scattering (SAXS) is useful for assessing the size of 'superstructures', such as the 10–50 nm long period of lamellar stacks in semicrystalline polymers.

Neutron scattering has in recent times shed new light on various problems in polymer science. The definite proof of the Flory theorem (that polymer molecules in the molten state are unperturbed; see Chapters 2, 3 and 4) was obtained by neutron scattering. The thermal neutrons used have a wavelength of approximately 0.45 nm, and by employing small-angle neutron scattering (SANS), the global molecular dimensions can be assessed. Wide-angle neutron scattering (WANS) provides information about more 'local' structures.

This chapter will present the different techniques separately. The basic elements of the methods are only briefly presented. The application of the methods in polymer physics is our main concern.

12.2 SPECTROSCOPY

12.2.1 VIBRATIONAL SPECTROSCOPY

Fundamentals

The atoms in molecules vibrate. This may be accomplished by changes in bond length, bond angle or torsion angle. The molecule consists of a set of harmonic oscillators. The disturbance of a molecule from its equilibrium causes a motion which is a combination of a number of simple harmonic vibrations. The latter are referred to as **normal modes**. The frequency of the motion of the atoms in each of the normal modes is the same and all atoms will pass the zero position simultaneously.

Certain frequencies of the radiation will be absorbed when the matter is irradiated by infrared light. The energy is transferred into the vibration, and either light may be re-emitted or the energy may be transferred to the surrounding molecules, i.e. converted to heat. The absorption frequencies correspond to the frequencies of the normal modes. This is the principle of infrared spectroscopy. It should be noted that not all of the normal modes are infrared-active. Only an oscillating dipole can absorb the infrared light, provided that the frequency is matching. The intensity of the absorption (I) is proportional to the square of the change in dipole moment (μ):

$$I \propto \left(\frac{d\mu}{dQ}\right)^2 \quad (12.1)$$

where Q is the displacement coordinate of the motion. Some vibrations of groups of atoms are not infrared-active despite the fact that the individual bonds participating in the vibration are dipoles. The symmetry of the vibrations counteracts in these cases the change in dipole moment associated with the individual bonds. The net change in dipole moment of the whole group becomes zero and hence the vibration is not infrared-active. A third requirement originates from the vector character of the transition moment ($\bar{\mathbf{M}}$) and the electrical field vector ($\bar{\mathbf{E}}$) associated with the infrared light:

$$I \propto (\bar{\mathbf{E}} \cdot \bar{\mathbf{M}})^2 = (|\bar{\mathbf{E}}| \cdot |\bar{\mathbf{M}}| \cdot \cos \theta)^2 \quad (12.2)$$

where θ is the angle between $\bar{E}$ and $\bar{M}$. The transition moment vector is oriented along the net change of dipole moment. The magnitude of absorption of the light depends on the state of polarization of the incoming light if the polymer chains have a preferential orientation. This is the basis for infrared dichroism, a method described in greater detail in Chapter 9.

Irradiation of matter also causes scattering. There will be both elastic scattering (Rayleigh scattering), involving no change in frequency of the scattered light, and inelastic scattering (Raman scattering). Only a few photons, one in 10^8, undergo Raman scattering. The energy is lost to vibrations (normal modes) and the change in frequency is equal to the frequencies of the normal modes. The strongest inelastic scattering (Stokes–Raman scattering), has a frequency which is lower than the frequency of the incoming light. Inelastic scattering with a higher frequency than the incoming light is called anti-Stokes–Raman scattering. Not all of the normal modes are Raman-active. The essential requirement for Raman scattering to occur is a change in polarizability accompanying the vibration. It is necessary to illuminate the sample under investigation with highly monochromatic (laser) light to detect the Raman scattering. Otherwise the weak Raman-scattered light is lost in the spectrum of the elastically scattered light.

Spectroscopy

Stretching:

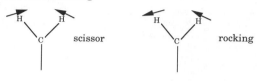

In-plane bending

Out-of-plane bending

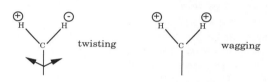

Figure 12.2 Vibrational modes in polyethylene.

For a molecule with n atoms, there are $3n - 6$ vibrational degrees of freedom. If this principle held for polymers there would be an enormous number of vibrational modes, but fortunately most of the vibrations are equivalent. The IR spectrum of a polymer is essentially independent of molar mass. The number of absorption peaks remains constant. Figure 12.2 shows some of the vibrational modes of polyethylene.

The Raman shift in frequency corresponds in many cases with the IR absorption spectrum. However, some vibrational modes appear only in the IR spectrum, whereas other modes appear only in the Raman spectrum. Bands with strong IR absorption originate from highly polar groups. The Raman scattering arises from non-polar groups. The strongest absorption bands in the IR spectrum of poly(ethylene terephthalate) are due to the carbonyl groups, whereas the C–C stretch of the aromatic groups gives rise to the strongest peaks in the Raman spectrum.

Figure 12.3 shows a summary of the absorption bands (group frequencies) from important groups. This

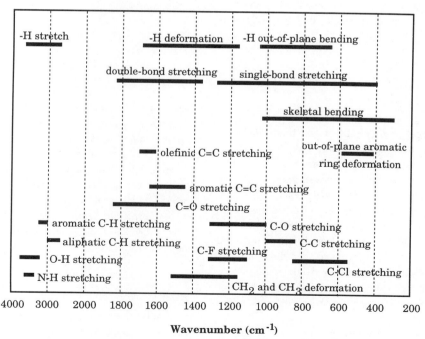

Figure 12.3 Group frequencies. Drawn after Bower and Maddams (1989).

scheme is very useful in chemical analysis. The presence of absorption bands in certain frequency regions suggests the presence of certain groups in the sample.

Instrumentation

IR instruments used to be dispersive and double-beam. Dispersive IR instruments use a mercury lamp as a light source, giving a continuous spectrum. They are equipped with an optical element consisting of prisms or gratings to disperse the IR radiation geometrically. A scanning mechanism passes the dispersed radiation over a slit system and isolates the frequency falling on the detector. Dispersive IR instruments are highly limited in resolution, because most of the energy does not reach the detector.

Currently used instruments are so-called Fourier transform infrared (FTIR) spectrometers using a Michelson interferometer. The light source is a mercury lamp giving a continuous spectrum. From the interferogram ($I(x)$), the frequency-domain spectrum ($G(v)$) is obtained by a mathematical procedure (Fourier transform) expressed in the following equation:

$$G(v) = \int_{-x}^{+x} I(x)\cos 2\pi v x \, dx \quad (12.3)$$

The advantage of the FTIR is the time efficiency. Dynamic processes can be monitored.

The main problem with the Raman spectrometers is the weakness of the Raman scattering. Only one in 10^8 incident photons is inelastically scattered. Powerful lasers (argon, krypton or helium–neon) are used. The laser beam has to be focused accurately on the specimen surface. The scattered light is recorded at 90° to the incident light beam. The discrimination of Raman scattering from Rayleigh scattering is accomplished by the use of a very efficient monochromator system, photomultiplier and amplifier.

Most infrared spectroscopy work is done in transmission mode. The path of the IR light is

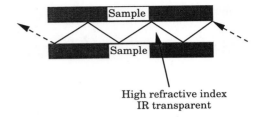

Figure 12.4 Schematic description of the working principle of ATR IR spectroscopy. The central waveguide is made of thallium bromoiodide (KRS5) or germanium.

through the sample. However, there are a number of reflection techniques available. Internal reflection spectroscopy (IRS), which is often called attenuated total reflection (ATR), is a useful technique for analysing samples with low transmission. It may also be considered as a surface characterization method since it 'samples' from the top micrometre of the specimen. Figure 12.4 shows the arrangement of the sample and how the wave is reflected many times from the surfaces.

The penetration depth (d) can be calculated from the following formula:

$$d = \frac{\lambda}{2\pi n_1 \left(\sin^2 \theta - \left(\frac{n_2}{n_1}\right)^2 \right)^{1/2}} \quad (12.4)$$

where λ is the radiation wavelength (in air), n_1 is the refractive index of the ATR crystal, n_2 is the refractive index of the sample and θ is the angle of incidence. A typical value for n_1 is 4.0 (germanium) and, for n_2, 1.5. The penetration depth is typically $\lambda/10$. It is important to emphasize that the penetration depth depends on the wavelength. This is one reason why the absorption peaks have a different weighing in the ATR spectrum than in the corresponding transmission spectrum.

Quantitive analysis

The infrared spectrum is conventionally presented in the form of a graph of transmittance (T) or absorbance (A) against frequency (λ) or wavenumber ($\bar{v}$ in cm^{-1}).

These quantities are defined as follows:

$$\bar{v}(\text{cm}^{-1}) = \frac{10^4}{\lambda(\mu\text{m})} \quad (12.5)$$

$$T = \frac{I}{I_0} \quad (12.6)$$

$$A = \log\left(\frac{I_0}{I}\right) \quad (12.7)$$

where λ is the wavelength, I is the intensity of the light at depth l in the medium and I_0 is the intensity of the incident light. Quantitive IR analysis is based on the Beer–Lambert law:

$$A = \varepsilon c l \quad (12.8)$$

where ε is the absorptivity or extinction coefficient, c is the concentration and l is the thickness of the specimen. The concentration of a given substance or fraction (group) is thus proportional to the absorbance values. The absorbance of a given peak is obtained by defining a base line in the A–v spectrum and by measuring the area under the peak or, simpler but less accurate, by just recording the height of the peak (Fig. 12.5). The base-line construction is often a difficult task, because it is not uncommon that nearby peaks overlap. In some cases, it is advisable to draw a sloping base line. In other cases, the composite spectrum has to be resolved into its components according to the following equation:

$$L(\bar{v}) = \sum_i L_i(\bar{v}) \quad (12.9)$$

where each $L_i(v)$ may be described by the Lorentzian distribution function:

$$L_i(\bar{v}) = \frac{I_{io}\left(\dfrac{w_i}{2}\right)^2}{(\bar{v} - \bar{v}_{io})^2 + \left(\dfrac{w_i}{2}\right)^2} \quad (12.10)$$

where $\bar{v}_{io}$ is the centre of the peak, I_{io} is the peak value and w_i is the breadth of the peak. In practice it is useful to know the $\bar{v}_{io}$ values of the absorption peaks involved. This can be accomplished by analysing simpler substances showing only the single absorption peaks. Equations (12.9) and (12.10) are thus fitted to the experimental data enabling determination of the adjustable parameters I_{io} and w_i.

Sample preparation

Liquids studied by IR spectroscopy are examined as thin films between two IR transparent plates or as solutions in IR transparent solvents (e.g. carbon tetrachloride, cyclohexane or chloroform). Solid polymers may be studied in solution using one of these solvents. The solvent may also cast films on a crystal of NaCl. The optimum thickness depends on the IR absorption of the polymer. Specimens of weakly absorbing polymers such as polyethylene can be relatively thick, approximately 100 μm, whereas strongly absorbing polar polymers have to be thinner, typically in the range 10–30 μm. It is also possible to prepare samples of suitable thickness by microtoming. Potassium bromide discs with approximately 1% (w/w) polymer are also frequently used. The grinding of the polymer has, for certain rubbers or other very tough polymers, to be made at cryogenic temperatures.

One of the advantages of Raman spectroscopy is the ease of sample preparation. Almost any sample shape or form is useful. Thin films should be oriented almost parallel to the incident beam to allow a long 'contact' length and hence a large surface which will Raman scatter (Fig. 12.6). Thicker transparent

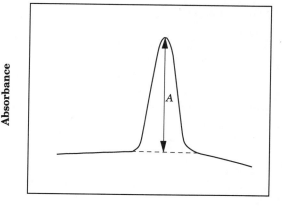

Figure 12.5 Graphical description of the assessment of the absorbance value of a well-separated peak.

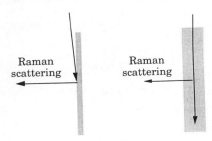

Figure 12.6 Arrangement of incident beam in Raman spectroscopy for a thin or thick specimen.

specimens can be illuminated from their end surface and in essence the specimen will act as a waveguide (Fig. 12.6). Powder samples and fibres are more difficult to study due to elastic scattering of the exciting radiation. Solutions and melts can be studied in glass containers.

Applications of vibrational spectroscopy in polymer physics

The list of useful applications of IR and Raman spectroscopy in polymer science and physics is almost endless. Characterization of chemical structure is one of the important uses. This is, however, not the main topic of this book and we will only briefly mention a few examples.

It is possible to measure the concentration of end groups, and hence assess the number average molar mass in polymers like poly(butylene terephthalate) (PBTP). The –COH end groups absorb at 3535 cm^{-1} and the –COOH end groups at 3290 cm^{-1}. End group analysis for the assessment of molar mass is, however, not applicable to high molar mass polymers, as for all other end-group analysis methods. One advantage with the IR method is that the polymer needs not to be dissolved prior to the analysis. This method is obviously less suited for branched polymers.

Stereoregularity (tacticity) can also be determined by IR spectroscopy for some polymers. The IR spectra of isotactic and syndiotactic polymethylmethacrylate are indeed different and the same observations have been made for polystyrene and polypropylene. Solid isotactic polypropylene shows several absorption peaks (805, 840, 898, 995 and 1100 cm^{-1}) which are absent in the spectrum of atactic polypropylene. However, the spectra of the two polymers in the molten state are identical. This suggests that these 'extra' absorption peaks present in semicrystalline isotactic polypropylene are due to vibrations of the regular 3_1 helix structure. The difference in configuration thus causes a difference in the conformational states, the latter being revealed by IR spectroscopy.

Tacticity, however, is ideally assessed by ^{13}C NMR spectroscopy (section 12.2.2). Chain branching in polyethylene is another popular field. The CH$_3$ groups, which in a polymer of medium to high molar mass are almost exclusively the end groups of the branches, absorb at 1379 cm^{-1} and their concentrations can be determined according to the equation:

$$\%\mathrm{CH}_3 = K\left(\frac{A_{1379}}{A_{1465}}\right) \quad (12.11)$$

where K is a constant which depends on the branch type, A_{1379} is the absorbance at 1379 cm^{-1} (methyl groups) and A_{1465} the absorbance at 1465 cm^{-1} (assigned to methylene groups). There are several methylene absorption bands interfering with the 1379 cm^{-1} peak, and these can be subtracted by taking the difference spectrum between the sample and a perfectly linear polyethylene. The proportionality constant K can be obtained by calibration using polyethylenes of known chain branching as assessed by ^{13}C NMR spectroscopy.

The concentration of unsaturated groups in polymers can be determined both by IR and Raman spectroscopy. These groups are susceptible to various chemical reactions, desirable in some cases and hazardous in other cases. The C=C stretch band appears at about 1650 cm^{-1} and a series of absorption peaks should appear at 965 cm^{-1} (*trans*-vinylene), 910 cm^{-1} (vinyl end group) and 730 cm^{-1} (*cis*-vinylene). These absorption bands refer to IR. Raman spectroscopy is, however, more suitable for quantitative analysis of unsaturation. The absorption peaks are stronger than for IR spectroscopy. Polybutadiene shows Raman peaks at 1650 cm^{-1}

(cis-vinylene), 1655 cm^{-1} (vinyl end group) and 1665 cm^{-1} (trans-vinylene).

Carbonyl-stretch vibration is commonly used for the assessment of oxidation of polyolefins. A number of overlapping absorption peaks essentially between 1700 and 1800 cm^{-1} appear in oxidized polyolefins. They may be assigned to various oxidation end products: ketones, aldehydes, esters and carboxylic acids. Quantitative analysis can only be made by resolution of the 'composite' spectrum and after proper calibration with pure low molar mass ketones, aldehydes, esters and carboxylic acids.

Let us now concentrate on topics dealt with in earlier chapters. Chain conformation analysis can be made with IR spectroscopy. Syndiotactic polyvinylchloride is a well-known example. The trans planar zigzag conformation in which the C–Cl bond is trans to the C–H bond shows stretching vibrations at 603 and 639 cm^{-1}, whereas both trans-gauche and gauche-gauche show absorption at 695 cm^{-1}. Spectral overlap is, however, a major problem in this kind of analysis. The absorption peaks at 850, 975, 1120, 1370 and 1470 cm^{-1}, all assigned to the ethylene group, are stronger in semicrystalline PETP than in amorphous PETP. It is known that the preferred conformation (i.e. the crystalline state) of this group is trans and hence these absorption peaks may be assigned to the trans conformation.

Hydrogen bonds are the strongest secondary bonds in polymers. They appear in several important polymers, e.g. polyamides. They play a vital role in certain miscible polymer blends, as is mentioned in Chapter 4. Their effect on the IR spectrum is well known. In polyamides, the vibrational stretching frequencies of hydrogen-bonded OH– and NH– groups are shifted to lower values compared to their non-hydrogen-bonded analogues. The vibration frequency of the NH– group drops from 3450 cm^{-1} to 3300 cm^{-1} (hydrogen-bonded). A similar shift of the absorption peaks in polymer blends (with reference to the frequencies of the pure components) may indicate specific interaction between the different polymers, which is an important reason for miscibility.

Specimens subjected to a mechanical load show a frequency shift of some absorption peaks. The shift is in many cases proportional to the applied stress. There is also a change in the shape of the frequency-shifted peaks from symmetric to asymmetric. The latter has been taken as evidence that different chain segments are subjected to different stresses, i.e. that the stress is unevenly distributed.

Another important application is infrared dichroism measurements for the assessment of chain (or group) orientation. This topic is treated in Chapter 9, and the discussion is not repeated here. It should be mentioned that Raman spectroscopy can also be used for the determination of chain orientation. Both IR and Raman spectroscopy are very useful for characterization of the physical structure of crystalline polymers. Assessment of the degree of crystallinity can be made by several methods. The preferred and internally consistent methods are X-ray diffraction, density measurements and calorimetry (DSC/DTA). This topic is described in detail in Chapter 7. However, both IR and Raman spectroscopy provide information about the crystallinity, although it is common for the actual crystallinity values obtained by these methods to deviate from values obtained by the three preferred methods: WAXS, density and DSC/DTA.

The major factor differentiating a chain in a crystal from a chain in the amorphous phase is that the former is in its preferred conformational state. An additional requirement is the presence of lateral order in the crystal. There are many reported values of crystallinity by IR or Raman spectroscopy which are based on absorption peaks characteristic of only preferred conformation. Zerbi, Ciampelli and Zamboni (1964) divided what they called regularity peaks into three categories.

The first group, being the 'weakest' indicator of crystallinity, refers to absorption bands assigned to certain 'single'-bond conformations. For example, the ratio of the 1450 (trans) and 1435 cm^{-1} (cis) bands of trans-1,4,-polybutadiene does indeed increase with increasing crystallinity, but it is equally clear that the trans conformer also is present in the amorphous component.

The second group is due to a group of adjacent chain segments with a specific chain conformation, e.g. the 3_1 helix of isotactic polypropylene. Absorption peaks at 805, 840, 898, 972, 995 and 1100 cm^{-1} are characteristic of this conformation. These peaks are replaced by a broad peak centring on 972 cm^{-1} in molten isotactic polypropylene.

There are several good arguments to account for the presence of the 3_1 helix conformation in the amorphous polymer but at a lower concentration. Thus, the intensity of these regularity peaks may not be proportional or even a direct measure of crystallinity.

The third group shows what might be called true crystallinity peaks. Their occurrence is a consequence of both intra- and intermolecular interaction typical of the crystal phase. These peaks disappear completely on crystal melting. Polyoxymethylene and syndiotactic polyvinylchloride are examples of polymers showing such regularity peaks.

One of the main uses of Raman spectroscopy in morphological analysis is the measurement of the longitudinal acoustic mode (LAM) in polyethylene which goes back to relatively early work (1949) on paraffins. The LAM frequency (v(LAM-n)) can be converted to all-*trans* chain length or crystal thickness (L_c) according to the following formula:

$$L_c = \frac{n}{2v} \cdot \sqrt{\frac{E}{\rho}} \qquad (12.12)$$

where n is the order of the vibration ($n = 1$ is practically the only value used), E is the Young modulus in the chain-axis direction and ρ is the density. The LAM ($n = 1$) frequencies are extremely low, 10–20 cm^{-1}. There are several complicating factors which may affect the results. The elastic modulus is not precisely known, although reasonably accurate estimates suggest values near 300 GPa for polyethylene. There is often a chain tilt in polyethylene crystals, i.e. the length of the all-trans chain is greater than the thickness of the crystal. There is nearly always a distribution of crystal thicknesses and it is not clear which type of average is assessed by Raman spectroscopy. Crystals contain defects and the exact effect of these on LAM frequency is not known. LAM frequency measurement for the assessment of crystal thickness has also been performed on a few other polymers, e.g. polyoxymethylene, polyethyleneoxide and polytetrafluoroethylene. The interpretation of these results is less straightforward than for polyethylene.

12.2.2 NUCLEAR MAGNETIC RESONANCE (NMR) SPECTROSCOPY

Fundamentals

NMR is based on the principle that nuclear spins under the influence of an external magnetic field are split into two energy levels. NMR-active nuclear spins are those having an odd number of either protons or neutrons. Hydrogen (^{1}H) is indeed NMR-active and is the subject of much work. Other nuclei subjected to studies are ^{13}C, ^{2}H and ^{19}F. These nuclei are present only in small concentrations in polymers. The nuclear spins are associated with angular momentum, characterized by the angular momentum vector ($\bar{\mathbf{I}}$):

$$|\bar{\mathbf{I}}| = \frac{h}{2\pi} \cdot \sqrt{I(I+1)} \qquad (12.13)$$

where h is Planck's constant and I is the nuclear magnetic spin quantum number. The latter is $\frac{1}{2}$ for ^{1}H, ^{13}C and ^{19}F, and 1 for ^{2}H. The magnetic moment ($\bar{\mathbf{\mu}}$) is proportional to $\bar{\mathbf{I}}$:

$$\bar{\mathbf{\mu}} = \gamma \bar{\mathbf{I}} \qquad (12.14)$$

where γ is the magnetogyric ratio. The magnetic moment is in scalar units given by:

$$\mu = g_N \beta_N I \qquad (12.15)$$

where g_N is a dimensionless constant and β_N is the nuclear magneton. Both these take different values for different nuclei. The magnetic quantum number (m_I) can, according to the quantum theory, only take specific values: $I, I-1, \ldots, -I$. For ^{1}H and ^{13}C, m_I can take the values $+\frac{1}{2}$ and $-\frac{1}{2}$. The magnetic moment can thus only take discrete values according to:

$$\mu = g_N \beta_N m_I \qquad (12.16)$$

The potential energy (E) of the isolated nucleus placed in a magnetic field (B_0) is given by:

$$E = -\mu B_0 \qquad (12.17)$$

and the energy difference (ΔE) between a nucleus with $m_I = +\frac{1}{2}$ and $-\frac{1}{2}$ is obtained by combining eqs (12.16) and (12.17):

$$\Delta E = h v_r = E_{-1/2} - E_{+1/2} = g_N \beta_N B_0 \qquad (12.18)$$

where v_r is the resonance frequency. Equation

(12.18) states the condition for resonance. The resonance frequencies for ^{1}H and ^{13}C are, at $B_0 = 1.0$ T, equal to 42.577 MHz and 10.705 MHz, respectively. The absorption of the radio frequency radiation will obviously only be detected if there is an excess of spins in the low-energy state. Equilibrium can only be obtained if a fraction of the excited spins are permitted to return to their ground state. The extra energy has to be transferred to the surrounding 'lattice'. This relaxation process occurs at a rate described by the spin-lattice relaxation time, T_1. There is a second relaxation mechanism, characterized by the spin-spin relaxation time, T_2.

NMR spectroscopy may be carried out by 'searching' for the resonance conditions either at constant magnetic field strength or at constant frequency. Both methods are used in practice.

NMR spectroscopy would not be particularly useful if it were not for the occurrence of the 'chemical shift'. The electrons surrounding a particular nucleus affect the effective magnetic field (B_{eff}) which is 'felt' by the nucleus:

$$B_{eff} = B_0(1 - \sigma) \quad (12.19)$$

where σ is the screening constant, whose magnitude depends on the chemical environment, which in turn affects the electron density around the nucleus. The resonance condition formula is then given by:

$$\Delta E = h v_r = g_N \beta_N B_0 (1 - \sigma) \quad (12.20)$$

The resonance condition depends on the field strength (B_0) which is undesirable. The accepted standard is to express chemical shift with reference to tetramethylsilane (TMS), which shows a single ^{1}H resonance peak, according to:

$$\delta(\text{in ppm}) = 10^6 \left(\frac{v_s - v_{TMS}}{v_{TMS}} \right) \quad (12.21)$$

The same substance is also used as reference for expressing chemical shift in ^{13}C NMR spectroscopy. Chemical shifts in ^{1}H NMR lie in the range 0–10 ppm, and in ^{13}C NMR the range is 0–250 ppm.

Nearby nucleus spins affect the local magnetic field by ΔB according to the following equation:

$$\Delta B = \pm \tfrac{3}{4} g_N \beta_N (1 - 3 \cos^2 \theta) r^{-3} \quad (12.22)$$

where θ is the angle between the magnetic field vector and the vector between the interacting spins and r is the distance between the nuclei. This so-called spin-spin interaction causes a broadening of the resonance line. However, if the mobility of the segments is high, as in a solution or a liquid, ΔB decreases markedly since $\langle \cos^2 \theta \rangle$ approaches values near to $\tfrac{1}{3}$. Solid samples show considerable broadening in their resonance lines due to spin-spin interaction. The problem can, however, be eliminated by rotating the sample at an angle of 54°44′ ($\cos^2(54°44') = \tfrac{1}{3}$) to the magnetic field. This angle is known as the 'magic angle'.

The 'coupling effect' refers to interaction between bonded nearby nuclei of the same kind. This so-called scalar through-bond interaction occurs due to polarization of the bonds by the nearby spins. This causes a splitting of the resonance lines into $2NI + 1$ lines, where N is the number of magnetically equivalent nuclei with spin I. The shifts are expressed in coupling constants given in hertz. The magnitude of coupling constants is essentially independent of magnetic field strength, but dependent on chemical structure. The relative intensities of the split lines are given by Pascal's triangle shown in Fig. 12.7. Starting at the top, $N = 0$, no interacting bonded nuclei and only one band appears, etc. This rule is only valid when the chemical shift differences are much greater than the coupling constants, which is normally the case.

Instruments and sample preparation

The NMR instrument consists of a high-field magnet (1.4–2.3 T), a radio frequency source, the NMR probe, a sweep system, and an amplification and recording system. Higher magnetic fields can be obtained by using superconducting solenoids. The main advantage

```
                    1
                 1     1
              1    2     1
           1    3     3    1
        1    4     6    4    1
     1    5    10    10    5    1
```

Figure 12.7 Pascal's triangle.

in using a high magnetic field is that both the magnitude of the chemical shift and the sensitivity increase.

Modern high-resolution instruments are Fourier transform (FT) spectrometers which record the response of the nuclei to strong radio frequency pulses, converting this by a mathematical procedure into a frequency spectrum. The advantage with the FT technique is that a spectrum is obtained in seconds and several hundred spectra can be added together, which greatly increases the signal-to-noise ratio. This is particularly important for ^{13}C NMR due to the low concentration of these nuclei in the samples.

There are a range of techniques employed to enhance spectral lines:

- The nuclear overhauser effect (NOE) enhances specified lines by changing the relaxation behaviour of one nuclear transition by saturation of other nuclear transitions. The coupling effect is only possible when the two nuclei are sufficiently close. NOE is used when recording ^{13}C spectra.
- Spin decoupling is used to simplify the analysis of complex spectra and is performed by modulation of the radio frequency radiation. The sample is irradiated with the resonance frequency of one nucleus while sweeping through the entire spectral region.
- Broadband decoupling is used when the resonance frequency of the decoupled nucleus occurs over a wider range. It is used in ^{13}C NMR work. The spectra are obtained with decoupling of the ^{1}H nuclei.

Application of NMR in polymer physics

NMR spectroscopy is the primary method for the assessment of the chemical structure of polymers. Characterization of tacticity and chain branching in polyethylene is best done by ^{13}C NMR. Assessment of meso and racemic dyads (Chapter 1) in tactic polymers is routine. Information about longer steric sequences, e.g. triads, is also available. A very detailed picture of the branch structure is obtained by ^{13}C NMR (Fig. 12.8). Each of the different carbons gives rise to separate lines and it is possible distinguish branches of six carbon atoms and shorter and also to

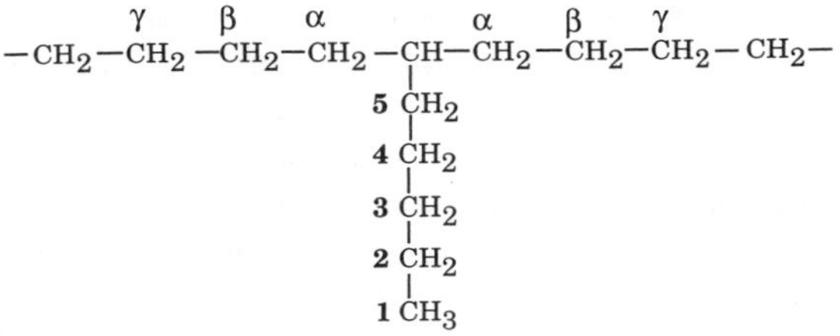

$$-CH_2-CH_2\overset{\gamma}{-}CH_2-CH_2\overset{\beta}{-}CH_2-\overset{\alpha}{CH}-CH_2\overset{\alpha}{-}CH_2-CH_2\overset{\beta}{-}CH_2-CH_2\overset{\gamma}{-}CH_2-$$
$$\underset{|}{5\ CH_2}$$
$$\underset{|}{4\ CH_2}$$
$$\underset{|}{3\ CH_2}$$
$$\underset{|}{2\ CH_2}$$
$$1\ CH_3$$

Figure 12.8 Designation of carbon atoms near a chain branch.

determine their concentrations. This method is clearly superior to IR spectroscopy. The disadvantages with the NMR method are that it is time-consuming and that relatively large amounts of sample are used for the analysis. Further details about the analysis can be found in Koenig (1992).

Head–tail configuration and sequence distribution of comonomers in copolymers are two good examples of problems solved by NMR.

Broad-line NMR line shapes provide information about both structure and segmental mobility. Broadening is mainly due to the spin-spin interaction which, when the spins (i.e. polymer segments) are immobile, leads to a considerable broadening of the resonance lines. Polymers of a more liquid-like nature exhibit more narrow resonance lines.

A semicrystalline polymer such as polyethylene consists of at least two components: solid crystals surrounded by an amorphous phase. The mobility of the chains in the orthorhombic crystal phase is negligible, whereas it is appreciable in the amorphous phase at room temperature. The crystal phase gives broad resonance lines, whereas considerably narrower lines are associated with the mobile amorphous phase. Detailed analysis of many polyethylene samples has shown that the NMR line profiles cannot be explained by these two components alone. A third component with intermediate mobility needs to be introduced in the model. Crystallinity data obtained by resolution of the proton spectrum or ^{13}C spectrum into these three components are in fair agreement with crystallinity data by X-ray diffraction and density measurements, and NMR can also assess an 'intermediate phase'. The latter is not revealed by WAXS or density measurements.

Thermal transitions and relaxation processes can be revealed by broad-line NMR. Detailed information about the mobility of individual carbon atoms obtained by ^{13}C NMR is indeed a valuable complementary method to the classical thermoanalytical techniques (see Chapter 10). NMR spectroscopy is useful in many other applications: assessment of miscibility (Chapter 4) and chain orientation (Chapter 9) to mention but two.

12.2.3 OTHER SPECTROSCOPIC METHODS

Electron spin resonance (ESR) spectroscopy provides information about the concentration and types of radicals. A radical has a spin which may be oriented in different ways, parallel or anti-parallel in a magnetic field. The external field causes an energy difference between the two states and resonance occurs when the energy of light is equal to this value (Fig. 12.9).

The surrounding atoms affect the ESR spectrum, which enables the structure of the radicalized chain to be assessed. ESR spectroscopy is useful in studies of radical reactions, e.g. polymerization and degradation. Chain scission accompanies mechanical fracture particularly in highly oriented polymers and in thermosets. ESR has provided useful results in this field. The chain dynamics can also be studied by using spin-labelling techniques, e.g. using nitroxide radicals.

Electron spectroscopy for chemical analysis (ESCA) is a surface analysis method. The sample is under high vacuum and is irradiated by monochromatic X-rays. The energy of the X-rays ($h\upsilon$) is greater than the binding energy (E_b) of the electrons, and by careful measurement of the kinetic energies (E_{kin}) of the excited (emitted) electrons useful information about the chemical composition of the outer 2 nm surface layer can be obtained:

$$E_b = h\upsilon - E_{kin} - \phi \qquad (12.23)$$

where ϕ is the so-called 'work function' of the spectrometer. The energy spectrum of the different emitted electrons provides direct information about the elemental composition of the surface layer. There is also a chemical shift of the binding energies due to the bonded surrounding to a given element. ESCA thus provides information about the elemental composition and chemical structure of the 2 nm top surface layer. However, no information is obtained about tacticity and physical structure of polymer.

UV-visible spectroscopy is primarily used to solve chemical problems. Absorption of visible and UV light is controlled by the chemical bonds. Sigma bonds absorb high-energy photons in the UV range whereas π electrons absorb at longer wavelengths. The groups which are of primary interest, however, are conjugated double bonds, phenyl groups, vinyl (vinylene) groups and carbonyl groups. This type of spectroscopy provides information about the chemical structure of a polymer, e.g. the remaining unsaturation in a thermoset. It has also proven valuable for determination of the concentration of certain additives (e.g. antioxidants).

12.3 SCATTERING AND DIFFRACTION METHODS

12.3.1 X-RAY METHODS

Fundamentals

The diffraction of X-rays by crystals is a particularly well-documented field and it is not necessary here to give a comprehensive presentation of the mathematics of diffraction. However, a few of the key formulae and the reasoning are presented.

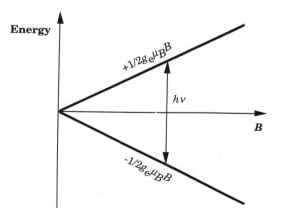

Figure 12.9 Principle behind ESR; B = magnetic field strength, g_e = free electron 'g-factor'; μ_B = Bohr's magneton.

Assume that a parallel and coherent X-ray beam (all waves are in phase) interacts with two atoms. The X-rays emitted from the atoms are scattered at all angles. Only scattered X-rays with the same wavelength as the incident X-rays are considered. Inelastically scattered X-rays (so-called Compton scattering) give rise to a diffuse background. The phase difference at a certain point is defined by the scattering vector ($\bar{K} = \bar{k}' - \bar{k}$, where $\bar{k}'$ and $\bar{k}$ are the vectors of the scattered and incident beams):

$$2\pi(\bar{k} - \bar{k}') \cdot \bar{r} = -2\pi \bar{K} \cdot \bar{r} \qquad (12.24)$$

where $\bar{r}$ is the vector between the two scattering centres. It can be shown that the total amplitude scattered from a group of n atoms is given by:

$$\psi_{tot} = \sum_n f_n(\theta) \exp(-2\pi i \bar{K} \cdot \bar{r}_n) \qquad (12.25)$$

where $f_n(\theta)$ is the so-called structure factor for the nth atom located at the position $\bar{r}_n$. The latter can be expressed in familiar lattice vectors $\bar{a}$, $\bar{b}$ and $\bar{c}$. We will use vector notation in the following equation:

$$\bar{r}_n = (n_x \bar{a} + n_y \bar{b} + n_z \bar{c}) + (p_x \bar{a} + p_y \bar{b} + p_z \bar{c}) \qquad (12.26)$$

where n_x, n_y and n_z are integers which effectively give the shift to the 'right' unit cell, and p_x, p_y and p_z are numbers smaller than 1 which account for the shift within the unit cell. The concept of reciprocal space is introduced in Chapter 7, and it is possible to relate the scattering vector to the reciprocal space vectors ($\bar{a}^*$, $\bar{b}^*$ and $\bar{c}^*$):

$$\bar{K} = u\bar{a}^* + v\bar{b}^* + w\bar{c}^* \qquad (12.27)$$

Equations (12.26) and (12.27) are inserted into eq. (12.25). The contributions from a single unit cell and from different appear as factors in the summation:

$$\psi_{tot} = \left\{ \sum_p f_n(\theta) \exp[-2\pi i(p_x u + p_y v + p_z w)] \right\}$$
$$\times \left\{ \sum_n \exp[-2\pi i(n_x u + n_y v + n_z w)] \right\} \qquad (12.28)$$

The so-called Laue conditions (constructive interference) are fulfilled when $u = h$, $v = k$ and $w = l$. The parameters h, k and l characterize the lattice planes (for details, see Chapter 7). Constructive interference occurs when:

$$\bar{K} = h\bar{a}^* + k\bar{b}^* + l\bar{c}^* = \bar{g}_{hkl} \qquad (12.29)$$

where $\bar{g}_{hkl}$ is a vector in the reciprocal space. The distance between the diffracting planes, $|\bar{d}_{hkl}|$, is equal to $1/|\bar{g}_{hkl}|$. Equation (12.29) is solved for several combinations of h, k and l values. It should be noted that not all hkl planes give rise to diffraction. Certain combinations are forbidden. This is known as the structure factor rule. Equation (12.29) can be rewritten as the well-known Bragg equation:

$$|\bar{K}| = \frac{2\sin\theta}{\lambda} = \frac{1}{|\bar{d}_{hkl}|} \qquad (12.30)$$

where λ is the wavelength of the X-ray and 2θ is the diffraction angle. An ideal crystal at 0 K would give rise to absolutely sharp diffraction points. However, thermal fluctuation, dislocations and, as is highly relevant for polymer crystals, the limited number of diffracting planes in a given crystal give rise to broadening of the diffraction peaks. This topic is dealt with in Chapter 7.

A few words about the X-ray sources. X-rays were discovered by Wilhelm Röntgen in 1895. Ideally the X-ray source should give a completely monochromatic radiation. An X-ray source gives a few sharp spectral lines corresponding to certain electron transitions in the target material and a superposed continuous spectrum. The Cu target, giving radiation of wavelength 0.1540 nm ($K\alpha_1$), 0.1544 nm ($K\alpha_2$) and 0.1392 nm ($K\beta_1$), can be regarded as standard for polymer studies. The $K\alpha_1$ radiation is the most commonly used of the three. Various filters and preferably crystal monochromators are used to eliminate the intensity at wavelengths other than that desired.

The so-called synchrotron source provides a very intense X-ray beam, which makes this configuration suitable for recording very rapid processes such as crystallization and deformation. It is used both for wide- and small-angle scattering work, sometimes in combination with DSC.

The intensity of the scattered radiation can be obtained by using photographic film or by using a counter. Both methods give good results and are in use. A microdensitometer can convert the blackness on the film into an intensity–2θ relationship. There are several counters, among them gas counters and scintillation counters. Detection is carried out by a 'moving' counter detector over the 2θ range. Another technique uses one-dimensional or two-dimensional detectors which permit simultaneous measurement of the scattered intensity at different 2θ. The 2θ range of these detectors is often limited, typically to $10°$.

Wide-angle X-ray scattering

Wide-angle X-ray scattering (WAXS) refers to studies of scattering at relatively large angles. Let us use the Bragg equation and assume that $\lambda = 0.15$ nm: $d = 0.075/\sin \theta$, where for large scattering angles (2θ) yields distances smaller than 1 nm. A scattering angle (2θ) of $36°$ corresponds to $d = 0.3$ nm. WAXS is used for the assessment of crystal unit cell structure, i.e. cell identification and the measurement of cell dimensions (a, b and c). Figure 12.10 shows a few wide-angle instruments.

Let us now briefly mention the important applications of WAXS in polymer physics. We have already presented a whole range of applications in Chapters 6, 7 and 9. First of all, WAXS provides direct evidence of the physical structure. The polymer can be considered as semicrystalline provided that sharp Bragg reflections are observed. A diffuse scattering pattern indicates that only short-range order is present. The polymer is probably fully amorphous. However, it is possible that it is liquid-crystalline with a nematic mesomorphism (see Chapter 6). WAXS is the standard method for the assessment of crystallinity. Details of the different X-ray methods used to determine crystallinity are presented in Chapter 7. X-ray diffraction is the major tool used in crystallographic work. The assessment of the unit cell of polymers with unknown crystal structure, and

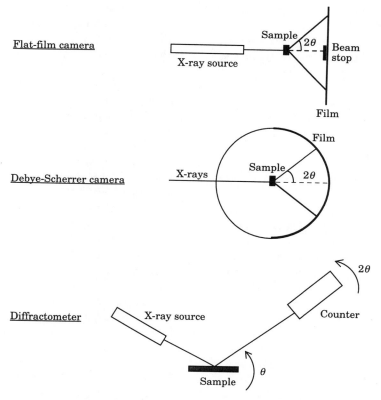

Figure 12.10 Wide-angle X-ray cameras and diffractometers.

the determination of the unit cell dimensions of polymers with known cell type (e.g. polyethylene) are important uses of WAXS data. WAXS is also used to record strains in the crystalline phase in stressed samples. The broadening of the diffraction peaks provides information about crystal size and internal disorder. This topic is treated in Chapter 7. WAXS is used together with small-angle X-ray scattering (SAXS) for the determination of mesophase structures (see Chapter 6). Characterization of c-axis orientation (crystalline chain orientation) is performed by recording the azimuthal angle dependence of the scattered intensity; for details see Chapter 9.

Small-angle X-ray scattering

Small-angle X-ray scattering (SAXS) is concerned with scattering phenomena occurring at small angles, typically less than a few degrees, which by application of eq. (8.30) can be converted into repeating distances 10–1000 times greater than those obtained by WAXS. The most important requirement is a very good collimation of the X-ray beam. There are several methods used: pinhole collimation and slit collimation. Three well-known instruments based on pinhole collimation are the Kiessig camera, the Rigaku Denki camera and the Statton camera. The Kratky camera is a slit camera with excellent collimation which also avoids diaphragm scattering, a problem which is present in the pinhole cameras.

SAXS is used for the assessment of superstructures in both crystalline and liquid-crystalline polymers. The so-called SAXS long period of semicrystalline polymers is the repeating distance in the stacks of lamellar crystals. The long period includes one crystal thickness and one amorphous interlayer thickness. It is thus possible to calculate the crystal thickness from the long period simply by multiplication with the volume crystallinity. Further details on this topic are presented in Chapter 7. Smectic polymers show a layered structure which is readily detected in a SAXS camera. The repeating distance is commonly in the range 2–5 nm. It is possible in a Statton camera to detect both the wide-angle reflections originating from the interchain spacing and the low-angle reflections arising from the layering. Further details of the structure of smectic polymers, including some X-ray patterns, are presented in Chapter 6. Microvoids created by fracture phenomena have also been assessed by SAXS.

12.3.2 OTHER SCATTERING METHODS

Three important scattering/diffraction methods will be mentioned only briefly: neutron scattering, light scattering and electron diffraction.

Neutron scattering typically uses thermal neutrons which have a wavelength of approximately 0.5 nm. The typical samples are mixtures of 'normal' hydrogenated (^{1}H) polymers and specially made deuterated (^{2}H) polymers. The requirement for neutron scattering to occur is that different molecules have different 'coherent scattering lengths': hydrogen has -0.374×10^{-12} m whereas deuterium has 0.667×10^{-12} m. Small-angle neutron scattering (SANS) provides direct information about the global structure of the molecules, i.e. the radius of gyration (s). Wide-angle neutron scattering (WANS) reveals more local structures. SANS instruments are only available at a few places world-wide. There are facilities in France (Laue-Langevin, Grenoble, 80 m instrument!), USA (Oak Ridge National Laboratory, 30 m instrument) and Germany (Jülich, 40 m instrument). Smaller instruments (less than 5 m) are available elsewhere. All the instruments are equipped with two-dimensional positron-sensitive arrays.

Neutron scattering has provided new data which have shed new light on several areas in polymer physics. The Flory theorem (see Chapter 2) was confirmed when SANS showed that the radius of gyration in the bulk state was indeed the same as in a theta solvent. The effect of deformation and swelling of rubbers was also studied by SANS, and the affine deformation model and the phantom network model were critically assessed (see Chapter 3). Crystallization of polyethylene (mixtures of hydrogenated and deuterated chains) was studied by both SANS and WANS. The radius of gyration showed only a weak molar mass dependence for solution crystals. This observation, together with WANS data, led to the proposal of the superfolding model (see Chapter 7). The radius of gyration of relatively rapidly cooled melt-crystallized samples (to avoid segregation of the deuterated molecules) is the same as that in the molten state prior to crystallization. It is not clear if this scheme also holds for samples crystallized at a lower supercooling. Neutron scattering has been used in a number of other fields (Wignall 1993).

Light scattering is a classical technique used to study solutions (with the necessary difference in refractive index between solvent and solute). This is described in Chapter 2. It is also used for characterization of superstructures (spherulitic, axialitic, etc.) in semicrystalline polymers. The technique is called small-angle light scattering (SALS) and information is obtained about type and size of superstructure (Chapter 7).

Electron diffraction provides information about crystal structure similar to that obtained by WAXS, but with the difference that the sample size (the beam size) can be made very small so that the diffraction pattern from isolated crystallites can be recorded. This is a great advantage for the electron diffraction technique. A sample analysed by WAXS contains many crystals which at best are oriented, but the diffraction pattern is not of the same 'quality' as that obtained by electron diffraction. The (electron) radiation sensitivity of polymers causes rapid deterioration of the electron diffraction patterns, which is obviously a drawback of this technique.

12.4 SUMMARY

Spectroscopy and scattering/diffraction methods play a central role in polymer science and physics. The enormous development of new techniques in the past decades, leading to improved sensitivity, has been a major force in the creation of new knowledge and new ideas in almost all the fields covered by this book.

12.5 EXERCISES

12.1. Suggest at least two different methods for the assessment of chain branching in polyethylene.

12.2. Suggest spectroscopy and diffraction methods by which the crystal thickness in a polyethylene sample can be determined.

12.3. Can miscibility of two amorphous polymers be judged by IR spectroscopy?

12.4. The following crystallinity values were obtained for poly(propylene-*stat*-ethylene) by X-ray diffraction using the Ruland method and by IR spectroscopy using the method of Trotignon:

Ethylene content (mol%)	X-ray crystallinity	IR crystallinity
0	49	67
3.6	36	49
6.6	33	48
9.3	37	47

Data supplied by S. Laihonen, Dept. of Polymer Technology, Royal Institute of Technology, Stockholm.

Suggest possible explanations for the results.

12.5. A fibre of a semicrystalline polymer is highly stressed. How can the deformation be recorded by spectroscopy and scattering techniques?

12.6. Polyolefins are readily oxidized at high temperatures, particularly when antioxidants are not present. IR spectroscopy is commonly used to determine the concentration of the oxidation products. What is the main problem with the IR analysis?

12.7. Chain orientation can be assessed by both spectroscopy and scattering methods. Write a short description of a few of the methods.

12.6 REFERENCES

Bower, D. I. and Maddams, W. F. (1989) *The Vibrational Spectroscopy of Polymers*. Cambridge University Press, Cambridge.

Koenig, J. L. (1992) *Spectroscopy of Polymers*. American Chemical Society, Washington, DC.

Wignall, G. D. (1993) Scattering techniques, in *Physical Properties of Polymers*, 2nd edn (J. E. Mark, ed.). American Chemical Society, Washington, DC.

Zerbi, G., Ciampelli, F. and Zamboni, V. (1964) *J. Polym. Sci., Polym. Symp.* **7**, 141.

12.7 SUGGESTED FURTHER READING

Baltá-Calleja, F. J. and Vonk, C. G. (1988) *X-ray Scattering of Synthetic Polymers*, Polymer Science Library 8. Elsevier, Amsterdam.

Campbell, D. and White, J. R. (1989) *Polymer Characterization: Physical Techniques*. Chapman and Hall, London and New York.

Klöpffer, W. (1989) *Introduction to Polymer Spectroscopy*. Springer-Verlag, Berlin.

Kroschwitz, J. I. (1990) *Polymers: Polymer Characterization and Analysis*, Encyclopedia Reprint Series. Wiley, New York.

SOLUTIONS TO PROBLEMS GIVEN IN EXERCISES

CHAPTER 1

1.1. Leo Baekeland made 'Bakelite' in 1905. Bakelite is a thermoset made from phenol and formaldehyde.

1.2. PE:

$$\begin{array}{c} H\ H \\ |\ \ | \\ -C-C- \\ |\ \ | \\ H\ H \end{array}$$

PP:

$$\begin{array}{c} H\ H \\ |\ \ | \\ -C-C- \\ |\ \ | \\ H\ CH_3 \end{array}$$

PMMA:

$$\begin{array}{c} H\ \ CH_3 \\ |\ \ \ \ | \\ -C-C- \\ |\ \ \ \ | \\ H\ \ \underset{\underset{O}{\|}}{C}-O-CH_3 \end{array}$$

PA 8:

$$-\underset{\underset{}{\|}}{\overset{O}{C}}-(CH_2)_7-\underset{\underset{}{|}}{\overset{H}{N}}-$$

1.3. **Configuration** refers to a permanent stereostructure that cannot be changed without a chemical reaction. **Conformation** refers to a stereostructure that can be changed by torsion about the sigma bonds and that involves much less energy than covalent-bond scission.

1.4. Atactic PS is irregular and the bulky side groups prevent it from crystallizing in this stereoform.

1.5. In **stepwise growth**, the system will at a given moment consist of a mixture of growing chains. The same kind of reaction is repeated (e.g. esterification) and the concentration of functional groups gradually decreases. A high molar mass polymer is only obtained if the yield is high.

In **chainwise growth**, the system at a given time consists of a mixture of monomer and polymer and very, very few growing chains. The growth of a chain from a few repeating units to a polymer is extremely rapid. The initiation of growth occurs sporadically and only a few chains grow simultaneously. High molar mass polymer is obtained also at low yields.

1.6. No.

1.7. Size exclusion chromatography (SEC) or, as it is also called, gel permeation chromatography (GPC).

1.8. Colligative properties only depend on the number of molecules: $\sum_i N_i = A$. The mass of polymer (B) in the sample is equal to $\sum_i N_i M_i$ and the number average molar mass is obtained from the following formula:

$$\bar{M}_n = \frac{\sum_i N_i M_i}{\sum_i N_i} = \frac{B}{A}$$

CHAPTER 2

2.1. For PE, the equation relating number of carbon atoms (n) and end-to-end distance (r) is given by:

$$\langle r^2 \rangle_0 \approx 6.85 n l^2$$

The number of carbons is $n = 10^7/14 \approx 7.14 \times 10^5$ g mol^{-1}; l is 0.127 nm:

$$\langle r^2 \rangle_0 = 7.892 \times 10^4 \text{ nm}^2 \Rightarrow \sqrt{\langle r^2 \rangle} \approx 281 \text{ nm}$$

The contour length of the chain is

$$r_c = nl \approx 7.14 \times 10^5 \times 0.127 \approx 90\,714 \text{ nm}$$

The ratio becomes

$$\frac{r_c}{\sqrt{r^2}} \approx 323$$

2.2. The possible conformations of n-hexane are generated by torsion about the three central C–C bonds, and here is a list of the different conformations:

States	Energy[a] (kJ mol^{-1})	Statistical weights[b]	
		20 K	400 K
TTT	0	1	1
TTG	2.1	3.2×10^{-6}	0.532
TTG'	2.1	3.2×10^{-6}	0.532
TGT	2.1	3.2×10^{-6}	0.532
TGG	4.2	1.0×10^{-11}	0.283
TGG'	14.5	1.2×10^{-38}	0.012 8
TG'G'	4.2	1.0×10^{-11}	0.283
TG'T	2.1	3.2×10^{-6}	0.532
TG'G	14.5	1.2×10^{-38}	0.012 8
TG'G'	4.2	1.0×10^{-11}	0.283
GTG	4.2	1.0×10^{-11}	0.283
GTG'	4.2	1.0×10^{-11}	0.283
GGG	6.3	3.3×10^{-17}	0.150
GGG'	16.6	3.7×10^{-44}	0.006 78
GG'G	26.9	4.2×10^{-71}	0.000 31
GG'G'	16.6	3.7×10^{-44}	0.006 78
G'TG	4.2	1.0×10^{-11}	0.283
G'GG'	26.9	4.2×10^{-71}	0.000 31
G'G'G'	6.3	3.3×10^{-17}	0.150

[a] Energy is calculated according to $E = \sum_{i=2}^{4} E_{\zeta\eta;i} = E_{\eta;2} + E_{\zeta\eta;3} + E_{\zeta\eta;4}$.
[b] Statistical weights is calculated from $u = e^{-E/RT}$.

Note the enormous difference in statistical weights between TTT and other states at 20 K and the more similar values obtained at 400 K.

2.3. See Fig. 13.1.
2.4. See Fig. 13.2.

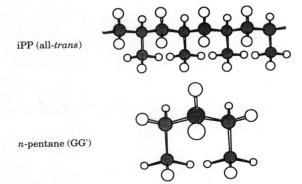

iPP (all-*trans*)

n-pentane (GG')

Figure 13.1 Illustration of similarities between iPP in all-*trans* state and *n*-pentane in a GG' state.

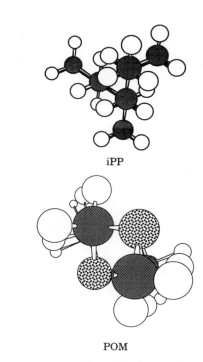

Figure 13.2 Views along helical axes of iPP and POM showing the preferred conformations.

2.5. The trans (x_t) and gauche (x_g, $x_{g'}$) contents assuming only **first order interactions** were calculated from their statistical weights, i.e.:

$$x_t = \frac{1}{1 + 2\exp(-E_g/RT)}$$

$$x_g = x_{g'} = \frac{\exp(-E_g/RT)}{1 + 2\exp(-E_g/RT)}$$

where E_g is the energy difference between trans and gauche state, which is set to 2100 J mol^{-1}. The end-to-end distance $\langle r^2 \rangle$, is calculated according to eq. (2.34):

$$\langle r^2 \rangle = nl^2 \left(\frac{1 + \cos(180 - \tau)}{1 - \cos(180 - \tau)} \right)$$
$$\times \left(\frac{1 + 2\exp(-E_g/RT)}{3\exp(-E_g/RT)} \right) = Cnl^2$$

The following values were obtained assuming only

first order interaction:

Temperature (K)	x_t	$x_g = x_{g'}$	C
20	1.000	0.000	409 651
100	0.862	0.069	17.35
200	0.639	0.181	5.38
300	0.537	0.232	3.76
400	0.485	0.258	3.17
600	0.432	0.284	2.70

The trans and gauche contents were calculated, assuming **second order interactions**, using the following equations:

$$x_t = \frac{1 - \lambda_1}{\lambda_1 - \lambda_2}$$

$$x_g = x_{g'} = \frac{1 - x_t}{2}$$

where λ_1 and λ_2 are defined in eq. (2.66) using $E_g = 2100 \text{ J mol}^{-1}$ and $E_{GG'} = 12\,400 \text{ J mol}^{-1}$. The following values were obtained:

Temperature (K)	x_t	$x_g = x_{g'}$
20	1.000	0.000
100	0.877	0.062
200	0.715	0.143
300	0.646	0.177
400	0.608	0.196
600	0.563	0.219

Flory showed on the basis of these data that C is equal to 6.87 at 400 K, which is consonant with experimental data (theta conditions).

2.6. The number of bonds (n) in a typical Gaussian amorphous chain sequence, which links the two adjacent crystallites separated by distance L_a (Fig. 13.3), is given by:

$$n = \frac{L_a^2}{Cl^2} \quad (13.1)$$

where C is a constant and l is the bond length.

The number of chain segments in a straight chain is:

$$n^0 = \frac{L_a}{l} \quad (13.2)$$

and the amorphous density, provided that all

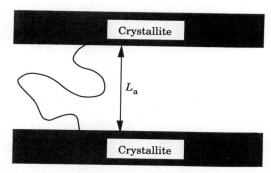

Figure 13.3 Simplified model of two lamellae and an amorphous interlayer showing a single Gaussian chain.

amorphous chain segments are random, is given by:

$$\rho_a = \rho_c \frac{L_a}{Cl} \quad (13.3)$$

Equation (13.3) produces an anomalously high ρ_a value, which can be reduced to more realistic values by postulating that a fraction (f_{fold}) of the polymer chains entering the amorphous phase must fold:

$$\rho_a = \rho_c \frac{L_a}{Cl}(1 - f_{fold})\cos\theta \quad (13.4)$$

where θ is the chain tilt angle. Equation (13.4) permits calculation of f_{fold} by insertion of the following values: $\rho_a/\rho_c = 0.85$; $L_a = 5$ nm; $l = 0.127$ nm; $C = 6.85$; $\theta = 30°$. The calculated value of f_{fold} is 0.83, i.e. 83% of all stems are expected to be tightly folded.

2.7. At equal eluation time conditions the following expression holds:

$$[\eta]_X M_X = [\eta]_{PS} M_{PS} \quad (13.5)$$

where X and PS are indices for the studied (X) polymer and polystyrene. Insertion of the Mark–Houwink equation yields:

$$K_X M_X^{a_X} M_X = K_{PS} M_{PS}^{a_{PS}} M_{PS} \Rightarrow M_X^{1+a_X} = \frac{K_{PS}}{K_X} \cdot M_{PS}^{1+a_{PS}}$$

which after taking logarithms and simplification gives:

$$\log M_X = \left(\frac{1}{1 + a_X}\right)\log\left(\frac{K_{PS}}{K_X}\right) + \left(\frac{1 + a_{PS}}{1 + a_X}\right)\log M_{PS} \quad (13.6)$$

CHAPTER 3

3.1. The general expression for the free energy is:

$$\Delta G = \frac{NRT}{2}(\lambda_1^2 + \lambda_2^2 + \lambda_3^2 - 3) \quad (13.7)$$

Biaxial stretching under constant volume yields the following deformation matrix:

$$\left(\lambda, \lambda, \frac{1}{\lambda^2}\right)$$

and

$$\Delta G = \frac{NRT}{2}\left(2\lambda^2 + \frac{1}{\lambda^4} - 3\right) \quad (13.8)$$

The force (f) is obtained by taking the derivative of ΔG:

$$f = \left(\frac{\partial \Delta G}{\partial \lambda}\right)\left(\frac{\partial \lambda}{\partial L}\right) = \frac{2NRT}{L_0}\left(\lambda - \frac{1}{\lambda^5}\right) \quad (13.9)$$

3.2. The diameter of the balloon increases with the factor α. The initial cross-sectional area (unstrained state) is A_0:

$$A_0 = \pi D_0 t_0 \quad (13.10)$$

where D_0 is the initial diameter and t_0 the initial wall thickness. The cross-sectional area of the expanded balloon is:

$$A = \pi D t = \pi D_0 \alpha \cdot \frac{t_0}{\alpha^2} = \frac{A_0}{\alpha} \quad (13.11)$$

The stress acting on the cross-sectional area becomes:

$$f = \sigma \cdot \frac{A_0}{\alpha} = \frac{2NRT}{L_0}\left(\alpha - \frac{1}{\alpha^5}\right) \Rightarrow$$

$$\sigma = \frac{2NRT}{V_0}\left(\alpha^2 - \frac{1}{\alpha^4}\right) = \frac{2\rho RT}{\bar{M}_c}\left(\alpha^2 - \frac{1}{\alpha^4}\right)$$

$$(13.12)$$

The stress acting on the balloon wall is related to the internal pressure by the following equation:

$$\sigma = \frac{pD}{4t} = \frac{pD_0 \alpha^3}{4t_0} \quad (13.13)$$

Insertion of eq. (13.13) into eq. (13.12) yields:

$$p = \left(\frac{8\rho RT}{\bar{M}_c}\right)\cdot\left(\frac{t_0}{D_0}\right)\left(\frac{1}{\alpha} - \frac{1}{\alpha^7}\right) \quad (13.14)$$

3.3. The deformation associated with the pressure maximum is obtained by:

$$\left(\frac{\partial p}{\partial \alpha}\right) \propto -\frac{1}{\alpha^2} + \frac{7}{\alpha^8} = 0 \Rightarrow \alpha^6 = 7 \Rightarrow \alpha \approx 1.38$$

This pressure maximum appears at a 38% increase in the diameter.

3.4. Is it possible to expand all parts of the balloon shown in Fig. 13.4? The important equation is eq. (13.14). The pressure is the same in all parts of the balloon. The pressure maximum which appears at $\alpha = 1.38$ will be greater for the small sphere. For the same wall thickness (t_0) it will be D_1/D_2 times as great in the small sphere as in the large sphere. The pressure required to expand the small-diameter sphere is not reached since the two spheres are at the same pressure and the small sphere will never exceed $\alpha = 1.38$. However, by keeping the ratio t_0/D_0 constant in all parts of the balloon, all parts expand to an equal level. The trick is thus to reduce the wall thickness in proportion to the diameter.

3.5. 'Physical' relaxation caused by disentanglement of chains. The entanglements can be considered as temporary crosslinks.

3.6. Prepare a dumb-bell-shaped specimen, heat it to a temperature well above the melting point (430–470 K), load the specimen with constant loads and record the extension (λ). The crosslink density expressed in $\bar{M}_c$ is obtained from the following equation:

$$\sigma = \frac{\rho RT}{\bar{M}_c}\left(1 - \frac{2\bar{M}_c}{M}\right)\left(\lambda^2 - \frac{1}{\lambda}\right) \quad (13.15)$$

where M is the number average molar mass of the polymer prior to the vulcanization. This equation assumes that the elastic force is entirely entropic, which is not true for polyethylene. The energetic

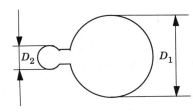

Figure 13.4 Simple model of balloon with small nose.

component (f_e) of the total force (f) is, according to the experimental data:

$$\frac{f_e}{f} \approx -0.45 \Rightarrow f_{entropic} = 1.45f$$

which can be inserted into eq. (13.15) as a correction factor.

3.7. The relevant equation for the 'modulus' is:

$$\frac{\rho RT}{\bar{M}_c} = 2\nu RT \quad (13.16)$$

where ν is the number of crosslinks per unit volume. The molar fraction of crosslinks is equal to the added mole fraction of peroxide (with reference to the number of carbon atoms), i.e.:

$$x_{crosslinks} = \frac{n_{peroxide}}{n_{CH_2}}$$

The number of repeating unit per unit volume is ρ/M_{rep}. Thus, the number of moles of peroxide is the molar fraction times the number of repeating units, i.e. $x\rho/M_{rep}$. The modulus of the rubber is $(2x\rho/M_{rep}) \cdot RT$.

3.8. The resulting temperature increase is given by the following equations:

$$\Delta T = -\frac{T}{c_p} \int_{l_0}^{l} \left(\frac{\partial S}{\partial l}\right) dl \quad (13.17)$$

$$\Delta S_N = -\frac{1}{2} nk\left(\lambda^2 + \frac{2}{\lambda} - 3\right)$$

$$= -\frac{NR}{2}\left(\frac{l^2}{l_0^2} + \frac{2l_0}{l} - 3\right) \quad (13.18)$$

$$\frac{\partial \Delta S_N}{\partial l} = -NR\left(\frac{l}{l_0^2} - \frac{2l_0}{l^2}\right)$$

CHAPTER 4

4.1. (a) The number of different ways of arranging 10 solute molecules in a lattice of 100 positions is given by:

$$P_1 = \frac{100!}{10!90!} = 17\,310\,309\,456\,440 \approx 1.73 \times 10^{13}$$

(b) The number of different ways of arranging an oligomer of 10 repeating units in a lattice of 100 positions is given by:

$$P_2 = 100Z(Z-1)^8$$

Assume that the coordination number is 6:

$$P_2 = 100 \times 6 \times 5^8 = 234\,375\,000 \approx 2.34 \times 10^8$$

(c) The combinatorial mixing entropy ($\Delta S_{mix,c}$) is much greater in the first case (L/L) than in the second case (L/P):

$$\Delta S_{mix,c} = k \ln P \quad (13.19)$$

$$\Delta S_{mix,c}(L/L) = k \ln P_1 = 30.48k$$

$$\Delta S_{mix,c}(L/P) = k \ln P_2 = 19.27k$$

4.2. The spinodal and binodal curves are obtained from the free energy–composition plot shown in Fig. 13.5. The phase diagram (T–x_1) is constructed from the data of Fig. 13.5. It shows a UCST.

4.3. The spinodal is obtained by setting the second derivative of $\Delta G_{mix}/N$ with respect to x_1 equal to zero. The equation to start with is:

$$\frac{\Delta G_{mix}}{N} = Bx_1 x_2 + RT(x_1 \ln x_1 + x_2 \ln x_2)$$

$$= Bx_1(1 - x_1) + RT[x_1 \ln x_1 + (1 - x_1)$$

$$\times \ln(1 - x_1)] \quad (13.20)$$

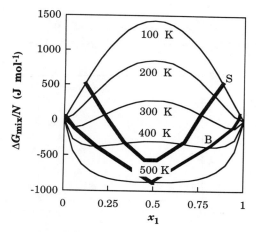

Figure 13.5 Free energy as a function of molar content (x_1) at different temperatures. The spinodal (S) and binodal (B) points are shown in the diagram.

Taking the first derivative:

$$\frac{d(\Delta G_{mix}/N)}{dx_1} = B - 2Bx_1$$
$$+ RT[\ln x_1 + 1 - \ln(1 - x_1) - 1]$$
$$= B - 2Bx_1 + RT[\ln x_1 - \ln(1 - x_1)] \quad (13.21)$$

and the second derivative:

$$\frac{d^2(\Delta G_{mix}/N)}{dx_1^2} = -2B + RT\left[\frac{1}{x_1} + \frac{1}{1-x_1}\right] = 0 \quad (13.22)$$

and the two spinodal concentrations are obtained as the solutions to eq. (13.22) are set equal to zero:

$$x_{1,spin} = \frac{1}{2} \pm \sqrt{\frac{RT}{B} + \frac{1}{4}} \quad (13.23)$$

Note that the spinodal points are located symmetrically around $x_1 = \frac{1}{2}$.

4.4. The critical parameters show the following x-dependences:

$$v_{2,c} = \frac{1}{1 + \sqrt{x}} \quad (13.24)$$

$$\chi_{12,c} = \frac{1}{2} + \frac{1}{2x} + \frac{1}{\sqrt{x}} \quad (13.25)$$

$$T_c = \frac{B}{R\chi_{12,c}} \quad (13.26)$$

The molar mass dependence of the critical parameters is shown in Figs 13.6–13.8. The T–v_2

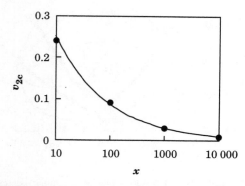

Figure 13.6 Critical volume fraction of polymer as a function of the number of mers (x).

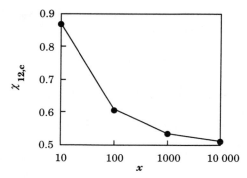

Figure 13.7 Critical interaction parameter as a function of the number of mers (x).

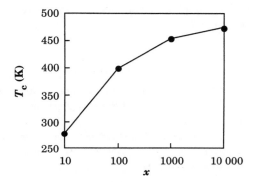

Figure 13.8 Critical temperature (UCST) as a function of the number of mers (x).

phase diagrams for polymers of different molar mass have the schematic forms shown in Fig. 13.9.

4.5. The threshold concentration relates to molar mass as follows:

$$c^* \propto M^{-4/5} \quad (13.27)$$

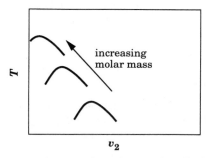

Figure 13.9 Schematic phase diagram showing the effect of molar mass.

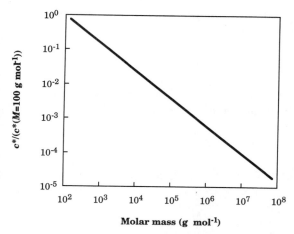

Figure 13.10 Reduced threshold concentration for molecular overlap as a function of molar mass.

which in a double logarithmic c^*–M plot looks like Fig. 13.10.

4.6. (i) $\langle r \rangle \propto M^{3/5}$; (ii) $\langle r \rangle \propto M^{1/2}$; (iii) $\langle r \rangle \propto M^{3/5 \to 1/2}$; (iv) $\langle r \rangle \propto M^{1/2}$ where $\langle r \rangle$ is the average end-to-end distance and M is molar mass.

4.7. Calculate first the distance between PVC and VC in three-dimensional solubility space; keep in mind that the dispersion values should be doubled:

$$R_A = \sqrt{4(18.2 - 15.4)^2 + (7.5 - 8.1)^2 + (8.3 - 2.4)^2}$$
$$= 8.2$$

$$\frac{R_A}{R_{AO}} = \frac{8.2}{3.5} > 1$$

It is thus predicted that PVC should be insoluble in VC.

4.8. There should be strong specific interaction (e.g. hydrogen bonding or charge-transfer complex formation between two groups of widely different polarities) between the two polymers. Infrared spectroscopy may detect a frequency shift of certain absorption bands caused by the specific interaction.

CHAPTER 5

5.1. **Polyethylene** has a regular chain structure (at least if we assume it to be unbranched). Its chain flexibility is high. This polymer crystallizes under normal cooling conditions. It can be quenched to a fully amorphous structure under extreme cooling conditions.

Isotactic polystyrene has a regular chain structure and can crystallize. However, the chain is very inflexible and this polymer is easily quenched to a fully amorphous glassy polymer.

Atactic polystyrene has an irregular chain structure and will not crystallize.

Atactic poly(vinyl alcohol) has an irregular chain structure but the hydroxyl groups are small, and this allows good packing of chains in the crystal. This polymer may crystallize to some extent. It can probably be quenched to a fully amorphous glassy polymer.

Atactic poly(vinyl acetate) has an irregular chain structure and does not crystallize.

Poly(ethylene-stat-propylene) 50/50 composition has an irregular chain structure and does not crystallize. Poly(ethylene-*stat*-propylene) 98/2 composition has long ethylene blocks and is crystallizable. This polymer can be quenched below the glass transition to a fully amorphous state by cooling at an enormously high cooling rate.

5.2. The experiment can be carried out in a dilatometer measuring the volume or in a differential scanning calorimeter recording changes in the enthalpy. Start at a temperature (T_1) well above T_g. Cool the sample at a constant rate to a temperature below T_g. Stop the cooling at a certain temperature (T_2) and hold the sample at that temperature while recording volume and enthalpy changes. The process may be so slow that the resolution of the DSC apparatus is not sufficient. In that case, the integrated evolved enthalpy between the starting time and a selected interruption time (t_1) can be measured by taking the difference between the integrated enthalpies in heating scans between T_2 and T_1 for the sample after ageing during the time period t_1 and for a sample that has not been 'aged' at T_2.

5.3. The formula to use is eq. (5.4) and by plotting T_g against $1/M$, T_g for $M = \infty$ is obtained as the intersection with the T_g axis, which occurs at 379.4 K (Fig. 13.11).

5.4. A similarity in the two polymers' T_g values gives rise to a single glass transition despite the fact that the polymers are immiscible as judged by

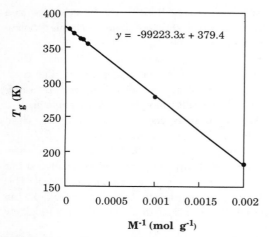

Figure 13.11 Glass transition temperature of atactic polystyrene as a function of the reciprocal of the molar mass.

their opaqueness. The second case can be given two different explanations. The first is that the polymers are immiscible but the similarity in refractive index of the two polymers results in only negligible light scattering, i.e. a transparent material. The second explanation is that the polymers are immiscible but the dispersion is considerably finer than the wavelength of light.

5.5. The following methods are suggested in order to increase T_g: crosslinking, increasing the polarity of the polymer by graft copolymerization with a polar monomer, or by making a miscible blend with a high-T_g polymer.

5.6. The relevant equation is eq. (5.6):

$$T_g = T_g^\infty - \frac{C_5}{M} + \frac{C_9}{\overline{M}_c}$$

The purpose is to obtain a polymer with a $T_g = T_g^\infty$, i.e.

$$\frac{C_5}{M} = \frac{C_9}{\overline{M}_c} \Rightarrow \overline{M}_c = M \frac{C_9}{C_5} \quad (13.28)$$

The number of crosslinks per unit volume (ν) is given by:

$$\nu = \frac{\rho}{2\overline{M}_c} = \frac{\rho C_5}{2MC_9} \quad (13.29)$$

which is equal to the number of peroxide molecules per unit volume, provided that each peroxide molecule yields one crosslink. The number of repeating units of polymer per unit volume (ν_{rep}) is:

$$\nu_{\text{rep}} = \frac{\rho}{M_{\text{rep}}} \quad (13.30)$$

The molar fraction of peroxide added per mole of repeating units is given by:

$$x_{\text{per}} = \frac{\nu}{\nu_{\text{rep}}} = \frac{C_5 M_{\text{rep}}}{2MC_9} \quad (13.31)$$

5.7. The main technique is wide-angle X-ray scattering (WAXS). The X-ray scattering pattern from an amorphous polymer is diffuse and liquid-like, whereas for a semicrystalline polymer the diffraction pattern also shows a number of sharp Bragg reflections. A preliminary, less exact assessment can be made by visual examination – the amorphous polymer is transparent, whereas crystalline polymers are opaque. Differential scanning calorimetry is another useful tool. An amorphous polymer displays only a glass transition, whereas a crystalline polymer shows a melting peak.

CHAPTER 6

6.1. (a) Relaxation (full equilibration of a molecule) is controlled by the monomeric friction coefficient and not by entanglements since only very few entanglements are present in a low molar mass polymer. Relaxation is rapid. (b) Entanglements are numerous and play a vital role. Relaxation is slow. (c) The shape of the chains is intact and relaxation is very slow.

6.2. The convergent flow causes axial orientation of the chains. The die-swell occurs in response to the axial orientation. During the flow through the narrow pipe section, relaxation of the oriented molecules occurs. The degree of die-swell decreases with increasing length of the narrow pipe section.

6.3. (a) > (b) > (c) > (d) > (e)

6.4. Figure 13.12, showing the director field, may be made from the polarized photomicrograph.

6.5. Melt-processable liquid-crystalline polymers should be nematic over an extensive temperature range. The melting point should be depressed with respect to the isotropization temperature. It is also important that both the transition temperatures, i.e. the isotropization and the melting temperature, are sufficiently low to avoid degradation. The molecular requirements can be formulated as follows: statistical copolymers with kinks or flexible units.

6.6. (a) Oriented smectic C; (b) oriented smectic A; (c) oriented nematic; (d) isotropic.

6.7. The order parameter can be calculated according to the equation:

$$\langle \cos^2 \phi_i \rangle = \frac{\int_0^\pi I(\phi)_i \cos^2 \phi \sin \phi \, d\phi}{\int_0^\pi I(\phi)_i \sin \phi \, d\phi} \quad (13.32)$$

where ϕ is the angle between the vertical director and the measured spot of the intermesogenic group spacing. The order parameter of the normals to these groups (f_i) is:

$$S_i = \frac{3\langle \cos^2 \phi_i \rangle - 1}{2} \quad (13.33)$$

The order parameter of the mesogens (S) is given by:

$$S = -2S_i \quad (13.34)$$

These equations can be applied to diffractograms of the type shown in Fig. 6.47 if orientation is uniaxial.

CHAPTER 7

7.1. The total thickness of amorphous layer (L_a) is given by:

$$\frac{L_a \rho_a}{L_c \rho_c + L_a \rho_a} = 0.15 \Rightarrow L_a = \frac{0.15 L_c \rho_c}{0.85 \rho_a} \quad (13.35)$$

where L_c is the crystal thickness, and ρ_a and ρ_c are the densities of the amorphous and crystalline material. The amorphous component is distributed on both sides of the single crystal and the thickness on each side (L_a^*) is:

$$L_a^* = \frac{0.15 L_c \rho_c}{1.7 \rho_a} \quad (13.36)$$

It may be assumed that the average end-to-end distance for a Gaussian amorphous chain is given by:

$$\langle r \rangle = x L_a^* = \frac{0.888\,24 x L_c \rho_c}{\rho_a} \quad (13.37)$$

where x is an (unknown) factor. The number of bonds (n) in a Gaussian chain of this length is given by:

$$n = \frac{\langle r \rangle^2}{C l^2} \quad (13.38)$$

where C is a constant and l is the bond length. The average number of bonds ($\langle n \rangle$) in all amorphous chains is:

$$\langle n \rangle = \frac{\langle r \rangle^2}{C l^2} (1 - f_{\text{fold}}) \quad (13.39)$$

where f_{fold} is the concentration of chain folds. The number of bonds for each entry is half of $\langle n \rangle$. The number of bonds (n_0) in a straight chain of length L_a^* is given by:

$$n_0 = \frac{L_a^*}{l} \quad (13.40)$$

The ratio of the amorphous to the crystalline density is given by:

$$\frac{\rho_a}{\rho_c} = \frac{\langle n \rangle}{2 n_0} = \frac{\langle r \rangle^2}{2 C l^2} (1 - f_{\text{fold}}) \frac{l}{L_a^*} = \frac{x^2 L_a^*}{2 C l} (1 - f_{\text{fold}}) \quad (13.41)$$

$$f_{\text{fold}} = 1 - \frac{2 C l \rho_c}{x^2 L_a^* \rho_a} \quad (13.42)$$

By insertion of eq. (13.36) into eq. (13.42):

$$f_{\text{fold}} = 1 - \frac{2 C l}{x^2 \cdot 0.888\,24 L_c} \quad (13.43)$$

By inserting the following values for linear polyethylene, $C = 6.85$; $l = 0.127$ nm; $x = 2$; $L_c = 10$ nm, we obtain $f_{\text{fold}} = 0.95$.

7.2. The following values were obtained:

Crystal thickness (nm)	Melting point (K)
5	364
10	390
50	409.9
100	412.5

7.3. A finite molar mass will modify eq. (7.23) as follows:

$$f_{\text{fold}} = 1 - \frac{\rho_a Cl}{\rho_c L_a \cos\theta f(M)} \quad (13.44)$$

where $f(M)$ is the fraction of the entries that are not chain ends. The latter can be expressed in simple terms:

$$f(M) = \frac{n_c - 2}{n_c} = \frac{M/M_{\text{crystal}} - 2}{M/M_{\text{crystal}}} \quad (13.45)$$

where n_c is the number of crystalline stems of a molecule with specified molar mass M, and M_{crystal} is the molar mass of a single stem. For simplicity let us assume that $L_c = 4L_a$:

$$f(M) = \frac{\dfrac{Ml}{4L_a \times 14} - 2}{\dfrac{Ml}{4L_a \times 14}} \quad (13.46)$$

which yields the final expression:

$$f_{\text{fold}} = 1 - \frac{\rho_a Cl}{\rho_c L_a \cos\theta \left(\dfrac{\dfrac{Ml}{4L_a \times 14} - 2}{\dfrac{Ml}{4L_a \times 14}} \right)} \quad (13.47)$$

Figures 13.12–13.14 plot f_{fold} against M, θ and L_a, respectively.

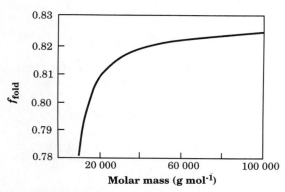

Figure 13.12 Concentration of tight folds as a function of molar mass.

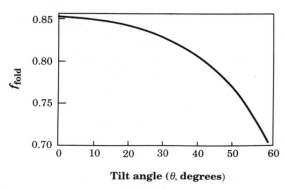

Figure 13.13 Concentration of tight folds as a function of tilt angle.

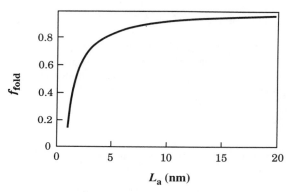

Figure 13.14 Concentration of tight folds as a function of thickness of amorphous layer.

7.4. Self-diffusion coefficients for polymers are in general very low. The segregation spacing, i.e. the distance which segregated species move during crystallization, is determined by the ratio between the crystallization rate and diffusion coefficient. A high diffusion coefficient means a short segregation spacing, i.e. a finely dispersed system of segregated low and early crystallizing high molar mass species.

7.5. By inserting a λ plate at an angle of 45° to the polarizer/analyser pair, quadrants 1 and 3 of the spherulites appear either blue or yellow, and quadrants 2 and 4 of the spherulites appear with the 'reverse' colour. This colour combination shows the sign of the spherulites. Negative spherulites, the common case, have a higher refractive index in the tangential plane than along the radius of the spherulite. Polymers with predominantly polarizable groups in the backbone chain normally form negative spherulites in the case where the chain-axis direction is in the tangential plane rather than along the spherulite radius. Polymers with polarizable pendant groups may show positive spherulites. Isotactic polypropylene sometimes shows positive spherulites due to the crosshatching and to the fact that the daughter lamellae are dominant in terms of volume fraction.

7.6. **Optical origin**. The regular variation in light intensity as a function of radial distance is due to a corresponding variation in orientation of the refractive index ellipsoid. The intrinsic birefringence is approximately uniaxial. The sample appears optically isotropic (dark on photomicrographs) at locations where the long axis of the refractive index ellipsoid is parallel to the light beam, i.e. the chain-axis director is parallel to the direction of propagation of the light. The transmitted light intensity is at a maximum at locations where the long axis of the refractive index ellipsoid is perpendicular to the light beam. This corresponds to the normal of the crystal lamellae being perpendicular to the light beam.

Underlying lamellar morphology. (a) An early suggestion was that continuously twisting lamellae radiating out from the centre of the spherulites led to rotation of the refractive index ellipsoid about the radius of the spherulite. The lamellar profiles (S or C) can, according to this view, result from slices that are at oblique angles (not parallel to [010]) through helicoids with straight profiles (from a view along [010]). (b) Samples that display banded spherulites also show C- and S-shaped lamellae. It is believed that banded spherulites consist of dominant lamellae whose profile, viewed down [010], is S- or C-shaped. Chain tilt and c-axis orientation in adjacent lamellae were found to be uniform. These uniform structures were untwisted along the radius of the spherulites about one-third of the band spacing. Changes in c-axis orientation occur sharply in screw dislocations of consistent sign and involve only two or three layers of spiral terrace.

7.7. The statistical distribution of the chain branches gives this polymer a multicomponent character. The cumulative distribution of linear chain segments of n carbon atoms, each terminated by two chain defects, is given by the equation:

$$W(n) = \int_0^\infty n(1-p)^2 p^{n-1} \, dn \quad (13.48)$$

where p is the probability that a main-chain carbon atom is not attached to a branch group. The cumulative distribution is shown in Fig. 13.15. The abscissa in this plot is transformed to an equilibrium melting-temperature scale by considering the Thompson–Gibbs equation:

$$T_m^0(n) = T_m^0(M = \infty)\left(1 - \frac{2\sigma}{\Delta h^0 \rho_c n l \cos\theta}\right) \quad (13.49)$$

where $T_m^0(M = \infty) = 418.1$ K, q is the specific fold surface free energy, ρ_c is the crystal density $= 1000$ kg m^{-3}, l is half the crystallographic unit $c = 0.127$ nm, and θ is the chain tilt angle which is set to 30°. Thus, the thickness of the equilibrium crystal is determined by the length of the linear segments and the equilibrium crystals are assumed to contain no chain branches.

7.8. The molecules are preferentially placed in the 'cylindrical plane' of the fibre. The molecules are aligned parallel to the fibre axis nearest to the fibre.

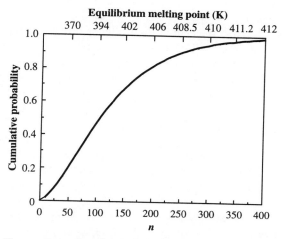

Figure 13.15 Cumulative distribution function $W(n)$, i.e. the mass fraction of linear chains shorter than n for polyethylene with 1.5% of chain defects.

7.9. The crystalline component is impermeable to small and intermediate molecules and all diffusion of these species occurs within the amorphous component. The width of the amorphous 'channels' decreases and their length increases with increasing crystallinity. This two-fold effect makes D a nonlinear function of crystallinity.

7.10. Mass crystallinity (w_c) is obtained from the following equations:

$$w_c(T_1) = \frac{\Delta h_{21} - \Delta h_{a21}}{\Delta h^0(T_1)} \quad (13.50)$$

$$\Delta h^0(T_1) = \Delta h^0(T_m^0) - \int_{T_1}^{T_m^0} (c_{pa} - c_{pc}) dT \quad (13.51)$$

The integral is first calculated. The specific heat data are fitted to polynomial expressions and the integral is calculated:

$$\int_{393}^{461} (c_{pa} - c_{pc}) dT = 19.18 \text{ cal mol}^{-1} \approx 6 \text{ J g}^{-1}$$

and $\Delta h^0(\alpha) = 200$ J g^{-1} and $\Delta h^0(\gamma) = 159$ J g^{-1}. The mass crystallinity is given by:

$$w_c(300 \text{ K}) = \frac{44}{0.5 \times 200 + 0.5 \times 159} = 0.24$$

which is distributed in 12% of α crystals and 12% of γ crystals.

7.11. The maximum in the scattering curve appears at $2\theta = 0.5°$. The Bragg equation can then be applied to obtain the long period $L = 20$ nm. The volume crystallinity is 70%, which gives a crystal thickness of 14 nm.

CHAPTER 8

8.1. The three curves are different parts of the bell-shaped crystallization rate–temperature curve.

8.2. The solidification time can be reduced by adding a nucleation agent.

8.3. The entropy is low and the free energy of the stretched rubber band is high. This favours any phase transitions, including crystallization.

8.4. The solution is presented in Fig. 13.16. The Avrami equation can be fitted to experimental data

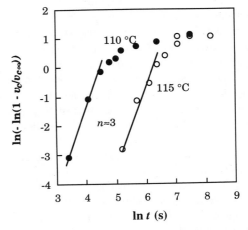

Figure 13.16 Avrami plot showing crystallization of a branched polyethylene crystallized at two different temperatures.

at low crystallinity levels. Pronounced deviation is obtained at the later stages of crystallization.

8.5. The LH view: change from regime I (high temperature) to regime II (low temperature). Point and Dosière (1989) suggested other causes for the slope change: molar mass segregation, a pronounced temperature dependence of the interfacial energies, viscosity effects, and temperature dependence of the nucleation processes.

8.6. Particles and impurities with nucleating power are present in all large-volume samples. The first step is to remove the majority of the particles by filtering a solution of the polymer. The polymer may be divided into very small droplets. The volume of each droplet is so small that an appreciable fraction of the droplets contain no particles and will nucleate homogeneously. The remaining particles contain nucleating particles and crystallization starts with a heterogeneous nucleation. The results of a droplet crystallization may be represented as shown in Fig. 13.17.

8.7. DSC, TEM and SAXS are useful in the assessment of segregation phenomena. The multimodality in the melting behaviour (DSC) and in the long period (SAXS) indicate segregation. TEM provides detailed information about the lamellar structure and can, in favourable cases, provide the most detailed and reliable information. Segregation in the molten state

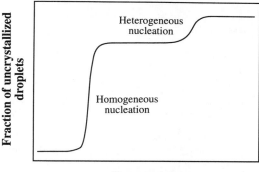

Figure 13.17 Schematic presentation of data from droplet experiment.

is best studied by rapid cooling from the molten state to prevent/minimize crystallization-induced segregation.

8.8. One example: low molar mass linear polyethylene cocrystallizes with intermediate to high molar mass branched polyethylene.

8.9. The fold surface free energy (σ) can be obtained from the slope coefficient in the plot of the melting temperature against the reciprocal of the crystal thickness:

$$\sigma = -\frac{\text{slope} \times \Delta h^0 \rho_c}{2T_m^0} \quad (13.52)$$

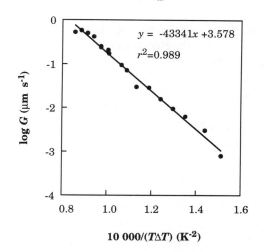

Figure 13.18 Linear growth rate as a function of crystallization temperature presented in an LH plot. Sample: branched polyethylene.

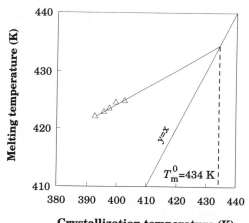

Figure 13.19 Melting peak temperature as a function of crystallization temperature. Sample: isotactic polypropylene.

where Δh^0 is the heat of fusion, ρ_c is the crystal density and T_m^0 is the equilibrium melting temperature. The product of σ and the surface free energy of the lateral surface (σ_L) can be obtained from linear growth rate (G) data:

$$G = \beta \exp\left(-\frac{Nb\sigma\sigma_L T_m^0}{\Delta h^0 kT_c \Delta Tf}\right) \quad (13.53)$$

where N is an integer according to LH theory.

8.10. The solution is presented in Fig. 13.18. The slope coefficient can be transformed to $K_g = 43\,341 \times 2.303 = 99\,800$ K^2.

8.11. The solution is presented in Fig. 13.19.

CHAPTER 9

9.1. It is important to note here that the problem is two-dimensional and that the two-dimensional orientation function is given by:

$$f = 2\langle\cos^2 \phi\rangle - 1 = \frac{2}{n}\sum_{i=1}^{n}\cos^2 \phi_i - 1 \quad (13.54)$$

The following f values were obtained:
(a) 0.6 (b) 1 (c) 0

9.2. The two polymers with a high positive Δn_0 (PETP and PC) have strongly axially polarizable phenylene groups in the backbone chain, whereas PS with a negative Δn_0 has the strongly polarizable phenyl group as pendant group. PVC constitutes an 'intermediate' case with a balance between axial and transverse polarizability.

9.3. The solution is as follows:

$$\langle \cos^2 \phi_{hko} \rangle = \frac{\int_0^{\pi/2} I_{hko}(\phi)\cos^2\phi \sin\phi \, d\phi}{\int_0^{\pi/2} I_{hko}(\phi)\sin\phi \, d\phi} = 0.0159$$

$$f_{hko} = \frac{3\langle \cos^2 \phi_{hko} \rangle - 1}{2} = -0.476$$

$$2f_{hko} + f_{ool} = 0 \Rightarrow f_{ool} = 0.95$$

The crystalline c-axis orientation is 0.95 (Hermans orientation function).

9.4. The amorphous orientation (f_a) is calculated as follows:

$$f = v_c f_c + (1 - v_c)f_a \Rightarrow f_a \frac{f - v_c f_c}{1 - v_c}$$

$$= \frac{0.8 - 0.5 \times 0.95}{0.5} = 0.65$$

9.5. The solution is given in Fig. 13.20 which shows the diffraction pattern of the uniaxially oriented sample taken at three different angular positions.

9.6. The solution is given in Fig. 13.21 which shows the diffraction pattern of the biaxially oriented sample taken at three different angular positions.

9.7. The variation in absorbance with angle ϕ is shown schematically in Fig. 13.22. The transition

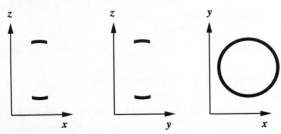

Figure 13.20 Schematic X-ray diffraction pattern of a uniaxially oriented sample, showing only the (00l) direction.

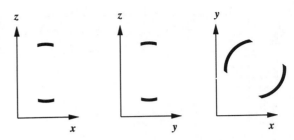

Figure 13.21 Schematic X-ray diffraction pattern of biaxially oriented sample, showing only the (00l) direction.

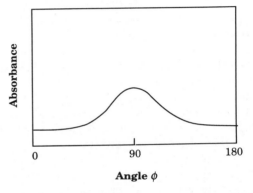

Figure 13.22 Infrared absorption associated with the carbonyl stretching band showing dichroism due to uniaxial orientation along the director (0°).

moment vector is in this case perpendicular to the chain axis and the Hermans orientation function can in this special case be expressed as follows:

$$f = 2\frac{(1 - R)}{(R + 2)} \quad (13.55)$$

where R is the dichroic ratio. The latter is defined according to:

$$R = \frac{A_\parallel}{A_\perp} \quad (13.56)$$

where $A_\parallel$ is the absorbance of light polarized parallel to the director and $A_\perp$ is the absorbance of light polarized perpendicular to the director.

9.8. The definition of the molecular draw ratio may be illustrated as in Fig. 13.23. The molecular draw ratio (λ) is defined, based on the notation used

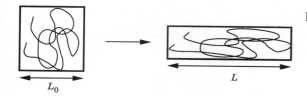

Figure 13.23 Deformation of single molecule.

in the figure, as follows:

$$\lambda = \frac{L}{L_0} \quad (13.57)$$

It may more strictly be defined as:

$$\lambda = \sqrt{\frac{\langle \bar{r}^2 \rangle_{\text{orient}}}{\langle \bar{r}^2 \rangle_0}} \quad (13.58)$$

where $\langle \bar{r}^2 \rangle_{\text{orient}}$ is the mean square of the end-to-end vector of the oriented material and $\langle \bar{r}^2 \rangle_0$ is the mean square of the end-to-end vector of the unoriented material. The molecular draw ratio may be determined by small-angle neutron scattering. A very simple, but still useful, experiment is to heat the oriented sample slightly above its melting point and then measure the shrinkage on melting. The macroscopic shrinkage ratio is approximately equal to the molecular draw ratio.

9.9. Polyamide 6 shows no crystalline α process and the crystals will not undergo plastic deformation due to the strong hydrogen bonds. Polyethylene shows a crystalline α process and the crystals, particularly at higher temperatures, readily undergo plastic deformation.

CHAPTER 10

10.1. Mass crystallinity (w_c) is obtained from the following equations:

$$w_c(T_1) = \frac{h - h_1}{\Delta h^0(T_1)} \quad (13.59)$$

$$\Delta h^0(T_1) = \Delta h^0(T_m^0) - \int_{T_1}^{T_m^0} (c_{pa} - c_{pc}) dT \quad (13.60)$$

The integral in eq. (13.60) is solved first. The specific heat data are calculated from the following polynomial expressions:

$$c_{pc} = 8.19697 \times 10^{-7} T^3 - 0.00075737 T^2 + 0.256479 T - 25.827302$$

$$c_{pa} = -3.80358 \times 10^{-9} T^3 - 3.683 \times 10^{-6} T^2 + 0.0066435 T + 5.380487$$

$$c_{pa} - c_{pc} = -8.235 \times 10^{-7} T^3 + 7.6105 \times 10^{-4} T^2 - 0.25011 T + 31.2078$$

$$\int_{370}^{418} (c_{pc} - c_{pc}) dT = 19.18 \text{ cal mol}^{-1} \approx 6 \text{ J g}^{-1}$$

$$\Delta h^0(370 \text{ K}) = 293 - 6 = 287 \text{ J g}^{-1}$$

The mass crystallinity is given by

$$w_c(370 \text{ K}) = \frac{140}{287} = 0.487$$

10.2. The entropy of fusion is equal to

$$\Delta s^0 = \frac{\Delta h^0}{T_m^0} = \frac{293 \times 14}{416} \text{ J mol}^{-1} \text{ K}^{-1}$$
$$\approx 9.86 \text{ J mol}^{-1} \text{ K}^{-1}$$

10.3. The enthalpy of fusion per main chain atom is greater for polyamide 6 than for polyethylene. The average intermolecular energy, expressed for example in CED, is much greater for polyamide 6 than for polyethylene.

10.4. Medium-density polyethylene is a blend of high-density and low-density polyethylenes. The two components crystallize in separate crystal lamellae with different melting points.

10.5. TG, by which the mass loss at 120°C is measured.

10.6. TG is used here. The polymer depolymerizes in an atmosphere of N_2 and the mass loss at 600–700°C is equal to the original polymer content. By changing the atmosphere from N_2 to O_2 at 750°C, oxidation of carbon black will take place, yielding gaseous CO_2. The remaining solid is CaO. The initial content of $CaCO_3$ can be calculated from the ash remaining at 750°C.

10.7. The liquid-crystalline polymer component (Vectra) is oriented in the blends and possesses an approximately invariant thermal expansion coefficient anisotropy.

10.8. The relationship between linear expansion coefficients ($\alpha_\parallel$ and $\alpha_\perp$) and the volume expansion coefficient (α_v) can be derived as follows:

$$V = L_0^3(1 + \alpha_\parallel)(1 + \alpha_\perp)^2$$
$$= L_0^3(1 + 2\alpha_\perp + \alpha_\perp^2 + \alpha_\parallel + 2\alpha_\parallel\alpha_\perp + \alpha_\parallel\alpha_\perp^2)$$
$$\approx L_0^3(1 + 2\alpha_\perp + \alpha_\parallel)$$

$$\frac{V_0(1 + \alpha_v)}{V_0} = 1 + 2\alpha_\perp + \alpha_\parallel \Rightarrow \alpha_v = 2\alpha_\perp + \alpha_\parallel$$

(13.61)

The following values are obtained:

Vectra content (%)	$\alpha_\parallel$	$\alpha_\perp$	α_v
0	6.5×10^{-5}	6.5×10^{-5}	19.5×10^{-5}
40	2×10^{-5}	6.6×10^{-5}	15.2×10^{-5}
80	-0.2×10^{-5}	7.5×10^{-5}	14.8×10^{-5}
100	-1.0×10^{-5}	8.0×10^{-5}	15.0×10^{-5}

10.10. First generate a reference sample by cooling at a specified rate (e.g. 80 K min^{-1}) to a temperature T_1 ($T_1 \ll T_g$). Heat the sample at 10 K min^{-1} to a temperature T_2 ($T_2 \gg T_g$) while recording the thermogram. This thermogram is referred to as REF.

Then start cooling from T_2 to a temperature T_3, hold the sample at that temperature for a specified time (t_3), cool the sample to T_1 and heat the sample from T_1 to T_2 at 10 K min^{-1}. The recorded thermogram should then be compared with REF and the difference in the enthalpy between the two is equal to the integrated enthalpy relaxation occurring during T_3. The experiment is repeated with different treatment times (t_3).

10.11. High temperature transition:
  nematic ↔ isotropic
Low temperature transition:
  smectic (or crystalline) ↔ nematic

10.12. OIT is more sensitive to changes in antioxidant concentration at high stabilization levels Fig. 13.24).

10.13. OIT varies between 50 and 100 min at 190°C. The change in OIT with temperature can be calculated from the Arrhenius equation using the activation energy 176 kJ mol^{-1}:

OIT(190°C)/OIT(220°C)
   $= \exp[176\,000/(8.31 \times 463)]/$
     $\exp[176\,000/(8.31 \times 493)]$
   $= 16.177$

OIT will vary between 3 and 6 min at 220°C.

OIT(190°C)/OIT(230°C)
   $= \exp[176\,000/(8.31 \times 463)]/$
     $\exp[176\,000/(8.31 \times 503)]$
   $= 38.003$

OIT will vary between 1.3 and 2.6 min at 230°C. The temperature dependence of OIT is illustrated in Fig. 13.25.

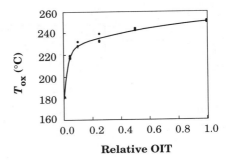

Figure 13.24 Oxidation temperature (from dynamic data) as a function of OIT for a medium-density polyethylene stabilized with a hindered phenol. Drawn after data from Karlsson, Assargren and Gedde (1990).

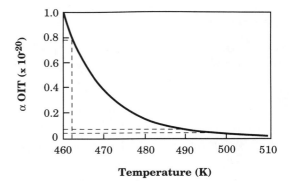

Figure 13.25 OIT as a function of temperature. Calculated curve from activation energy data.

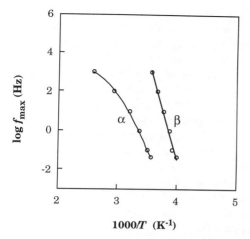

Figure 13.26 Arrhenius diagram showing the temperature dependence of the two relaxation processes.

10.14. The Arrhenius diagram of the two relaxation processes is shown in Fig. 13.26. The activation energies are slope coefficient × ln 10 × R.

The low temperature (β) process exhibits a very high activation energy, 202 kJ mol^{-1}, whereas the high temperature (α) process obeys WLF-temperature dependence.

CHAPTER 11

11.1. Prepare thin films (<5 μm) by solvent casting. Try to make the samples large, preferably 5–10 mm in diameter. The polymer with large spherulites can be studied by polarized microscopy and, with the lambda plate inserted in the beam, the optical sign of the spherulites can be assessed. The small size of the spherulites of the other sample makes direct assessment using polarized microscopy impossible. Small-angle light scattering is the preferred method. It can be conducted in the microscope. Use the cross polarizers; close the field and aperture irises and focus the back focal plane of the objective by using the Bertrand lens. A clover-leaf pattern will appear from which the spherulite diameter can be calculated.

11.2. PVC/NR: use osmium tetroxide; it will react with NR.

ABS: use osmium tetroxide; it will react with the polybutadiene.

PE/PS: solvent etching; use chloroform, which will dissolve PS.

PETP/LCP: etching with n-propylamine; it removes PETP.

PMMA/PVC: PMMA is very sensitive to the electron beam and good contrast is obtained in the microscope.

11.3. Fractography and SEM.

11.4. The typical view is along the b axis [010].

11.5. Electron diffraction.

11.6. The dark field for OM means that the specimen is illuminated from the side, i.e. at a large angle to the optical axis. The bright parts of the image are due to the scattered (deflected) light. A dark field for TEM is accomplished by adjusting the objective aperture so that it transmits only diffracted/scattered electrons. The bright parts of the image in this case are also due to scattered electrons.

11.7. Scanning electron microscopy due to its large depth of field. It has also the capacity to reveal foreign particles often present in the initiation region of fracture.

CHAPTER 12

12.1. The first method is infrared spectroscopy: the 1379 cm^{-1} absorption band is assigned to methyl groups whereas the 1465 cm^{-1} band is due to methylene groups. Methyl concentration (i.e. branch concentration) is determined by comparison of the two absorption peaks (internal standard) after suitable calibration with polymers of known chain branching. There is some interference between the 1379 cm^{-1} band and adjacent methylene bands. This 'background' can be removed by taking the difference spectrum between the sample spectrum and the spectrum of a branch-free polymer. The second method is ^{13}C NMR, by which both the type (branches up to six carbon atoms can be distinguished) and their concentrations can be determined.

12.2. (a) Small-angle X-ray scattering (SAXS) measuring the long period (L) combined with wide-angle X-ray scattering determining the volume crystallinity (v_c). The average crystal thickness (L_c) is calculated by the formula $L_c = Lv_c$.

(b) Raman spectroscopy by measuring the frequency (v) of the longitudinal acoustic mode

(LAM):

$$L_c = \frac{n}{2v}\sqrt{\frac{E}{\rho}}$$

where n is the 'order' of the vibration, E is the Young's modulus in the chain-axis direction and ρ is the density of the crystal component.

(c) A third method, possibly less attractive than the former two, is to measure the width of the crystalline wide-angle reflections (WAXS) and to calculate the crystal thickness perpendicular to the diffracting planes by using the Scherrer equation.

12.3. It may be possible when strong specific interaction between the two polymers occurs. Hydrogen bonding may cause a frequency shift of vibrations of the participating groups, e.g. the carbonyl stretch in polyesters.

12.4. X-ray scattering measures the crystallinity considering only crystals thicker than 2–3 nm. IR measures the total content of helix structures. The very thin crystals (<2–3 nm) are included in the IR crystallinity assessment. It is also possible that the helices are present in the amorphous phase.

12.5. X-ray diffraction records changes in the unit cell parameter. IR spectroscopy records frequency shift of certain absorption bands which, in addition to a shift in the peak frequency, will also turn asymmetric due to the variation in stress among the different segments. Small-angle neutron scattering of labelled (deuterated) guest chains provide information about deformation of whole molecules.

12.6. Oxidation of polyolefins leads to the formation of a spectrum of different carbonyl-containing compounds – ketones, aldehydes and esters – all with carbonyl-stretching vibrations at slightly different frequencies. The assessment of their concentrations can only be made by a resolution of the broad carbonyl band into its basic components (ketone-, aldehyde- and ester-carbonyl bands).

12.7. Infrared dichroism measurement provides information about the orientation of groups. If the angle between the transition moment vector and the chain axis is known, it is possible to determine a Hermans orientation parameter for the chain axis. The c-axis orientation (crystal-phase orientation) can be obtained from wide-angle X-ray scattering. The orientation of an amorphous polymer can be assessed by IR.

REFERENCES

Karlsson, K., Assargren, C. and Gedde, U. W. (1990) *Polymer Testing* **9**, 421.

Point, J. J. and Dosière, M. (1989) *Polymer* **30**, 2292.

INDEX

Absorbance, infrared spectroscopy 263
Acrylonitrile-butadiene-styrene plastics, glass transition temperatures 81
Aliphatic polyesters, history 16
Alpha process, relaxation, molecular criteria for 163–4
Anelastic behaviour 89, 90
Anti-plasticization 82
Anti-Stokes–Raman scattering 260
Argand plot 225, 226
Arrhenius equation, dielectric relaxation 226, 227
Atacticity 3
Attenuated total reflection 262
Atomic force microscopy 240
ATR, *see* Attenuated total reflection
Avrami equation
  analysis of crystallization kinetics data 178
  athermal nucleation 175, 176
  exponent values for special growth cases 177
  general 176
  limitations for polymers 177
  primary crystallization 177
  secondary crystallization 178
  thermal nucleation 176
Axialite 154

Bakelite, *see* Phenol-formaldehyde resin
Bimodal melting 229
Bingham plastics 101
Binodal points 57, 58
Birefringence
  assessment, compensator 204–5
  biaxial orientation 208
  form 204
  semicrystalline polymers 204
  uniaxial orientation 202, 203
Boltzmann superposition principle 90
Bonds
  dissociation energy 2
  length 2
Bragg equation 270
Bravais lattices 131, 132
Burger's model 91–2

C-axis orientation 205–6
CED, *see* Cohesive energy density
Cellulose nitrate, history 16
Chain branching, polyethylene, crystal structure 137
Chain entanglements
  general 99
  rheological properties 106
Chain folding 138, 139, 145, 150, 151
Chain of hindered rotation
  independent torsion angle potentials 26–9
  interdependent torsion angle potentials 29–32
Chain orientation
  assessment, infrared spectroscopy 206–7
  assessment, sonic modulus 207
  biaxial orientation 208
  effect on properties 211–14
  end-to-end vector orientation 200–1
  methods to orient polymers 208–11
  segmental orientation 200
  semicrystalline polymers 207–8
  uniaxial orientation 203
Chain with preferred conformation 35–6
Chlorosulphonation, staining 248, 249
Cholesteric phase 114, 115
Clausius–Mesotti equation 223
Cohesive energy density 67
Cold-drawing 208–9
Cole–Cole equation 225
Cole–Cole plot, *see* Argand plot
Cole–Davidson equation 226
Compton scattering 270
Configuration
  definition 2
  head-to-tail 4
  tacticity 2
Conformation
  anti-gauche 19
  definition 4
  eclipsed position 19
  energy map of butane 20
  energy map of ethane 20

  gauche 19, 20
  staggered position 19
  trans 19, 20
Conoscopy 240
Constitutional repeating unit 1
Copolymer
  alternating 5
  block 5
  definition 5
  graft 5
  random 5
  statistical 5
  terpolymer 5
Crankshaft motions 95
Crystal lamella
  chain folding 138, 139, 145, 150, 151
  crystal thickening 142, 143, 144
  equilibrium shape 142
  fundamentals 138–40
  history 137–8
  hollow pyramid, polyethylene 139
  initial crystal thickness 143
  lateral habit, other polymers, solution-grown 141
  lateral habit, polyethylene 139, 140
  melt-grown 147–51
  melt-grown, fold surface 149–51
  methods for assessment of structure 145–6
  polyethylene, annealing of monolayer crystals 142, 143
  roof-shaped, polyethylene 148
  rough lateral surfaces 140, 141
  sectorization, polyethylene 139
  solution-grown 138–41
Crystal lamella stack
  dominant 149
  general 147–9
  subsidiary 149
Crystal systems 131, 132
Crystal thickening 142, 143, 144, 227
Crystallinity
  assessment
    by calorimetry 158–9, 229
    by density measurements 158
    by wide-angle X-ray scattering 157–8

Crystallinity *contd*
  definition 157
  infrared spectroscopy 265, 266
  nuclear magnetic resonance
    spectroscopy 268
  polyethylene, dependence on
    molecular structure 159–60
    thermal treatment 160
  polyethylene, effect on
    mechanical properties 161
    transport properties 161
Crystallite orientation 206
Crystallization, metastable
    phases 188–9
Crystallization kinetics, assessment by
    differential scanning
    calorimetry 229–30

Degree of polymerization 11
Dendrimer 6
des Cloiseaux' law 66
DETA, *see* Dielectric thermal analysis
Diamond, maximum modulus 212
Die-swell 103
Dielectric α process 163
Dielectric constant 225
Dielectric loss 225
Dielectric thermal analysis 223–6
Differential scanning calorimetry
  amorphous polymers 230–1
  degree of vulcanization 234
  fundamentals 218, 219, 220
  liquid-crystalline polymers 231,
    232
  methods to obtain high accuracy in
    data 220
  molecular fractionation,
    polyethylene 193, 194
  polymer degradation 232, 233
  semicrystalline polymers 226–30
Differential thermal analysis 218,
    219, 220
Diffraction 259
Dilatant liquid 101
Dilatometry 221–2
Dimension of Gaussian chains
  light scattering 22
  long-range effects 22, 33
  small-angle neutron scattering 22
  theta conditions 22
  viscometry 22
Disclination 112, 113, 114
Dislocation
  edge 136
  screw 136
DMTA, *see* Dynamic mechanical
    thermal analysis
Doi theory 123
Doi–Edwards theory 108, 109

Doolittle's viscosity equation 85, 87
Droplet experiment 286
DSC, *see* Differential scanning
    calorimetry
DTA, *see* Differential thermal analysis
Dyad 3
Dynamic mechanical thermal
    analysis 222–3

Elasticity
  compliance, dependence on chain
    orientation 213
  energy-driven 39, 40
  entropy-driven 39, 40
  theoretical 212
Elastomer 39
Electrically conductive polymers 17
Electron diffraction 96, 273
Electron microscopy
  history 239
  permanganic acid etching 146, 147
  solution-grown single crystals 145
  staining of semicrystalline
    polymers 146
Electron spectroscopy for chemical
    analysis 269
Electron spin resonance
    spectroscopy 269
Elongational flow 100
End-to-end distance, chain
  definition 21
  experimental methods 21–2
  point distribution function 35
  radial distribution function 35
Enthalpy of isotropization 120
Entropic crystal growth theory, *see*
    Sadler–Gilmer theory
Entropy of isotropization 120
Equation-of-state theories 68–9
Equilibrium crystallization growth
    theories 178
Equilibrium melting temperature
  dependence on chain
    structure 172–3
  diluted systems 174
  from Hoffman–Weeks plot 172
  from small-molecule data 172
  from Thompson–Gibbs
    equation 171
  molar mass dependence 173–4
Equivalent chain model 32
Erhrenfest equation 89
ESCA, *see* Electron spectroscopy for
    chemical analysis
ESR spectroscopy, *see* Electron spin
    resonance spectroscopy
Etching, electron microscopy 249–51
Extended chain crystals,
    polyethylene 174

Fingerprint texture 115
Flory temperature, *see* Theta
    temperature
Flory theorem 21, 23
Flory theory for liquid crystals 123–6
Flory–Huggins theory
  general 58–63
  mixing enthalpy 61–2
  mixing entropy 59–61, 63
  rubber materials 48
  upper critical solution
    temperature 62
Fluids 99
Focal-conic fan texture 116
FOVE model 68
Fox equation 70, 82
Frank constants 114
Free volume 79, 80
Freely jointed chain 24–5
Freely rotating chain 25–6

Gauche state 19, 20
Gaussian chain, models 21, 24–35
Gel permeation, chromatography, *see*
    Size exclusion chromatography
Gibbs–DiMarzio theory 88–9
Glass transition
  mechanical properties 14
  semicrystalline polymers 162, 163
Glass transition temperature
  effect of
    cohesive energy density 79
    crosslink density 81
    molar mass 80
    molecular architecture 80
    pressure 82
    repeating unit structure 78–9
  polymer blends 70, 72, 81, 82
Glass transition theory
  free volume theories 87–8
  kinetic theories 85–7, 88
  thermodynamic theory 88–9
Glassy amorphous polymer
  mechanical behaviour 89–90, 92–5
  molecular factors 77, 78
  structure 95–6
  thermal factors 77, 78
Göler equation 176
GPC, *see* Size exclusion
    chromatography
Gutta-percha 3

Havriliak–Negami equation 226
Hermans orientation function
  definition 202
  optical birefringence 201–2
Higher order interaction 21
Hildebrand solubility parameter, *see*
    Solubility parameter

History of polymers 15–18
Homopolymer 5
Hyperbranched polymer 6

Infrared dichroism 206–7, 260
Infrared spectroscopy
  chain conformation 265
  chain orientation 265
  chemical structure analysis 264–5
  crystallinity 265, 266
  dichroism 206–7, 260
  Fourier transform
    spectrometers 262
  fundamentals 260, 261
  instrumentation 262
  miscibility of polymer 265
  polymer blends 71, 72
  quantitative analysis 262–3
  sample preparation 263
Injection moulding 211
Interaction parameter 61
Interlamellar tie chains 151
Internal reflection spectroscopy 262
iPP, see Polypropylene, isotactic
IR spectroscopy, see Infrared
    spectroscopy
IRS, see Internal reflection
    spectroscopy
Isotacticity 2, 3
Isothermal volume recovery
  memory effect 84
  phenomenology 83, 84
Isotropization temperature
  definition 113
  effect of mesogenic group 117–19
  effect of spacer group 120

KAHR model, see
    Kovacs–Aklonis–Hutchinson–
    Ramos model
Kevlar 17, 110
Kinetic crystallization growth
  theories 178–88
Kink 137
Kovacs–Aklonis–Hutchinson–Ramos
  model 86

LAM, see Longitudinal acoustic mode
Lattice plane index 134–5
Laue condition 270
Lauritzen–Hoffman theory
  experimental data 183–4
  general 178–86
  recent criticism 185–6
  recent developments 185
  regime I growth 181–2
  regime II growth 182–3
  regime III growth 183
  Z test 184

Leslie–Ericksen theory 123
LH theory, see Lauritzen–Hoffman
    theory
Light scattering 22, 70, 72, 273
Liquid crystal
  history 109
  definition 109
  discotic 111
  distortion 114
  X-ray diffraction patterns 116
Liquid crystalline polymers
  chemical structure 110
  crystal structure, effect of
    copolymerization 121
  history 17, 18, 110–11
  main-chain 110, 111
  melting point, effect of
    copolymerization 121
  mesogenic group 110
  microscopy 253, 254
  morphology of injection-moulded
    specimen 211
  phase diagrams 119, 126
  physical structure 111–17
  rheology 121–3
  side-chain 110, 111
  spacer group 110
  theory 123–6
  thermal behaviour 117
Long-chain branch 6
Longitudinal acoustic mode 266
Lower critical solution
    temperature 63, 64, 72
Lyotropic liquid 109

Macromolecular concept, history 16
Maier–Saupe theory 126
Maxwell model
  dynamic mechanical behaviour 92
  static mechanical behaviour 91
Meander model 96
Mechanical α process 163
Melt-spinning 211
Meso dyad 3
Mesomorphic material 109
Mesophase
  enantiotropic 117
  monotropic 117
  virtual 117
Michel–Levy chart 204, 244
Microscopy, different techniques 240
Miller's index 134, 135
Mixtures of branched and linear
    PE 191–3
Molar mass
  colligative properties 10–11
  end-group analysis 10
  mass average 9
  methods for assessment 10–11

number average 9
polydispersity index 10
viscosity average 9
Z average 9
Molar mass distribution functions 12
Molar mass segregation 149
Molecular architecture 6
Molecular fractionation
  general 189–93
  methods for assessment 193–4
  theory 190
Monomer, definition 1
Mooney–Rivlin theory 51

Natural rubber
  energetic stress ratio 44
  history 15, 16
  structure 39
  tacticity 3
Necking, cold-drawing 210
Nematic phase 112–13
Network polymer 6
Newtonian liquid 101
NMR spectroscopy, see Nuclear
    magnetic resonance
    spectroscopy
Non-periodic layer crystallite 121
Nuclear magnetic resonance
    spectroscopy
  applications 268–9
  chemical shift 267
  chemical structure 268
  coupling effect 267
  crystallinity 268
  fundamentals 266–7
  instrumentation 267–8
  nuclear overhauser effect
    (NOE) 268
  polymer blends 71, 72
  segmental mobility 268
  thermal transitions 269
Nucleation, crystal
  fundamentals 169, 170
  heterogeneous 170
  homogeneous 170
  molecular 190
  primary 170
  secondary 170
  tertiary 170

OIT, see Oxidation induction time
Oligomer 1
Optical microscopy
  bright-field microscopy 241
  condenser 242
  conjugate planes 242, 243
  dark-field microscopy 241
  differential interference-contrast
    microscopy 244

296 Index

Optical microscopy *contd*
  fundamentals 241–3
  history 239
  numerical aperture 241, 242
  objective 242
  phase-contrast microscopy 244
  polarized light microscopy 243–4
  polymer blends 71, 72
  resolution 241
Order parameter
  definition 112
  domain, temperature dependence 113
  see also Hermans orientation function
Orientation-induced crystallization 194–5
Osmium tetroxide staining 247–8, 249
Osmosis 62–3
Osmotic pressure
  concentration dependence 65–6
  second virial coefficient 64
Ostwald–de Waele equation 102
Oxidation induction time
  general 232, 233
  temperature dependence 233

PA 6, *see* Polyamide 6
PA 6,10, *see* Polyamide 6,10, repeating unit
PA *n*, *see* Polyamide *n*
PAN, *see* Polyacrylonitrile
Paracrystalline model 147
PBTP, *see* Poly(butylene terephthalate)
PE, *see* Polyethylene
Pentad 3
PEO, *see* Polyethyleneoxide
Permanganic acid etching 250, 251
PETP, *see* Poly(ethylene terephthalate)
Phantom chain, *see* Unperturbed chain
Phenol-formaldehyde resin 16
Physical ageing 82–5
Plasticizer 82
PMMA, *see* Polymethylmethacrylate
Point groups 132
Polarization
  atomic 224
  dipolar 224
  electronic 224
Polarized light microscopy
  fundamentals 243–4
  molecular fractionation, polyethylene 194
  supermolecular structure 155, 156
Poly(1,4-butadiene)
  crystallization via metastable phase 188
  repeating unit 7

Poly(1-butene), crystal lamella, solution-grown 141
Poly(4-methyl-1-pentene)
  crystal lamella, solution-grown 141
  repeating unit 7
Poly(butylene terephthalate) 8
Poly(dimethyl siloxane) 44
Poly(ethylene terephthalate)
  crystal lamella, solution-grown 141
  etching techniques for electron microscopy 251
  maximum modulus 212
  repeating unit 8
Poly(*n*-alkylacrylate) 8
Poly(*n*-alkylmethacrylate) 8
Poly(vinyl alcohol) 7
Poly(vinylidene dichloride) 8
Poly(vinylidene difluoride) 8
Polyacrylonitrile 7
Polyalcohols, staining for TEM 249
Polyaldehydes, staining for TEM 249
Polyamide 6
  crystal lamella, solution-grown 141
  maximum modulus 212
  repeating unit 8
Polyamide 6,6
  crystal lamella, solution-grown 141
  equilibrium melting data 173
  supermolecular structure 155
Polyamide *n* 8
Polyamide 6,10, repeating unit 8
Polyamides
  etching techniques for electron microscopy 251
  repeating unit 8
  staining for TEM 249
Polybutadiene 4
Polydienes, staining for TEM 249
Poly(etherether ketone)
  etching techniques for electron microscopy 251
Polyethylene
  Avrami exponent data 177–8
  crystal lamella, solution-grown 138–40
  crystal structure 133
    temperature dependence 134
  crystallization kinetics data, LH theory 184–5
  elastic modulus as function of draw ratio 214
  energetic stress ratio 44
  equilibrium melting data 173
  etching techniques for electron microscopy 251

  heating rate dependence of melting point 227, 228
  hexagonal phase 188
  history 16, 17
  maximum modulus 212
  molecular fractionation 190, 191, 192
  polymerization 13
  preferred conformation 36
  Raman spectroscopy, crystal thickness 147
  repeating unit 7
  staining for TEM 249
  statistical weight matrix 29
  supermolecular structure 152, 153, 154, 155
  transmission electron microscopy 252–3, 254
Polyethyleneoxide
  crystal lamella, solution-grown 141
  equilibrium melting data 173
  repeating unit 8
Polyisoprene
  repeating unit 7
  tacticity 3
Polymer
  definition 1
  names
    source-based 6
    structure-based 6
Polymer blends
  electron microscopy 254–5
  glass transition temperatures 70, 72
  infrared spectroscopy 71, 72
  light scattering 70, 72
  methods for the assessment of morphology 70–1
  molecular interpretation of miscibility 72
  nuclear magnetic resonance spectroscopy 71, 72
  optical microscopy 71, 72
  phase diagrams 72
  scanning electron microscopy 71, 72
  small-angle X-ray scattering 70, 72
  transmission electron microscopy 71, 72
  turbidity measurements 70
  wide-angle X-ray scattering 70, 72
Polymer crystallography 131–7
Polymer solutions
  concentration regimes 65–6
  dilute 65, 66
  general 58–68
  semi-dilute 66

Polymerization
  anionic 13
  cationic 13
  chain-growth 12–13
  coordination 13
  definition 1
  history 16–18
  radical 12
  step-growth 12
Polymethylmethyacrylate
  etching techniques for electron microscopy 251
  relaxation behaviour 93, 94
  repeating unit 7
Polymorphism
  polyethylene 133
  polypropylene 133, 134
Polyoxymethylene
  crystal lamella, solution-grown 141
  equilibrium melting data 173
  maximum modulus 212
  preferred conformation 36
  repeating unit 8
Polypropylene isotactic
  cross-hatching 149
  crystal lamella, solution-grown 141
  crystal structure 133, 134
  equilibrium melting data 173
  etching techniques for electron microscopy 251
  infrared spectrum, conformational analysis 265–6
  maximum modulus 212
  preferred conformation 36
  staining for TEM 249
  supermolecular structure 155
Polypropylene, repeating unit 7
Polystyrene
  glass transition temperature, molar mass dependence 81
  relaxation behaviour 95
  repeating unit 7
Polystyrene, isotactic
  crystal lamella, solution-grown 141
  etching techniques for electron microscopy 251
Polytetrafluoroethylene
  crystal lamella, solution-grown 141
  equilibrium melting data 173
  repeating unit 8
Polyvinylacetate, repeating unit 7
Polyvinylchloride, repeating unit 7
POM, see Polyoxymethylene
PP, see Polypropylene, isotactic
Primary normal stress coefficient 100

PS, see Polystyrene, isotactic
Pseudoplastic liquid 101
PTFE, see Polytetrafluoroethylene
PVAc, see Polyvinylacetate, repeating unit
PVAL, see Poly(vinyl alcohol)
PVC, see Polyvinylchloride, repeating unit
PVDC, see Poly(vinylidene dichloride)
PVDF, see Poly(vinylidene difluoride)

Racemic dyad 3
Radius of gyration, definition 21
Raman spectroscopy
  chemical structure analysis 264
  crystal thickness 147
  fundamentals 260, 261
  sample preparation 263–4
Random chain, see Gaussian chain
Random-flight analysis 33–5
Random lamellar structure 154
Random switch-board model 145
Reciprocal space 135–6
Regular solution model
  general 55–8
  mixing enthalpy 56
  mixing entropy 56
  upper critical solution temperature 58, 62
Relaxation processes
  molecular interpretation 93–5
  semicrystalline polymers 162–4
Relaxation time
  Arrhenius temperature dependence 93
  dielectric measurements 225, 226
  Maxwell model 91
  melts, according to reptation theory 108
  melts, molar mass dependence 105–6
  terminal 104
Relaxation time spectrum 92
Reptation model 107–9
Retardation time, Voigt–Kelvin model 91
Rheology
  definition 99
  liquid-crystalline polymers 121–3
  methods for assessment of properties 104–5
Rheopexi 102
Rotational isomeric state approximation 19–20
Rotational partitioning function 27
Rouse model 107
Row-nucleated structure 154
Rubber elastic theory, history 16
Rubber elasticity, statistical mechanical theory 44–8
Rubber-plateau 14

Rubbers
  energetic stress ratio 43, 44
  structure 41
  swelling in solvents 48
  thermodynamics 41–4
  thermo-elastic behaviour 41
Ruthenium tetroxide, staining 248, 249

Sadler–Gilmer theory 186–8
SALS, see Small-angle light scattering
SAXS, see Small-angle X-ray scattering
Scanning electron microscopy
  depth of field 245
  fundamentals 244, 245
  polymer blends 71, 72
  preparation of samples 247
  resolution 245
  supermolecular structure 156–7
Scattering 259
Scherrer equation 147
Schlieren texture 112, 113
Schultz distribution 12
Schultz–Flory distribution 12
SEC, see Size exclusion chromatography
Secondary normal stress coefficient 100
Second order interaction 19
Sectioning 249, 250
Sectorization, see Crystal lamella
SEM, see Scanning electron microscopy
Sheaf structure, see Axialite
Shear modulus
  complex 101
  loss 101
  storage 101
Shear viscosity 100
Shish-kebab 154
Short-chain branch, definition 6
Short-fibre composites, assessment of fibre orientation 255
Single-crystal, see Crystal lamella, solution-grown
Size exclusion chromatography 11
Small-angle light scattering 156
Small-angle neutron scattering
  deformation of rubbers 51
  general 22, 260, 272
  glassy amorphous polymers 96
  polyethylene, melt-grown crystals 150–1
Small-angle X-ray scattering
  crystal thickness 146
  instrumentation 272
  molecular fractionation, polyethylene 194
  polymer blends 70, 72

Smectic phase 115–16
Solid-state extrusion 209
Solubility parameter
  definition 67
  Hansen's three-dimensional 67, 68
  methods for assessment 67
Solution, definition 55
Solution-spinning 211
Solvent etching 250, 251
Space groups 132
Spectroscopic methods, spectral ranges 259
Spectroscopy, definition 259
Spherulite
  banded 152–3
  general 152, 153, 155, 156
  negative 152, 156
  non-banded 152
  positive 152, 156
Spinodal decomposition 56
Spinodal points 56, 57, 58
Staining, electron microscopy 247–9
Statistical weight matrix 29–30
Steady elongational viscosity 100
Steady simple shear flow, constitutive equations 96–100
Steady-state recoverable shear compliance
  definition 103
  melts, according to reptation theory 108
  molar mass dependence 106
Stokes–Raman scattering 260
Subglass processes
  amorphous polymers 93–5
  semicrystalline polymers 162, 163
Super-folding model 145
Superheating, crystal 144, 228
Supermolecular structure
  general 151–5
  methods of assessing 155–7
Swiss cheese structure 142
Syndiotacticity 3

Tacticity 2
TEM, see Transmission electron microscopy
Tetrad 3
TG, see Thermogravimetry
Theory of rubber elasticity
  affine deformation model 44–7
  deviations from classical treatment 48–50

Mooney–Rivlin theory 51
non-Gaussian statistics 50–1
phantom network model 44, 45, 47, 48
Thermal analysis
  definition 217
  methods, summary 217
Thermal conductivity, dependence on chain orientation 213
Thermal mechanical analysis 221–2
Thermal optical analysis 223
Thermo-elastic inversion 41, 42
Thermogravimetry
  content of volatile species 233
  filler content 233–4
  fundamentals 221
  oxidation induction time 233
Thermoplastic elastomers 41
Thermoplastics, definition 15
Thermosets, definition 15
Thermotropic liquid 110
Theta condition 21, 23
Theta temperature 64, 65
Thixotropy 102
Thompson–Gibbs equation 144–5, 147, 171
TMA, see Thermal mechanical analysis
TOA, see Thermal optical analysis
Torsion pendulum 222, 223
Total enthalpy method 158–9
Trans state, definition 19, 20
Trans(1,4-polyisoprene), energetic stress ratio 44
Trans-crystalline structure 154
Transmission electron microscopy
  artificial structures 251–2
  dark-field microscopy 246
  depth of field 245
  fundamentals 244, 245, 246
  molecular fractionation, polyethylene 194
  origin of contrast 246
  polymer blends 71, 72
  preparation of samples 247–51
  resolution 245
Triad 3

Ultra-oriented polymers, history 17
Ultraviolet microscopy 240
Unit cell, crystal 131
Universal calibration procedure, SEC 11
Unperturbed chain
  definition 23

  dimensions 23, 24, 65
  temperature coefficients 24, 28
Upper critical solution temperature 58, 72
Utracki–Jukes equation 70
UV-visible spectroscopy 269

Vectra 121
Vibrational spectroscopy
  fundamentals 260–2
  group frequencies 261
Viscosity
  complex 101
  dynamic 101
  Miesowisz 122
  pressure dependence 107
  shear 100
  steady elongational 100
  temperature dependence 107
  zero-shear-rate 101, 102
Viscosity measurement 22
Voigt–Kelvin model 91
Vulcanization methods 41

Wavenumber 263
WAXS, see Wide-angle X-ray scattering
Wide-angle neutron scattering 260, 272–3
Wide-angle X-ray scattering
  crystal thickness 147
  glassy amorphous polymers 95
  instrumentation 271
  polymer blends 70, 72
William–Landel–Ferry equation
  dielectric relaxation 226, 227
  general 87, 88, 89, 92, 93
WLF equation, see William–Landel–Ferry equation

X-ray scattering
  fundamentals 269–71
  synchrotron source 270

Young's modulus 213

Zero-shear-rate viscosity
  definition 101, 102
  melts, according to reptation theory 108
  molar mass dependence 106
Zimm plot 22